Rectangular Dielectric Resonator Antennas

Rajveer S. Yaduvanshi · Harish Parthasarathy

Rectangular Dielectric Resonator Antennas

Theory and Design

Springer

Rajveer S. Yaduvanshi
Department of ECE
Ambedkar Institute of Advanced Communication Technologies and Research
New Delhi, Delhi
India

Harish Parthasarathy
Department of ECE
Netaji Subhas Institute of Technology
New Delhi, Delhi
India

ISBN 978-81-322-2499-0 ISBN 978-81-322-2500-3 (eBook)
DOI 10.1007/978-81-322-2500-3

Library of Congress Control Number: 2015942632

Springer New Delhi Heidelberg New York Dordrecht London

Printed on acid-free paper

Springer (India) Pvt. Ltd. is part of Springer Science+Business Media (www.springer.com)

Preface

Microwave dielectric resonator antenna (DRA) materials or ceramics were demonstrated by Richtmeyer in 1939. Richtmeyer showed that these dielectric ceramics can resonate. Theory of DRA was expanded by Okaya and Brash in 1960. More experimental work on DRAs, done by Long in 1980, proved that DRAs can become efficient radiators and can be used as antennas. S.A. Long experimentally implemented DRAs of different shapes and sizes as a low-profile antenna.

Analysis and studies on characteristic equation, radiation patterns, and excitation methodology made DRAs popular by providing a new avenue compared to traditional patch antennas suffering from low gain and low bandwidth. Aldo Petosa made DRAs a very successful candidate as functional antennas. Both the limitations of low gain and low bandwidth in patch antennas can be eliminated by the use of a rectangular dielectric resonator antenna (RDRA) operating in higher modes and hybrid modes.

The modes theory of RDRA gives an important analysis on current distribution, impedance, and radiation patterns of an antenna. Modes form a real, orthogonal basis function for currents on the antenna. These are defined by boundary value problems using eigenvalues and eigenvectors. The scope of this book has been restricted to RDRAs, however, the concept can be extended to other geometries. In RDRAs, once the excitation is given, the total distributed current on the antenna structure becomes a weighted sum of eigen currents or a superposition of various modes at any instant of time.

Resonant modes in RDRAs can be classified as dominant and higher modes. Dominant modes correspond to lowest resonant frequency. These are called as TE, TM, and HEM modes. *E* and *H* field formats inside the RDRA at any instant of time at a known frequency are termed as resonant modes. Modes excitation is directly related to the surface current densities of the structure due to applied RF current. This current gets converted into modal fields based on Maxwell's equations. These fields are restricted by RDRA boundary conditions. Reflection and refraction of electromagnetic waves takes place because of dielectric interface at the boundary.

The generation of higher modes generally depends on RF excitation, device dimensions, permittivity of dielectric material and coupling techniques used in design of the antenna. The higher-order modes and hybrid modes have much flexibility and design space in RDRA for different applications, but the excitation techniques are complex. Rectangular DRA has a high degree of design flexibility due to two aspect ratios (*a*/*d* and *b*/*d*), low cost, simplicity, and ease of fabrication. It can retrofit to the existing patch antenna technology for gain improvements.

Researchers have long felt the need for a rigorous theoretical analysis on resonant modes of RDRA, and resonators have become a demanding field for industry and academia. This is because knowledge of resonant modes gives physical insight to the antenna designer, based on which input impedance and radiation characteristics can be predicted. We hope that this book will help to fill the gap.

The investigations and theory developed are based on applying waveguide theory models. Propagation of electromagnetic fields has been taken along z-axis, i.e., $\exp(-\gamma z)$. Initially, these are exploited via the Maxwell's curl equations and then manipulating them to express the transverse components of the fields in terms of the partial derivatives of the longitudinal components of the fields w.r.t. x and y (i.e., the transverse coordinates).

Waveguide models of four different boundary conditions filled with homogeneous as well as inhomogeneous dielectric materials with linear and nonlinear permittivity, permeability, and conductivity have been developed to determine TE and TM propagating electromagnetic fields. These have resulted in different sine–cosine combinations. TE modes generation required H_z fields as longitudinal fields and E_x, E_y, H_x, and H_y fields as transverse fields.

If input excitation is applied along x-axis as partial fields, y-axis will have fixed variation and z-axis will have desired variation in propagating fields. For example, TE δ_{13}. Similar cases can be developed for TM modes so as to propagate E_z fields as longitudinal and E_x, E_y, H_x, and H_y as transverse fields. H_z field will get vanished because of boundary conditions.

An equivalent but computationally simpler way to pass on from waveguide physics to resonator physics is to just replace (γ) by $(-\frac{\partial}{\partial z})$ in all the waveguide formulae that express the tangential field components in terms of the longitudinal components. This is done after solving the full 3D Helmholtz equations using separation of variable as x, y, z.

$$\left(\nabla^2 + \frac{\omega^2}{c^2}\right)\begin{pmatrix} E_z \\ H_z \end{pmatrix} = 0$$

The discrete modes $\omega(mnp)$ enable us to visualize the resonator as collection of L, C oscillators with different L, C values. The outcome of all this analysis enables us to write down the $\underline{E}$ and $\underline{H}$ fields inside the resonator, as superposition of four and three vector-valued basis functions.

$$\underline{E}(x,y,z,t) = \sum_{mnp=1}^{\infty} \mathrm{Re}\left\{C(mnp)e^{j\omega(mnp)t}\underline{\psi}^{E}_{mnp}(x,y,z)\right\} + \sum_{mnp=1}^{\infty} \mathrm{Re}\left\{D(mnp)e^{j\omega(mnp)t}\underline{\bar{\phi}}^{E}_{mnp}(x,y,z)\right\}$$

and

$$\underline{H}(x,y,z,t) = \sum_{mnp=1}^{\infty} \mathrm{Re}\left\{C(mnp)e^{j\omega(mnp)t}\underline{\psi}^{H}_{mnp}(x,y,z)\right\} + \sum_{mnp=1}^{\infty} \mathrm{Re}\left\{D(mnp)e^{j\omega(mnp)t}\underline{\bar{\phi}}^{H}_{mnp}(x,y,z)\right\}$$

We note that there are only two sets $\{C(mnp)\}$ and $\{D(mnp)\}$ of linear combination of coefficients using from the E_z and H_z *expansions*. The vector-valued complex functions are as follows:

$$\underline{\psi}^{E}_{mnp}, \underline{\bar{\phi}}^{E}_{mnp}, \underline{\psi}^{H}_{mnp}, \underline{\bar{\phi}}^{H}_{mnp} \in R^3$$

where R is autocorrelation and contain components $\{\cos,\sin\} \otimes \{\cos,\sin\} \otimes \{\cos,\sin\}$ functions and hence for $(m',n',p') \neq (m,n,p)$, each function of the set, where m, n, p are integers.

$$\left\{\underline{\psi}^{E}_{mnp}, \underline{\bar{\phi}}^{E}_{mnp}, \underline{\psi}^{H}_{mnp}, \underline{\bar{\phi}}^{H}_{mnp}\right\}$$

is orthogonal to each function of the set:

$$\left\{\underline{\psi}^{E}_{m'n'p'}, \underline{\bar{\phi}}^{E}_{m'n'p'}, \underline{\psi}^{H}_{mnp}, \underline{\bar{\phi}}^{H}_{m'n'p'}\right\}$$

w.r.t. the measure of dx dy dz over surface of RDRA [0, a] × [0, b] × [0, d], where a, b, and d are RDRA dimensions. The exact form of the function $\underline{\bar{\phi}}^{E}, \underline{\bar{\phi}}^{H}, \underline{\psi}^{E}, \underline{\psi}^{H}$ depends on the nature of RDRA boundaries.

Excitation of RDRA plays very important role for modal analysis. To calculate the amplitude coefficients $\{C(mnp)\}$ and $\{D(mnp)\}$, we assume that at $z = 0$, an excitation $E_x^{(e)}(x,y,t)$ or $E_y^{(e)}(x,y,t)$ is applied for some time say t ε [0, T] and then removed, as usually is done in L, C oscillators. Then, the Fourier components in this excitation corresponding to the frequencies $\{\omega_{mnp}\}$ are excited, and their solutions are the oscillations for $t > T$. The other Fourier components decay within the resonator.

$\{C_{mnp}, D_{mnp}\}$ magnitude components can be determined based on principle of orthonormality:

$$\sum_{mnp} \mathrm{Re}\left(C(mnp)e^{j\omega(mnp)t}\psi^{E}_{mnp\,X}(x,y,0)\right) + \mathrm{Re}\left(D(mnp)e^{j\omega(mnp)t}\underline{\bar{\phi}}^{E}_{mnp\,X}(x,y,0)\right) = E^{(e)}_{x}(x,y,t)$$

and

$$\sum_{mnp} \mathrm{Re}\left(C(mnp)e^{j\omega(mnp)t}\psi^{E}_{mnp\,Y}(x,y,0)\right) + \mathrm{Re}\left(D(mnp)e^{j\omega(mnp)t}\underline{\bar{\phi}}^{E}_{mnp\,Y}(x,y,0)\right) = E^{(e)}_{y}(x,y,t)$$

By using orthogonality of $\{\psi^{E}_{mnp\,X}(x,y,0),\ \underline{\bar{\phi}}^{E}_{mnp\,X}(x,y,0)\}$; for different (m, n), we write p fixed and likewise of $\{\psi^{E}_{mnp\,Y}(x,y,0), \underline{\bar{\phi}}^{E}_{mnp\,Y}(x,y,0)\}$; in addition, we need to use Kolmogorov–Arnold–Moser (KAM) type of time averaging to yield:

$$C(mnp)\psi^{E}_{mnp\,X}(x,y,0) + D(mnp)\underline{\bar{\phi}}^{E}_{mnp\,X}(x,y,0) = \lim_{T\to\infty} \frac{1}{2T}\int_{-T}^{T} E^{(e)}_{x}(x,y,t)e^{-j\omega(mnp)t}\mathrm{d}t$$

and likewise

$$C(mnp)\psi^{E}_{mnp\,Y}(x,y,0) + D(mnp)\underline{\bar{\phi}}^{E}_{mnp\,Y}(x,y,0) = \lim_{T\to\infty} \frac{1}{2T}\int_{-T}^{T} E^{(e)}_{Y}(x,y,t)e^{j\omega(mnp)t}\mathrm{d}t$$

In this book, RDRA resonant modes theoretical as wells as practical aspects have been investigated along with rigorous mathematical analysis for TE, TM, and HEM. Higher modes generation and control of resonant modes have been experimented. Shifting of dominant mode toward higher modes and vice versa is desired phenomenon for reconfigurability, merging of neighboring resonant modes have been exploited with simulation results. Use of higher modes for practical applications in antennas has been described. Merging of neighboring modes significantly increased antenna bandwidth. The device miniaturization using high-permittivity materials has been described. The devising control on modes has imparted reconfiguration of operating frequency, beam pattern, beam width, polarization, gain, and bandwidth. Higher modes radiation pattern, sensitivity analysis by changing dimensions, and permittivity analysis by changing permittivity have been mathematically modeled, and each is supported with simulated and experimental results. Selecting and cancelling a particular resonant mode has also been described. The concept of modes has been supported with practically implemented case studies. Devising control on resonant modes in RDRA can be used for software-defined radios and military applications, where frequent change of antenna parameters is operational requirement. For automation on modes control, microcontrollers equipped with lookup table can be used.

The modes have been modeled by R, L, C networks. Antenna far fields patterns and impedance have been computed and measured. Analysis on hybrid modes in RDRA has been discussed. Hybrid modes are complex to determine. Their mathematical formulations have been described. These modes are diversified.

The excitation of hybrid modes is complex, and their effective control can revolutionize the antenna technology. Detailed study of mathematical modeling of hybrid modes has been described. Hybrid modes are more popular for azimuthally field variations. The transcendental equation and characteristic equation for RDRA modes are used for determining propagation constants and then resonant frequency.

The solution of resonant modes can be obtained using the following:

(a) H_z and E_z fields are expressed as $u_{mnp}(x, y, z)$, $v_{mnp}(x, y, z)$ and ω_{mnp} based on solving Maxwell's equations with given boundary conditions.
(b) At $z = 0$, surface (x, y) excitation with applied surface current density is

$$\left(J_{sx}(x, y, t), J_{sy}(x, y, t)\right)$$

(c) Surface current density is equated with generated magnetic fields

$$\left\{J_s(x, y, \delta) = \left(J_{sx}, J_{sy}\right) = (\hat{z} \times \underline{H}) = \left(-H_y, H_x\right)\right\};$$

at $z = 0$; amplitude coefficients (D_{mnp} and C_{mnp}) are obtained on expansion of H_z is terms D_{mnp}, and E_z terms as C_{mnp}.
(d) Equate tangential component of E_z at boundary, i.e., $\underline{E}_y|_{z=0}$ to zero, and compute the coefficients D_{mnp} for H_z and C_{mnp} of E_z.
(e) Excited by ω_{mnp} and arbitrary feed position in xy plane $(x_0, y_0)(\phi_0, \theta_0)$

$$\mathrm{H}_{\perp} = \left[\sum_{mnp} \mathrm{Re}\left\{\tilde{D}_{mnp} e^{j\omega(mnp)t}\right\} \nabla_{\perp} \tilde{u}_{mnp}(x, y, z)\right] - \sum_{mnp} \mathrm{Re}\left\{\tilde{C}_{mnp} e^{j\omega(mnp)t}\right\} \nabla_{\perp} \tilde{u}_{mnp}(x, y, z)$$

and similarly $\mathrm{E}_{\perp}$.

Depending on the boundary conditions, four cases have been developed. In RDRA, these four walls are assumed as perfect magnetic conductors and top and bottom walls are taken as perfect electric conductors.

$$u_{mnp} = \sin \sin \sin = E_z$$

$$v_{mnp} = \cos \cos \cos = H_z$$

Sidewalls and top walls all are perfect electric conductors

$$u_{mnp} = \sin\sin\cos = E_z$$

$$v_{mnp} = \sin\sin\sin = H_z$$

Sidewalls and top walls all are perfect magnetic conductors

$$u_{mnp} = \cos\cos\sin = E_z$$

$$v_{mnp} = \sin\sin\cos = H_z$$

Top and bottom walls are perfect magnetic conductors, and all four sidewalls are PEC

$$u_{mnp} = \cos\cos\cos = E_z$$

$$v_{mnp} = \sin\sin\sin = H_z$$

Transcendental equation is used to solve propagation constants, i.e., k_x, k_y, and k_z. The propagation constant gives rise to resonant frequency with the help of characteristic equation. These wave numbers k_x, k_y, and k_z are in x, y, and z-directions, respectively. The free space wave number is k_0. The resonant frequency can be determined from combined solution of transcendental equation and characteristic equation of rectangular DRA. Time-averaged electric energy = time-averaged magnetic energy

$$\begin{aligned} \varepsilon_r k_0^2 &= k_x^2 + k_y^2 + k_z^2 \\ \varepsilon_0 k_0^2 &= k_x^2 + k_y^2 + k_z'^2 \\ k_z' &\neq p\pi/d \\ \tan(k_z d) &= \frac{k_z}{\sqrt{k_0^2(\varepsilon_r - 1) - k_z^2}}; \end{aligned}$$

the final result of transcendental equation is thus achieved.

The contents of this book are the outcome of our research work on RDRA higher-order resonant modes. In this book, analyses have been restricted to rectangular resonators higher modes, however, the concept can be extended to other geometry resonators, such as cylindrical, conical, and hemispherical. With this book, we hope to fill the gap for rigorous theoretical analysis on RDRA resonant modes. The work is supported with live projects data and their case studies. This book should be very useful for antenna designers, both in research and development and for practical implementations. This book is written in a simple and reader friendly manner and can be easily understood with an initial knowledge of basic electromagnetic theory. All the chapters are self-reliant, and no initial specialization is required to understand the contents. We hope that this book will help open the design space for a new class of antenna implementations.

This book is organized into 12 chapters including rigorous theoretical analysis of modes along with case studies and design data annexure. Introduction along with history of RDRA is given in Chap. 1. Introduction of resonant modes is explained in Chap. 2. Mathematical derivations for modes and the generation of TE/TM modes have been discussed in Chap. 3. Chapter 4 presents the derivation of RDRA transcendental equations. In Chap. 5, mathematical description of amplitude coefficients of even and odd modes is presented. Chapter 6 contains radiation parameters and mathematical explanations of RDRA. Chapter 7 describes derivations of higher-order resonant modes and their applications for high-gain antenna designs. Chapter 8 explains the effect of angular variation on excitation to produce various types of radiation patterns to meet military requirements. Chapter 9 discusses sensitivity analysis and mathematical modeling of radiation pattern solutions in RDRA. Chapter 10 presents the excitation of hybrid modes in RDRA and their possible applications. Chapter 11 covers inhomogeneous solution along with measurements. Basic RDRA resonant frequency formulations, materials required, and their sources are given in the annexures. Complete and detailed solutions of RDRA have been explained in case studies. Design data are provided in the annexures. Chapter 12 discusses case studies.

New Delhi, India

Rajveer S. Yaduvanshi
Harish Parthasarathy

Acknowledgments

We would like to express special thanks to reviewers and editors for sparing their valuable time in reviewing this manuscript. We acknowledge the support and cooperation extended by all our family members, without whom it would not have been possible to complete this book.

We in person acknowledge the healthy discussion with our students and colleagues. Special thanks to Sujata, R. Gupta, Promod, Vipin, and Chander P. for imparting their help and support in preparing this manuscript, i.e., typing work and proofreading. We extend special thanks to Springer for publishing this book.

Rajveer S. Yaduvanshi
Harish Parthasarathy

Contents

About the Authors

Dr. Rajveer S. Yaduvanshi is working as an associate professor in the Department of ECE at Ambedkar Institute of Advanced Communication Technologies and Research Delhi for the last 7 years. Before that, he was working as a senior scientific officer in Ministry of Defence, Government of India. He has a total of 31 years of experience in teaching and research. He is the author of a book on MHD Antenna, Design and Applications and has published 27 papers in reputed journals and conferences. He has supervised 151 B.Tech. projects and 11 M.Tech. projects and is currently supervising seven Ph.D. students. He holds an M.Tech. degree from NIT Allahabad and a Ph.D. degree from NSIT, Delhi (Delhi University). His research interests include design of Dielectric resonator antennas, MHD embedded antennas and analysis of higher modes in DRA.

Dr. Harish Parthasarathy is working as a professor in the Department of ECE at Netaji Subhas Institute of Technology. He is the author of more than 11 books and has guided several Ph.D. students. He holds his Ph.D. and B.Tech. degrees from IIT Delhi. His research involves antenna and signal processing with specialization in mathematical modeling.

Chapter 1
Rectangular DRA Fundamental Background

Abstract This chapter introduces rectangular dielectric resonator antenna. Working mechanism of rectangular DRA (RDRA) has been explained. Survey work along with citations on related works based on the available literature has been described. RDRA as a new candidate in the field of antennas whose comparison has been made with existing patch antennas. Their advantages have been listed. Mathematical solution of one-dimensional resonator has been derived.

Keywords RDRA (rectangular DRA) · Working mechanism · Survey · Characteristics · Advantages · One-dimensional resonator

1.1 Introduction

Antenna is usually visualized as metallic device for radiating and receiving electromagnetic waves. It is an interface (transducer) between space and communication device. For wireless communication system or radar system, antenna is used to couple radio energy from transmitter to space in transdirection, and space to receiver in receive direction. Antennas are frequency dependent. The design of antenna corresponds to specific bandwidth and resonant frequency. These are purely designed as per requirements. The antenna rejects all signals beyond their bandwidth. An antenna is an integral part of any wireless communication. Hence, its development must be in synchronization with communication system. There have been revolutionary developments in communication systems since last decades. The emergent requirements are being felt in antenna development. The Gigabytes of data transmission at very high speeds are today's communication requirements. To match today's advanced communications requirements, rectangular DRA (RDRA) is the most suitable candidate. Rectangular dielectric resonator antenna is new kind of antenna, which is different from traditional metal or patch antenna. The patch or metal antennas generally suffer from low bandwidth, high conducting loss and low gain. RDRA has high gain and wide bandwidth antenna.

R.S. Yaduvanshi and H. Parthasarathy, *Rectangular Dielectric Resonator Antennas*, DOI 10.1007/978-81-322-2500-3_1

1.2 History of DRA

Dielectric resonator antenna is a microwave antenna consists of block of ceramics material having permittivity greater than 10 F/m. In 1939, R.D. Richtmeyer showed that non-metalized dielectric material objects can resonate and function as an antenna, these are called as dielectric resonator antenna [1, 2]. There were no practical applications of these DRAs until 1960. Dielectric resonator antenna was first introduced by S.A. Long in 1980 [3]. Since then, vast research has been carried out for analysis of DRA material properties, and their effective use as DRA. Various shapes and excitation methods for DRA have been developed. Many research papers have been published in reputed journals by researcher such as Kishk [4], Lee [5], Leung [6], Luk [7], Mongia [8], Shum [9], Junker [4], Antar [10], and Petosa [11] till date. No rigorous theoretical analysis for RDRA is available in the literature so for. It is felt that if good literature along with sound mathematical analysis on RDRA is made available, it can benefit the society in large. Only few books are available on introduction of DRA, but no book is available for sound theoretical analysis supported with mathematical computations of RDRA.

1.3 Working Mechanism of RDRA

The electromagnetic waves were generated by rapid oscillations of electrons in atoms causes acceleration or deaccelerations which become electromagnetic wave radiation. The radio waves are introduced into ceramics forming resonator as shown in Fig. 1.1 from RF transmitter circuits. These RF waves bounce back and forth between resonator walls, thus forming standing waves, hence stores electrical energy. Oscillating current introduces oscillating magnetic fields, *H* fields, and oscillating electric fields, *E* fields. The time-varying field radiates away from antenna into space due to accelerrating currents. The walls of ceramic formed partially transparent magnetic walls, and the magnetic energy leaks through these transparent walls due to fringing effect. Thus, radio power is radiated into space.

Let RDRA having dimensions a, b, d lengths is excited by external electric fields $E_{ix}(x, y)$ and $E_{iy}(x, y)$ in x, y plane. The equivalent circuit as shown in Fig. 1.2 is

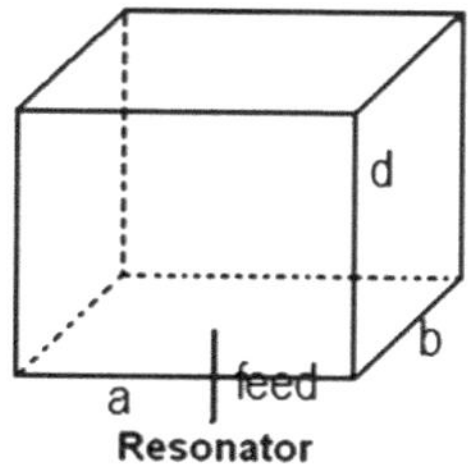

Fig. 1.1 Ceramics rectangular DRA with a, b, and d dimensions

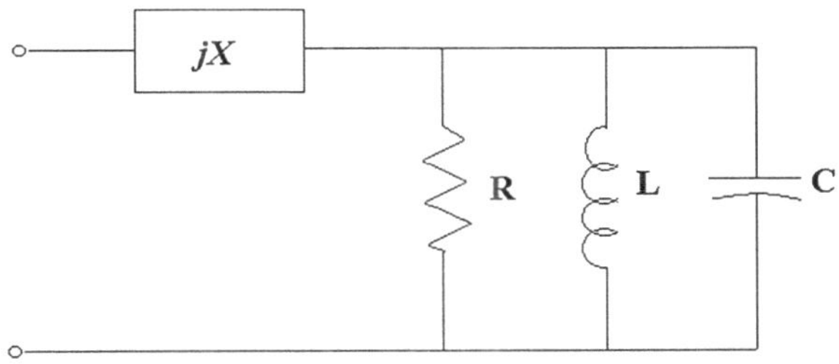

Fig. 1.2 RDRA equivalent RLC circuit

drawn based on the electrical properties of this ceramic RDRA. This results into longitudinal E_z and H_z fields. The probe currents are equated with RDRA radiating currents as per principle of conservation of energy. In other words, time-average (KAM) electric energies inside the RDRA are equated with time-average magnetic energies. Figure 1.2 presents RDRA equivalent RLC circuit for computing quality factor of RDRA.

$$\int_V |E|^2 \mathrm{d}V = \int_V |H|^2 \mathrm{d}V; \tag{1.1}$$

$$h_{mn}^2 = \gamma^2 + h^2; \tag{1.2}$$

where $\gamma = \frac{j\pi p}{d}$.

These fields are computed using Helmholtz equations, taking into account of source and RDRA boundaries. Mathematical solution of transverse and longitudinal fields is obtained by half-wave and full-wave Fourier analysis, taking inside medium and outside medium into consideration.

Figure 1.3 is shown as RDRA placed on infinite ground plane. The image theory can be applied to this RDRA for reducing its height. Varying sinusoidal in time, energy flow in particular direction can be treated as power radiated per unit solid angle (energy per unit area per unit time).

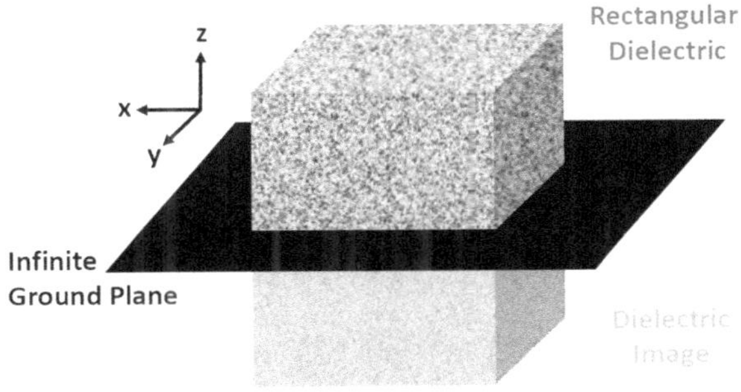

Fig. 1.3 RDRA with ground plane

1.4 Antenna Radiation Parameters

Antenna radiation parameters are as follows:

- Antenna radiation pattern
- Power
- Gain
- Polarization
- Impedance
- Efficiency radiation

1.5 Advantage of RDRA

- Lower conduction losses due to use of dielectric material;
- Most suitable at microwaves and millimeter waves;
- Compact in size and portable;
- Dimensions of RDRA are of the order of $\frac{\lambda g}{\sqrt{\varepsilon_r}}$; choosing higher ε_r RDRA size can be reduced significantly.
- Ease of fabrication;
- No frequency drift due to change in temperature;
- High-power handling capability;
- High gain and high bandwidth;
- Can be integrated with MIMC;
- RDRA has advantage of two aspect ratios. Hence, various modes can be generated by varying any of the aspect ratio;
- Simple coupling schemes;
- Bandwidth can be variable by choosing dielectric constant; and
- High Q factor.

1.6 Resonant Modes

In RDRA, resonant modes represent the radiating phenomena with the help of E and H field patterns. These fields inside RDRA are presented azimuthally. With the knowledge of modes, radiation characteristics of an antenna can be predicted. The designer can get insight of antenna design and hence can provide correction in the antenna design. Resonant modes are real current vectors. These modes are found by orthogonal Fourier basis functions. These are generated based on the current distribution on the surface of antenna due to field perturbations. These can be classified as TE or TM modes. The loss tangent (δ) introduced is due permittivity of the material. The principle conservation of energy is applied, in which, time-average electrical energy is equated to magnetic energy at any instant of time to compute radiated fields. RDRA is excited by input radio frequency currents at proper impedance match at input port. The transverse components are defined in terms of longitudinal components and vice versa. The modal field equations are developed using Fourier basis functions of cosine or sine terms appearing based on the RDRA boundary conditions, i.e., six walls of RDRA can be PMC, PEC, or any combination of these PMC and PEC walls. Hence, resonant modes bring physical insight into the radiating phenomena taking place inside the RDRA. The resonant modes form a set of orthogonal functions to compute total current on the surface of RDRA.

Figure 1.4 shows the resonant modes configuration generated into RDRA. Wave can only propagate if wave vector $k > k_c$, where k_c is cutoff frequency. The lowest resonance is called dominant mode.

This is solved and the solution consists of a superposition of a source (particular solution) term and a homogeneous term (i.e., general solution of the homogeneous

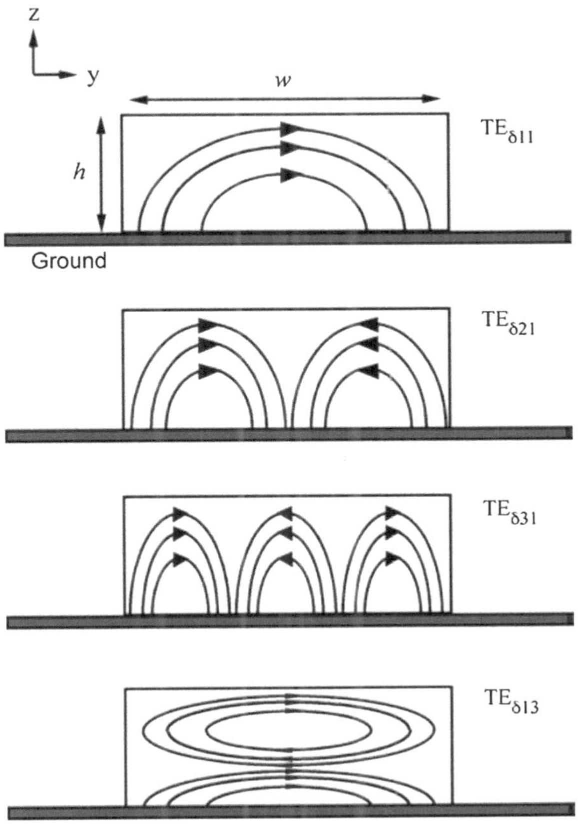

Fig. 1.4 Resonant modes in *yz* plane

part). Two constants in the homogeneous part are determined by applying the vanishing boundary conditions on H_z and E_z at top and bottom surfaces, i.e., at $z = 0, d$.

1.7 Characterization of Resonant Modes

The radiation can be identified as magnetic dipoles. Any function can be decomposed or separated by projecting that function into basis function, i.e., inner product with basis function

$$\langle F_1, F_2 \rangle = \int_a^b F_1(\vec{r}) F_2(\vec{r}) \mathrm{d}R; \tag{1.3}$$

The resonance modes are E and H field patterns inside the RDRA. Figure 1.5 has shown that electric fields are always associated with magnetic fields and vice versa. These can be three types, i.e., TE, TM, and HEM modes. The amplitude coefficients and phase of RDRA are C_{mnp}, ψ_{mnp} and D_{mnp}, ϕ_{mnp}. E_z and H_z fields are based on the orthonormality. These can be determined by applying principle of orthonormality. The characteristics equations of RDRA are given as follows:

$$\varepsilon_r k_0^2 = k_x^2 + k_y^2 + k_z^2; \tag{1.4}$$

where k_0 is free space wave and k_x, k_y, k_z are propagation constants in x-, y-, and z-directions, respectively. Also, $k_0^2 = \omega_0^2 \mu_0 \varepsilon_0$; hence, resonant frequency in free space can be determined based on the free space wave number. To determine propagation constants, i.e., k_x, k_y and k_z, knowledge of transcendental equation is required. The transcendental equation is developed for RDRA when fields are propagating in z-direction and given below as

$$k_z \tan\left(k_z \frac{d}{2}\right) = \sqrt{(\varepsilon_r - 1)k_0^2 - k_z^2}; \tag{1.5}$$

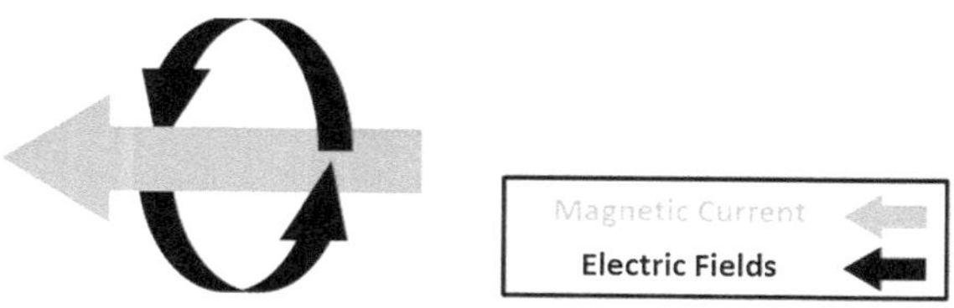

Fig. 1.5 Electric fields and magnetic fields are associated

The solution resonant frequency of RDRA can be determined depending upon the resonant mode generated, i.e., $TE_{111}, TM_{111}, TE_{11\delta}, TE_{1\delta1}$, and $TE_{11\delta}$. Fields are expanded into summation of their modal functions, which may be by C_{mnp} and D_{mnp} amplitude coefficients. Applying continuity equation across regional interfaces tangential fields, current distribution along surfaces of an antenna can be computed as J_e, i.e., current density. Input impedance Z_{in} and radiation pattern P_{rad} can be computed based on the current distribution. Eigenvectors or Eigen functions are formed as characteristic modes. Modes are orthogonal over source region. E_z electric fields produced by J_n characteristic currents on the surfaces. These modes are mainly dependent on the RDRA boundary and excitation.

Electrical walls of RDRA:

$$E_{\text{tan}} = n \times E = 0; \tag{1.6}$$

$$H_{\text{nor}} = n \cdot H = 0; \tag{1.7}$$

Magnetic walls of RDRA:

$$H_{\text{tan}} = n \times H = 0; \tag{1.8}$$

$$E_{\text{nor}} = n \cdot E = 0; \tag{1.9}$$

The solution of resonant modes shall vary in terms of sine and cosine as these are dependent on PEC and PMC walls of RDRA. E_z and H_z fields can be determined as linear combinations of these functions sin or cosine in xy plane and z-component of source, to get these propagation constant. Propagating fields in particular direction x or y or z is assumed to be continuous inside and outside the RDRA. While taking into account inside the resonator both, reflected as wells as propagating fields are available, outside the RDRA only outgoing field components are taken and reflected component is negated. This solves the transcended equation for RDRA. The modal characteristics of antenna give rise to fields, i.e., resonant modes. These are also known as eigenvector and eigenvalues. Eigenvectors are current amplitudes C_{mnp} and D_{mnp}, and eigenvalues are resonant frequencies ω_{mnp}. The resonant frequency can be given as follows:

$$(f_r)m, n, p = \frac{c}{2\pi\sqrt{\varepsilon\mu}}\sqrt{\left(\frac{m\pi}{a}\right)^2 + \left(\frac{n\pi}{b}\right)^2 + \left(\frac{p\pi}{d}\right)^2}; \tag{1.10}$$

This book contents are lucid, simple, and pedagogical.

1.8 Magnetic Dipole Moment

The radiation in RDRA is taking place due to short magnetic dipole formation.

$$d = \sum e \cdot r; \tag{1.11}$$

where

d dipole moment
e charge
r distance between two charges

$$d = \frac{\mathrm{d}}{\mathrm{d}t} \sum e \cdot r = \sum e \cdot v; \tag{1.12}$$

$$d^{\cdot\cdot} = \frac{\mathrm{d}}{\mathrm{d}t} \sum e \cdot v; \tag{1.13}$$

Hence, charges can radiate only if they move with acceleration. There will be no radiation even if they move with fixed or uniform velocity.

1.9 Spring Resonator of Length L

As shown in Fig. 1.6, one single string of AB length is applied with external excitation to produce oscillations. These oscillations will give rise to resonant frequency of the resonator.

$$x''(t) + \omega_0^2 x(t) = C_2 e^{j\omega t}; \tag{1.14}$$

$$x(t) = C_1 e^{j\omega t}$$

$(\omega_0^2 - \omega^2)C_1 = C_2$, where C_1 and C_2 are constants and f is the function of length L.

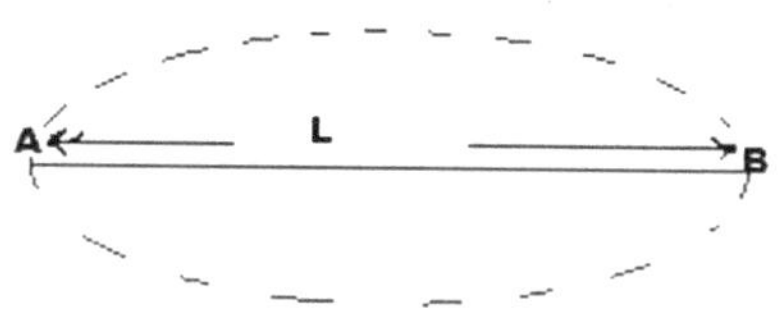

Fig. 1.6 Simple spring resonator

Hence, $C_1 = \frac{C_2}{\omega_0^2 - \omega^2}$

$x(t) = \frac{C_2 e^{j\omega t}}{\omega_0^2 - \omega^2}$, if $\omega_0 = \omega$; then, $x(t)$ will be ∞.

Now, $\omega = \omega_0 + \delta$; when δ is small deviation,

$$= \frac{C_2 e^{j\omega t}}{(\omega_0 + \omega)(\omega_0 - \omega)}; \tag{1.15}$$

Hence, the solution of spring resonator in one dimension is given as follows:

$$= \frac{C_2 e^{j\omega t}}{\delta(2\omega_0)}$$

$$\left(\frac{\partial^2}{\partial x^2} - \frac{1}{c^2}\frac{\partial^2}{\partial t^2}\right) f(x,t) = 0; \text{ at boundaries}$$

$$f(0,t) = 0 \quad \text{and} \quad f(L,t) = 0$$

Taking Fourier Transform of above equation,

$$\left(\frac{\partial^2}{\partial x^2} + \frac{\omega^2}{c^2}\right) \hat{f}(x,\omega) = 0; \tag{1.16}$$

Writing above terms in sine and cosine form, we have

$$C_1 \sin\left(\frac{\omega x}{c}\right) + C_2 \cos\left(\frac{\omega x}{c}\right) = 0 \text{ or } \hat{f}(0,\omega) = 0 \text{ or } \hat{f}(L,\omega) = 0$$

$\sin\left(\frac{\omega L}{c}\right) = 0$. Hence, $kL = n\pi$; sine values to be zero.

$$\omega = kc = \frac{n\pi c}{L}, \text{ when } n = 1,2,3\ldots \text{ where } k = \omega/c; \tag{1.17}$$

String length $2L$, dominant frequency ω_1, length L, dominant frequency is $2\omega_1$. Length is $2L/3$, dominant frequency $3\omega_1$; Eqs. (1.1)–(1.17) used in this chapter presented the mathematical concept of topic.

References

1. Harrington RF (1961) Time harmonic electromagnetic fields. Wiley, New York
2. Richtmyer RD (1939) Dielectric resonator. J Appl Phys 10:391–398
3. Wakino K, Tamura H, Sudo T (1987) Dielectric resonator materials and their applications. Microw J 6:133–148
4. Junker GP, Kajfez D, Kishk AA, Lisson AW (1995, May 11) Effect of aperture filling on slot coupled dielectric resonator antennas operating in HEM_{11} mode. Electron Lett 31(10):774–775

5. Lee RQ, Simons RN (1994) Bandwidth enhancement of dielectric resonator antennas. In: IEEE antennas and propagation society international symposium, Seattle, WA, June 1994, pp 1500–1503
6. Luk KM, Leung KW, Chow KY (1997) Bandwidth and gain enhancement of a dielectric resonator antenna with the use of stacking element. Microw Opt Technol Lett 14(4):215–217
7. Leung KW, Chow KY, Luk KM, Yung EKN (1997) Excitation of dielectric resonator antenna using a soldered through probe. Electron Lett 33(5):349–350
8. Mongia RK, Ittipiboon A, Cuhaci M, Roscoe D (1994) Radiation Q-factor of rectangular dielectric resonator antennas theory and experiment. In: IEEE antennas and propagation society international symposium, Seattle, WA, pp 764–767, June 1994
9. Shum SM, Luk KM, Leung WK, Wa K (1994) Mutual impedance of hemisphere dielectric resonator antennas. IEEE Trans Antennas Propagat 42(12):1652–1654
10. Antar YMM, Chang D, Sequin G, Henry B, Keller MG (1998, Oct 5) Modified wave guide model (MWGM) for rectangular dielectric resonator antennas. Microw Opt Technol Lett 19 (2):158–160
11. Petosa A, littipiboon A, Cuhaci M, Larose R (1996) Bandwidth improvement for a micro strip fed series array of dielectric resonator antennas. Electron Lett 32(7):608–609

Chapter 2
Rectangular DRA Resonant Modes and Sources

Abstract Basics of resonant modes have been described. Their mathematical analysis for generation of different resonant modes have been presented in this chapter. Realization of resonant modes based on MATLAB has also been worked. Modes are generated by applying voltage source. Various types of resonant modes have been described along with all possible applications.

Keywords Cavity resonator · Resonant modes · Type of modes · Wave guide analysis · Mathematical description of resonant modes · Simulated work

2.1 Introduction

In the early 1960s, Okaya and Barash [1] reported the first ever DRA in the form of a single-crystal TiO_2. Since then, no rigorous theoretical analysis has been developed so far in the literature to evaluate the resonant modes in Rectangular DRA. Based on Cherenkov principle of radiations, an external electric field brings the charges of the molecules of the dielectric into a certain ordered arrangement in space and creates acceleration phenomenon in dielectric material itself. The dielectric polarization P is equal to the total dipole moment induced in the volume of the material by the electric fields. In most cases, the magnitude of polarization is directly proportional to the intensity of the electric field at a given point of a dielectric. The relative permittivity is related to the dielectric susceptibility. A dielectric resonator is defined as "object of dielectric material which *functions as a resonant cavity by means of reflections at the dielectric air interface*." The discontinuity of the relative permittivity at the resonator surface allows a standing electromagnetic wave to be supported in its interior at a particular resonant frequency, thereby leading to maximum confinement of energy within the resonator.

Certain fields distribution or modes will satisfy Maxwell's equations and boundary conditions. Resonant modes are field structures that can exist inside the DRA. Modes are the pattern of motion which repeat itself sinusoidally. Infinite number of modes can excited at same time. Any motion is superposition or

R.S. Yaduvanshi and H. Parthasarathy, *Rectangular Dielectric Resonator Antennas*, DOI 10.1007/978-81-322-2500-3_2

weighted sum of all the modes at any instant of time by combining amplitudes and phases. As in the case of all resonant cavities, there are many possible resonant modes that can be excited in dielectric resonators. The boundary conditions are $n \cdot H = 0$; where H denotes the electric field intensity and n denotes the normal to the surface of the resonator.

And, $n \times E = 0$, is not necessarily satisfied at all the surfaces of the RDRA by all the modes. Different resonant modes have distinct electromagnetic field distributions within the DRA, and each mode may provide a different radiation pattern.

Operation of DRA is based on the process that if excitation is applied, then a high magnetic field is created inside the dielectric object placed on a ground plane. Phenomena which occur like a charge particle passing to the field create the physical environment like any metal ball passing through liquid. Thus, there will be change in the field, contraction, and expansion, which causes fringing effect. This way dielectric object starts to radiate. Another phenomenon that occurs is that there might be reflection of the field from sidewalls of dielectric object due to change in the refractive index of the medium. The dielectric object acts as an oscillator.

Theory of characteristic modes can be applied in the design of antenna or DRA. These modes give insight into physical phenomenon taking place inside device in terms of current vectors as maxima and minima. This helps to locate the feeding point and desired dimension of RDRA.

In 1968, modes were introduced by Garbacz and later by Harrington. Inagaki gave simpler theory on modes for radiation mechanism in an antenna. It requires lot of computation, for loading quality factor, double feeding to improve bandwidth, and circular polarization. Characteristic modes are current modes or eigenvectors, which are the solution of characteristic equation. These are orthogonal functions that can predict total current on surfaces of body of the antenna. Also, desired mode can be excited for specific radiating pattern. Excitation of mode mainly depends on feeding arrangement, geometry of the device, and dielectric material used. In time domain, varying electric field can produce magnetic fields and vice versa. By applying RF, excitation currents in RDRA get converted into surface current density distributed over the surfaces, i.e., RF excitation with proper impedance match can generates J. This probe current produced magnetic vector potential "A." The radiated magnetic fields are presented in the form E-electric field intensity using Lorentz gauge transformation. An antenna can propagate electromagnetic fields, if wave vector $k > k_c$. The cutoff wave vector k_c determines the cutoff frequency. There can be dominant resonant frequency or higher-order resonant frequency. The propagation takes place along x-axis if propagation constant $k > \frac{n\pi}{b}$. There will not be any propagation if $k \leq \frac{n\pi}{b}$, as it will lead to formation of standing waves. Similar conditions persist for propagation along y-axis and z-axis. Maxwell's equations define the behavior of electromagnetic wave propagation, while the solution of Maxwell equation is defined by Helmholtz equation. The radiated power is given by Parseval's power theorem. Half-wave Fourier analysis is used to determine the time domain behavior of antenna radiations. The magnitude and phase of the radiated field is given by Poynting vector ($S = E \times H$). The image theorem can be applied for antenna size reduction. It can be implemented by extending ground plane to an isolated RDRA. Resonance in RDRA is created due to formation of standing waves inside the device. Frequency ω_{mnp} is the spectral solution of an antenna, and this can determine the base half-wave Fourier analysis. Principle of orthonormality is used to determine radiation parameters by equating electric time average energy equal to magnetic time average energy by KAM (Kalmogorov–Arnold–Moser).

At any instant of time, n number of modes exist. The particular mode can be excited by increasing weighted amplitude of desired mode. More than one mode can also be excited into RDRA. Blocking modes can take place if E_z and H_z fields of same frequency are available in RDRA at any instant of time. Hence, mode spectrum will result into corresponding resonant frequency generation. Wave propagation can be defined by Helmholtz equation. The Maxwell's equation describes the behavior of electromagnetic fields and forms the basis of all EM classical phenomena. The size of antenna can be reduced to half by image theorem, converting isolated cavity into infinite ground plane. Dielectric resonator antenna is formed with high permittivity substrate. The abrupt change in permittivity due to change in medium forms standing waves. These waves establish resonance as they bounce back and forth in-between two walls due to fields perturbation. Modes are spectral resolution of electromagnetic fields of waves radiated by the RDRA. Modal excitation mainly depends on:

(a) Position of probe;
(b) Magnitude of probe current; and
(c) Phase of input current.

To compute resonant modes, vector principle of orthonormality on half-wave Fourier analysis has been applied, i.e., radiated magnetic energies time averaged are compared with applied electrical energies time averaged in the case of resonator antennas. More number of modes along z-axis in RDRA can be generated either by increasing electrical height "d" of RDRA or by increasing resonant frequency of DRA. Figure 2.1 depicts the prototype RDRA with moat under fabrication.

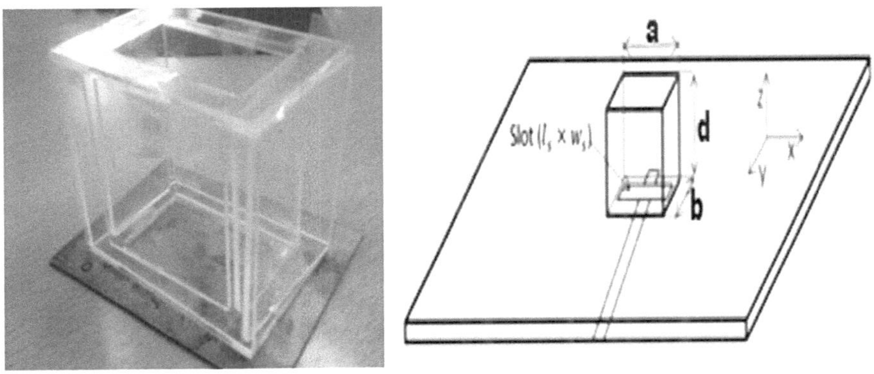

Fig. 2.1 RDRA prototype with neat sketch

Depending on the nature of the surfaces, different linear combinations of the $\pm\gamma$ modes are formed. The propagation constant (γ) itself is taking discrete values. This forces the natural frequencies of the field oscillations to take discrete values (*mnp*) indexed by three positive integers, namely m, n, and p. The solutions of the waveguide problem yield discrete values of γ, i.e., $\gamma(m, n, \omega)$ for a given frequency ω by applying boundary conditions to the electromagnetic fields on the sidewalls. The corresponding field amplitudes are solutions to the 2-D Helmholtz equations corresponding to the transverse Laplacian $\nabla_{\perp}^{2}$. These amplitudes are called "the waveguide modes" and are of the form given below in Sect. 2.2.

2.2 Type of Modes (TE, TM, HEM)

EM waves are of four types given below:
Transverse electric and magnetic (TEM) mode

- Transverse electric (TE) mode
- Transverse magnetic (TM) mode
- Hybrid electric and magnetic (HEM) or HE odd and EH even mode

Modes propagation depends mainly on following configuration:

1. Excitation
2. Dimensions
3. Coupling
4. Medium
5. Point of excitation
6. Input impedance

Cross-polarization solution can be the outcome of modes. They can be merged, separated and mixed depending upon the requirements. Half Fourier analysis can be used to describe modes of propagation and excitation. Even and odd modes can be studied. They can be analyzed with magnetic dipole moments. They help to predict far field radiation patterns. Modulated bandwidth and gain control can be achieved. High gain at higher modes can be used for hardware implementations. Device dimensions can be minimized by proper selection of modes for resonant frequencies. In case of milli metric (mm) wave, device size can be enlarged for easy hardware development or hardware implementation. The solution is based on waveguide method when boundaries have been all six electrical walls. The solution is based on solution of Maxwell's equations and then restricted to given boundary conditions for confined modes of EM waves.

2.3 Solutions of Helmholtz Equation

Helmholtz equation solution with source

$$\nabla \times H = J + \frac{\partial D}{\partial t} \tag{2.1}$$

$$D = \epsilon E$$

$$B = \mu H = \nabla \times A$$

$$H = \frac{1}{\mu}(\nabla \times A)$$

Considering the sources to be natural time harmonic

$$E = E_m e^{j\omega t} \tag{2.2a}$$

$$H = H_m e^{j\omega t} \tag{2.2b}$$

Now,

$$\nabla \times E = -\frac{\partial B}{\partial t} \tag{2.3}$$

or

$$\nabla \times E = -j\omega\mu H = -j\omega(\nabla \times A)$$

$$\nabla \times (E + j\omega A) = 0 \tag{2.4}$$

Using vector identity

$$\nabla \times (-\nabla ø_e) = 0$$

$$E + j\omega A = -\nabla ø_e$$

$$E = -j\omega A - \nabla ø_e$$

Using the vector identity

$$\nabla \times (\nabla \times A) = \nabla(\nabla \times A) - \nabla^2 A \tag{2.5}$$

Now,

$$\nabla \times (\mu H) = \nabla(\nabla \times A) - \nabla^2 A$$

or

$$\mu(\nabla \times H) = \nabla(\nabla \times A) - \nabla^2 A$$

From Maxwell's fourth equation,

$$\nabla \times H = J + j\omega\epsilon E$$

or

$$\nabla \times \frac{B}{\mu} = J + j\omega\epsilon E$$

or

$$\nabla \times B = \mu J + j\omega\epsilon\mu E$$

or

$$\nabla \times (\nabla \times A) = \mu J + j\omega\epsilon\mu E$$

or

$$\nabla(\nabla \times A) - \nabla^2 A = \mu J + j\omega\epsilon\mu E$$

$$\nabla(\nabla \times A) - \nabla^2 A = \mu J + j\omega\epsilon\mu(-j\omega A - \nabla\o_e)$$

or

$$\nabla(\nabla \times A) - \nabla^2 A = \mu J + \omega^2\epsilon\mu A - j\omega\mu\epsilon(\nabla\o_e)$$

or

$$\nabla^2 A + k^2 A = -\mu J + j\omega\mu\epsilon(\nabla\o_e) + \nabla(\nabla \times A)$$

or

$$\nabla^2 A + k^2 A = -\mu J + \nabla(j\omega\mu\epsilon\o_e + (\nabla \times A))$$

where $k^2 = \omega^2\mu\epsilon$.

Using Lorentz condition, i.e.,

$$\nabla \times A = -j\omega\varepsilon\mu\text{ø}_e$$

or

$$\text{ø}_e = -\frac{1}{j\omega\mu\varepsilon}(\nabla \times A)$$

$$\nabla^2 A + k^2 A = -\mu J + \nabla(j\omega\mu\epsilon\text{ø}_e - j\omega\varepsilon\mu\text{ø}_e) \tag{2.6}$$

Hence, $\nabla^2 A + k^2 A = -\mu J$.

2.4 Rectangular Waveguide Analysis

Propagation in waveguide has been taken along z-axis, and all the four sidewalls of waveguide are PEC; the fields computed are as follows:

$$\begin{matrix} H_x & E_x \\ H_y & E_y \\ H_z & E_z \end{matrix}$$

$$E_z(x,y,z) = \sum_{m,n=1}^{\infty} C(m,n)\sin\left(\frac{m\pi x}{a}\right)\sin\left(\frac{n\pi y}{b}\right)\exp(-\gamma_{mn}z); \tag{2.7a}$$

$$E_{z,y} = \frac{\partial E_z}{\mathrm{d}y};$$

$$E_{z,x} = \frac{\partial E_z}{\mathrm{d}x}$$

$$H_z(x,y,z) = \sum_{m,n=1}^{\infty} D(m,n)\cos\left(\frac{m\pi x}{a}\right)\cos\left(\frac{n\pi y}{b}\right)\exp(-\gamma_{mn}z) \tag{2.7b}$$

Form Maxwell's equations

$$\text{Curl } \underline{E} = \nabla \times \underline{E} = -j\omega\mu\underline{H} = -B,t \tag{2.8a}$$

$$\text{Curl } \underline{H} = \nabla \times \underline{H} = -j\omega\epsilon\underline{E} = J + D,t \tag{2.8b}$$

Solution of above equations is based on separation of variables solving LHS of both sides first

$$\nabla \times E = \begin{vmatrix} i & j & k \\ \frac{\partial}{\partial x} & \frac{\partial}{\partial y} & \frac{\partial}{\partial z} \\ E_x & E_y & E_z \end{vmatrix} = i\left(\frac{\partial E_z}{\partial y} - \frac{\partial E_y}{\partial z}\right) - j\left(\frac{\partial E_z}{\partial x} - \frac{\partial E_x}{\partial z}\right) + k\left(\frac{\partial E_y}{\partial x} - \frac{\partial E_x}{\partial y}\right) \tag{2.9a}$$

$$\nabla \times H = \begin{vmatrix} i & j & k \\ \frac{\partial}{\partial x} & \frac{\partial}{\partial y} & \frac{\partial}{\partial z} \\ H_x & H_y & H_z \end{vmatrix} = i\left(\frac{\partial H_z}{\partial y} - \frac{\partial H_y}{\partial z}\right) - j\left(\frac{\partial H_z}{\partial x} - \frac{\partial H_x}{\partial z}\right) + k\left(\frac{\partial H_y}{\partial x} - \frac{\partial H_x}{\partial y}\right) \tag{2.9b}$$

Comparing with RHS in both equations and getting value of H_x, H_y, H_z from (2.9a) and E_x, E_y, E_z from (2.9b) we get

$$H_x = \frac{1}{-j\omega\mu}\left(\frac{\partial E_z}{\partial y} - \frac{\partial E_y}{\partial z}\right); \quad E_x = \frac{1}{j\omega\epsilon}\left(\frac{\partial H_z}{\partial y} - \frac{\partial H_y}{\partial z}\right); \tag{2.10a}$$

$$H_y = \frac{1}{j\omega\mu}\left(\frac{\partial E_z}{\partial x} - \frac{\partial E_x}{\partial z}\right); \qquad E_y = \frac{1}{-j\omega\epsilon}\left(\frac{\partial H_z}{\partial x} - \frac{\partial H_x}{\partial z}\right); \tag{2.10b}$$

$$H_z = \frac{1}{-j\omega\mu}\left(\frac{\partial E_y}{\partial x} - \frac{\partial E_x}{\partial y}\right); \quad E_z = \frac{1}{j\omega\epsilon}\left(\frac{\partial H_y}{\partial x} - \frac{\partial H_x}{\partial y}\right); \tag{2.10c}$$

$$E_{z,y} + \gamma E_y = -j\omega\mu H_x$$

$$\gamma E_x + E_{z,x} = j\omega\mu H_y$$

$$H_{z,y} = \frac{\partial H_z}{\mathrm{d}y}$$

$$H_{z,x} = \frac{\partial H_z}{\mathrm{d}x}$$

Similarly,

$$H_{z,y} + \gamma H_y = j = j\omega\epsilon E_x$$

$$\gamma E_x + H_{z,x} = -j\omega\epsilon E_y$$

These above equations can be placed in matrix form

$$\begin{bmatrix} j\omega\epsilon & -\gamma \\ \gamma & -j\omega\mu \end{bmatrix} \begin{bmatrix} E_x \\ H_y \end{bmatrix} = \begin{bmatrix} H_{z,y} \\ -E_{z,x} \end{bmatrix}$$

and

$$\begin{bmatrix} j\omega\epsilon & x \\ \gamma & -j\omega\mu \end{bmatrix} \begin{bmatrix} E_y \\ H_x \end{bmatrix} = \begin{bmatrix} -H_{z,x} \\ E_{z,x} \end{bmatrix}$$

On manipulating them

$$\begin{bmatrix} E_y \\ H_x \end{bmatrix} = \begin{bmatrix} j\omega\mu & -\gamma \\ -\gamma & j\omega\epsilon \end{bmatrix} \begin{bmatrix} -H_{z,x} \\ E_{z,y} \end{bmatrix}$$

Hence, on simplification

$$E_y = \frac{j\omega\mu}{h^2_{m,n}} H_{z,x} - \frac{\gamma}{h^2_{m,n}} E_{z,y}$$

$$\begin{aligned} E_x &= \sum_{m,n} j\omega\mu \frac{\left[D(m,n)\left(\frac{n\pi}{b}\right)\cos\left(\frac{m\pi x}{a}\right)\sin\left(\frac{n\pi y}{b}\right)e^{(-\gamma_{mn}z)}\right]}{h^2_{m,n}} \\ &\quad + \sum \gamma \frac{\left[C(m,n)\left(\frac{m\pi}{a}\right)\cos\left(\frac{m\pi x}{a}\right)\sin\left(\frac{n\pi y}{b}\right)e^{(-\gamma_{mn}z)}\right]}{h^2_{m,n}}; \\ &= \sum_{m,n} j\omega\mu \frac{\left[D(m,n)\left(\frac{n\pi}{b}\right) + \gamma C(m,n)\left(\frac{m\pi}{a}\right)\right.}{h^2_{m,n}} \cos\left(\frac{m\pi x}{a}\right)\sin\left(\frac{n\pi y}{b}\right)e^{(-\gamma_{mn}z)}\Big) \end{aligned} \tag{2.11}$$

Similarly, we can compute

$$\gamma^2_{m,n} + k^2 = h^2_{m,n}; \quad k^2 = \mu\epsilon\omega^2$$

$$h^2_{m,n} = \left(\frac{m\pi}{a}\right)^2 + \left(\frac{n\pi}{b}\right)^2$$

$$\gamma_{mn} \rightarrow \text{propagation constant}$$

$$\begin{aligned} E_{ix}(m,n) &= \frac{2}{ab}\int_0^a\int_0^b E_{ix}(x,y)\cos\left(\frac{m\pi x}{a}\right)\sin\left(\frac{n\pi y}{b}\right)\mathrm{d}x\mathrm{d}y \\ &= \frac{j\omega\mu\left(\frac{n\pi}{b}\right)D(m,n) + \gamma_{m,n}C(m,n)\left(\frac{m\pi}{a}\right)}{h^2_{m,n}}; \end{aligned} \tag{2.12a}$$

$$\begin{aligned} E_{iy}(m,n) &= \frac{2}{ab}\int_0^a\int_0^b E_{iy}(x,y)\sin\left(\frac{m\pi x}{a}\right)\cos\left(\frac{n\pi y}{b}\right)\mathrm{d}x\mathrm{d}y \\ &= -\left(\frac{j\omega\mu}{h^2_{m,n}}D(m,n)\left(\frac{m\pi}{a}\right) + \frac{\gamma_{m,n}C(m,n)\frac{n\pi}{b}}{h^2_{m,n}}\right); \end{aligned} \tag{2.12b}$$

$$\begin{bmatrix} E_{ix}(m,n) \\ E_{iy}(m,n) \end{bmatrix} = \begin{bmatrix} \frac{m\pi}{a}\frac{\gamma_{m,n}}{h_{m,n}^2} & \frac{n\pi}{b}\frac{j\omega\mu}{h_{m,n}^2} \\ -\frac{\gamma_{m,n}}{h_{m,n}^2}\frac{n\pi}{b} & -\frac{j\omega\mu}{h_{m,n}^2}\frac{m\pi}{a} \end{bmatrix} \begin{bmatrix} C(m,n) \\ D(m,n) \end{bmatrix}$$

Hence, $C(m,n), D(m,n)$ amplitude coefficients can be computed, when the boundary conditions are given as:

$$\left\{ \begin{array}{c} x = 0, a \\ y = 0, b \\ z = 0, d \end{array} \right\}$$

Incident waves at input of waveguide are $E_{ix}(x,y)$, $E_{iy}(x,y)$

$$E_{ix}(x,y) = \sum_{m,n} \frac{j\omega\mu D(m,n)\left(\frac{n\pi}{b}\right) + \gamma_{m,n}C(m,n)\left(\frac{m\pi}{a}\right)}{h_{m,n}^2}\left[\cos\left(\frac{m\pi x}{a}\right)\sin\left(\frac{n\pi y}{b}\right)\right]; \tag{2.13a}$$

$$E_{iy}(x,y) = \sum_{m,n} \frac{j\omega\mu D(m,n)\left(\frac{m\pi}{a}\right) + \gamma_{m,n}C(m,n)\left(\frac{n\pi}{b}\right)}{h_{m,n}^2}\left[\sin\left(\frac{m\pi x}{a}\right)\cos\left(\frac{n\pi y}{b}\right)\right]. \tag{2.13b}$$

2.5 Two-Dimensional Resonator

Solution is obtained by the application of Helmholtz equation.

$$\frac{\partial^2\psi(x,y,t)}{\partial x^2} + \frac{\partial^2\psi(x,y,t)}{\partial y^2} - \frac{1}{c^2}\frac{\partial^2\psi(x,y,t)}{\partial t^2} = 0 \tag{2.14}$$

Applying boundary conditions in rectangular plane,

$$\psi(0,y,t) = \psi(a,y,t) = 0$$

$$\psi(x,0,t) = \psi(x,b,t) = 0$$

Let input excitation be some tension T

$$\sigma \mathrm{d}x\mathrm{d}y\frac{\partial^2\psi}{\partial t^2} = \frac{\partial}{\partial x}\left(T\cdot \mathrm{d}y\frac{\partial\psi}{\partial x}\right)\mathrm{d}x + \frac{\partial}{\partial y}\left(T\cdot \mathrm{d}x\frac{\partial\psi}{\partial y}\right)\mathrm{d}y \tag{2.15}$$

$$\frac{Y''}{Y} = -k_y^2; \qquad \frac{X''}{X} = -k_x^2;$$

$$\frac{\partial^2 \psi}{\partial t^2} - c^2 \nabla^2 \psi = 0 \tag{2.16}$$

Using separation of variables:

$$\psi(x, y, t) = X(x)Y(y)T(t) \tag{2.17}$$

$$-\omega^2 = \frac{T''(t)}{T(t)} = c^2 \left(\frac{X''(x)}{X(x)} + \frac{Y''(y)}{Y(y)} \right)$$

let

$$X(x) = \sin(k_x x)$$

$$Y(y) = \sin(k_y y)$$

$$k_x^2 + k_y^2 = \frac{\omega^2}{c^2}$$

where k_x and k_y can be written as:

$$k_x = \frac{m\pi}{a}; \quad k_y = \frac{n\pi}{b} \tag{2.18}$$

Frequency can be written as: $\omega(mn) = c\pi\sqrt{\left(\frac{m}{a}\right)^2 + \left(\frac{n}{b}\right)^2}$.

2.6 Basic Mathematical Representation of Resonant Modes

$$\nabla^2 A_x + k^2 A_x = 0; \tag{2.19}$$

$k_r \gg 1$ far field pattern
$k_r \ll 1$ near field pattern

where A_z is the magnetic vector potential and k is the wave vector or wave number along z-axis.

$$\begin{aligned} A_z = {} & (C_1 \cos(k_x x) + C_1 \sin(k_x x))(C_3 \cos(k_y y) \\ & + C_4 \sin(k_y y))(C_5 \cos(k_z z) + C_6 \sin(k_z z)); \end{aligned}$$

$$A_z = \frac{\mu}{4\pi}\int J(z')\frac{e^{jkR}}{R}d^3z'; \tag{2.20}$$

$k_c = \frac{2\pi}{\lambda}$, $k_0 = \sqrt{k_x^2 + k_y^2 + k_z^2} = \omega^2\epsilon\mu$, where k is the wave number.

The wave number can be defined as rate of change of phase w.r.t. distance in the direction of propagation. Resonant frequency $\omega = \omega_{mnp}$ in RDRA and its mathematical expression is given below:

$$(f_r)m, n, p = \frac{c}{\sqrt[2\pi]{\varepsilon_r}}\sqrt{\left(\frac{m\pi}{a}\right)^2 + \left(\frac{n\pi}{b}\right)^2 + \left(\frac{p\pi}{d}\right)^2}; \tag{2.21}$$

where m, n, p are the half-wave field variations along x, y, z directions.

$H = \nabla \times A$

$E = -\nabla ø - \frac{\mathrm{d}A}{\mathrm{d}t}$; scalar and magnetic vector potential from Lorentz Gauge conditions.

$S = (E \times H^*)$; S is Poynting vector (energy flow or flux).

$Z = \frac{P_{\text{rad}}}{|I|^2}$ = input Impedance.

$E_x, E_y, E_z, H_x, H_y, H_z$ are electric and magnetic fields

$$\mathcal{L}\oint\left\{\cos\left(\frac{n\pi x}{a}\right), \sin\left(\frac{n\pi x}{a}\right)\right\} \otimes \mathcal{L}\oint\left\{\cos\left(\frac{m\pi y}{b}\right), \sin\left(\frac{m\pi y}{b}\right)\right\}; \tag{2.22}$$

where $\mathcal{L}$ denotes linear components. It turns out that depending on the nature of wall surfaces (PEC or PMC), four possible linear combinations can appear ($\cos \otimes \sin, \sin \otimes \cos, \sin \otimes \sin$, and $\cos \otimes \cos$).

In rectangular DRA, we've got to applying in additional boundary conditions on top and bottom surfaces to be the linear combinations as compared to waveguide.

$$C_1\exp\{(-\gamma(m, n, \omega)z\} + C_2\exp\{+\gamma(m, n, \omega)z\}$$

and these cases are $\gamma(m, n, \omega) = \frac{\pi p}{d}$, when $p = 1, 2, 3\ldots$ and have two possible linear combinations of $\sin\left(\frac{\pi p z}{d}\right)$ and $\cos\left(\frac{\pi p z}{d}\right)$.

Thus, the possible frequencies ω obtained by solving $\gamma(m, n, \omega) = \frac{\pi p}{d}$; then comes out to be:

$$\omega(m, n, p) = \pi\left[\frac{m^2}{a^2} + \frac{n^2}{b^2} + \frac{p^2}{d^2}\right]^{1/2}. \tag{2.23}$$

An equivalent but computationally simpler way to pass on from waveguide physics to resonator physics is to just replace γ by $-\frac{\partial}{\partial z}$ in all the waveguide formulae that express the tangential field components in terms of the longitudinal components. This is done after solving the full 3-D Helmholtz equations using separation of variable in x, y, z.

$$\left(\nabla^2 + \frac{\omega^2}{c^2}\right)\begin{pmatrix} E_z \\ H_z \end{pmatrix} = 0 \tag{2.24}$$

The discrete modes ω_{mnp} enable us to visualize the resonator as collection of L, C oscillators with different L, C values. The outcome of all this analysis enables us to write down the $\underline{E}$ and $\underline{H}$ fields inside the resonator, as superposition of four, three vector-valued basis functions.

$$\begin{aligned} \underline{E}(x, y, z, t) = & \sum_{mnp=1}^{\infty} \mathrm{Re}\left\{C_{mnp} e^{j\omega(mnp)t} \underline{\psi}^{E}_{mnp}(x, y, z)\right\} \\ & + \sum_{mnp=1}^{\infty} \mathrm{Re}\left\{D_{mnp} e^{j\omega(mnp)t} \underline{\bar{\phi}}^{E}_{mnp}(x, y, z)\right\}; \end{aligned} \tag{2.25}$$

and

$$\begin{aligned} \underline{H}(x, y, z, t) = & \sum_{mnp=1}^{\infty} \mathrm{Re}\left\{C(mnp) e^{j\omega(mnp)t} \underline{\psi}^{H}_{mnp}(x, y, z)\right\} \\ & + \sum_{mnp=1}^{\infty} \mathrm{Re}\left\{D(mnp) e^{j\omega(mnp)t} \underline{\bar{\phi}}^{H}_{mnp}(x, y, z)\right\}; \end{aligned} \tag{2.26}$$

We note that there are only two sets $\{C_{mnp}\}$ and $\{D_{mnp}\}$ of linear combination of coefficients from the E_z and H_z *expansions*. The vector-valued complex functions are $\underline{\psi}^{E}_{mnp}$, $\underline{\bar{\phi}}^{E}_{mnp}$ $\underline{\psi}^{H}_{mnp}$ $\underline{\bar{\phi}}^{H}_{mnp}$ ϵR^3 (where R is autocorrelation) and contains components $\{\cos, \sin\} \otimes \{\cos, \sin\} \otimes \{\cos, \sin\}$, functions and hence for $(m'n'p') \neq (mnp)$, each function of the set:

$$\left\{\underline{\psi}^{E}_{mnp}, \underline{\bar{\phi}}^{E}_{mnp}, \underline{\psi}^{H}_{mnp}, \underline{\bar{\phi}}^{H}_{mnp}\right\};$$

is orthogonal to each functions of the set:

$$\left\{\underline{\psi}^{E}_{m'n'p}, \underline{\bar{\phi}}^{E}_{m'n'p}, \underline{\psi}^{H}_{mnp}, \underline{\bar{\phi}}^{H}_{m'n'p}\right\};$$

w.r.t. the measure of dx dy dz over $[0, a] \times [0, b] \times [0, d]$.

The exact form of the function $\underline{\bar{\phi}}^E$, $\underline{\bar{\phi}}^H$, $\underline{\psi}^E$, $\underline{\psi}^H$ depends on the nature of the boundaries. The next problem addressed can be on excitations of RDRA. To calculate the amplitudes' coefficients $\{C_{mnp}\}$ and $\{D_{mnp}\}$, we assume that at $z = 0$, excitations $E_x^{(e)}(x, y, t)$ or $E_y^{(e)}(x, y, t)$ are applied for some time say $t \in [0, T]$ and then removed. Then, the Fourier components in this excitation corresponding to the frequencies $\omega\{(mnp)\}$ are excited and their solutions are the oscillations, while the waveguide for $t > T$. The other Fourier components decay within the resonator.

$\{C_{mnp}, D_{mnp}\}$ are the components of the form:

$$E_x^{(e)}(x, y, t) = \sum_{mnp} \mathrm{Re}\left(C(mnp)e^{j\omega(mnp)t}\psi^E_{mnp\,x}(x, y, 0)\right) + \mathrm{Re}\left(D(mnp)e^{j\omega(mnp)t}\underline{\bar{\phi}}^E_{mnp\,x}(x, y, 0)\right) \tag{2.27}$$

and

$$E_y^{(e)}(x, y, t) = \sum_{mnp} \mathrm{Re}\left(C(mnp)e^{j\omega(mnp)t}\psi^E_{mnp\,y}(x, y, 0)\right) + \mathrm{Re}\left(D(mnp)e^{j\omega(mnp)t}\underline{\bar{\phi}}^E_{mnp\,y}(x, y, 0)\right); \tag{2.28}$$

By using, orthogonality of $\{\psi^E_{mnp\,x}(x, y, 0), \underline{\bar{\phi}}^E_{mnp\,x}(x, y, 0)\}$. For different (m, n), we write p to be fixed and likewise of $\{\psi^E_{mnp\,y}(x, y, 0), \underline{\bar{\phi}}^E_{mnp\,y}(x, y, 0)\}$.

In addition, we need to use KAM type of time averaging to yield field components:

$$C(mnp)\psi^E_{mnp\,x}(x, y, 0) + D(mnp)\underline{\bar{\phi}}^E_{mnp\,x}(x, y, 0) = \lim_{T \to \infty} \frac{1}{2T} \int_{-T}^{T} E_x^{(e)}(x, y, t)e^{-j\omega(mnp)t}\mathrm{d}t. \tag{2.29}$$

and likewise

$$C(mnp)\psi^E_{mnp\,y}(x, y, 0) + D(mnp)\underline{\bar{\phi}}^E_{mnp\,y}(x, y, 0) = \lim_{T \to \infty} \frac{1}{2T} \int_{-T}^{T} E_y^{(e)}(x, y, t)e^{j\omega(mnp)t}\mathrm{d}t. \tag{2.30}$$

2.7 Voltage Source Model

This method of excitation can be compared with connecting a voltage or current source to an LC circuit for sometimes and then switching it off. After a sufficiently long time, all frequencies in the LC circuit decay away except the frequency $\frac{1}{\sqrt{LC}}$. We can more generally compare a resonator with the material medium having non-zero conductivity. Thus, the medium is characterized by the triplet $(\epsilon, \beta, \sigma)$ which corresponds to an array of (*C*, *L*, *R*) = RLC circuits.

Such a resonator is analyzed in the same way replacing ϵ by $\epsilon' = \epsilon - \frac{j\sigma}{\omega}$, i.e., complex permittivity depending on frequency. The resonant frequencies $\omega(mnp)$ now have a non-zero imaginary part corresponding to decay of the field with time. Their frequencies and fields may also be determined by applying separation of variables with boundary conditions to the Helmholtz equations.

$$[\nabla^2 - j\omega\mu(\sigma + j\omega\epsilon)]\begin{bmatrix} E_z \\ H_z \end{bmatrix} = 0; \tag{2.31}$$

To have sustained oscillations in such a resonator, we must never switch off the excitation. We may for example apply a surface current source at $z = \delta_0$, where $0 < \delta_0 \ll d$. Letting $J_{sx}(x, y, \omega)$ and $J_{sy}(x, y, \omega)$ be this surface current excitations in the Fourier domain, the current density corresponds to this is given as:

$$\underline{J}_e(x, y, z, \omega) = (J_{sx}(x, y, \omega)\hat{X} + J_{sy}(x, y, \omega)\hat{Y}\,\delta(z - \delta_0); \tag{2.32}$$

This current is computed by substituting into the Maxwell curl equations

$$\text{Curl}\,\underline{E} = -j\omega\mu\,\underline{H},$$

$$\text{Curl}\,\underline{H} = \underline{J}_e + (\sigma + j\omega\epsilon)\underline{E},\ \text{div}\,\underline{H} = 0$$

The method of solution is to express it as the sum of a general solution to the homogeneous equations, i.e., with $\underline{J}_e = 0$ and a particular solutions for $\underline{J}_e \neq 0$. The general solutions to the homogeneous problem are the same as earlier explained, i.e., containing only the frequencies $\{\omega(mnp)\}$. Particular solution to the $\underline{J}_e \neq 0\,(inhomogenous)$ problem is obtained by taking the curl of the second equation and substituting the fields into third equation to obtain

$$\nabla^2\underline{H} = -\underline{\nabla} \times \underline{J}_e + j\omega\mu(\sigma + j\omega\epsilon)\underline{H}; \tag{2.33}$$

We express a particular solution to this equation by setting

$$\sum_{m,n=1}^{\infty}(\mathcal{L}_1(m,n,\omega)\exp(-\gamma(mn\omega)z)\underline{u}_{mn}(x,y,\omega) \\ + \beta_1(m,n,\omega)\exp(\gamma(mn\omega)z)\underline{v}_{mn}(x,y,\omega) \quad \text{for } d \geq z > \delta;\ (2.14) \tag{2.34a}$$

$$\underline{H}_p(x,y,z,\omega) = \sum_{m,n=1}^{\infty}(\mathcal{L}_2(m,n,\omega)\exp - \gamma(mn\omega)z)\underline{u}_{mn}(x,y,\omega) \\ + \beta_2(m,n,\omega)\exp(\gamma(mn\omega)z)\underline{v}_{mn}(x,y,\omega) \quad \text{for } 0 \leq z < \delta;\ (2.15) \tag{2.34b}$$

Here, ω is a continuous variable, unlike $\{\omega(mnp)\}$, $\underline{u}_{mn}(x,y,\omega)$ and $\underline{v}_{mn}(x,y,\omega)$ are multiples (ω-dependent) of

$$\mathcal{L}\left\{\cos\left(\frac{m\pi x}{a}\right), \sin\left(\frac{n\pi x}{a}\right)\right\} \otimes \mathcal{L}\left\{\cos\left(\frac{m\pi y}{p}\right), \sin\left(\frac{m\pi y}{p}\right)\right\}$$

To meet the boundary conditions on the sidewalls, if $z = 0$, d; if the walls are PEC, $H_{pz} = 0$; when $z = 0$, d. That gives use

$$H_{pz}(x,y,z,\omega) = \sum_{m,n}\mathcal{L}(m,n,\omega)\sin h\{\gamma(m,n,\omega)(z-d)\}\underline{u}_{mnz}(x,y,\omega), \quad \delta < z \leq d; \tag{2.35}$$

and

$$H_{pz}(x,y,z,\omega) = \sum_{m,n}\beta(m,n,\omega)\sin h\{\gamma(m,n,\omega)z)\}\underline{u}_{mnz}(x,y,\omega), \quad 0 \leq z < \delta; \tag{2.36}$$

The fields $H_{p\perp}(x,y,z,\omega)$ are easily determined from these equations in the region $z > \delta$ and $z < \delta$ by differentiating them w.r.t. x, y, z; wherever γ comes in the multiple w.r.t. $\exp(-\gamma z)$, we replace it by $-\frac{\partial}{\partial z}$ etc.

In this way, we get

$$H_{px}(x,y,z,\omega) = \sum_{m,n=1}^{\infty}\mathcal{L}_1(m,n,\omega)\psi_{mnx}(x,y,z,\omega), \quad \text{for } z > \delta; \tag{2.37a}$$

and

$$H_{py}(x,y,z,\omega) = \sum_{m,n=1}^{\infty}\mathcal{L}_2(m,n,\omega)\psi_{mny}(x,y,z,\omega), \quad \text{for } z < \delta; \tag{2.37b}$$

where, $\psi_{mnx}(x, y, z, \omega)$ and $\psi_{mny}(x, y, z, \omega)$ are obtained by differentiating

$$\underline{u}_{mnz}(x, y, \omega) \sin h\{\gamma(m, n, \omega)(z - d)\} \quad \text{w.r.t. } x, y, z.$$

Likewise for $z < \delta$, we have expression of the form

$$H_{px}(x, y, z, \omega) = \sum_{m,n=1}^{\infty} \beta(m, n, \omega) \underline{\bar{\phi}}^{E}_{mnp\,x}(x, y, z, \omega); \tag{2.38a}$$

and

$$H_{py}(x, y, z, \omega) = \sum_{m,n=1}^{\infty} \beta(m, n, \omega) \underline{\bar{\phi}}^{E}_{mnp\,y}(x, y, z, \omega); \tag{2.38b}$$

The coefficients $\mathcal{L}(m, n, \omega)$ and $\beta(m, n, \omega)$ are obtained from the boundary conditions

$$\hat{z} \times \left(H|_{z=\delta+} - H|_{z=\delta-}\right) = J_s|_{z=\delta+}.$$

Hence, current density

$$J = J_{sx}(x, y, \omega)\hat{X} + J_{sy}(x, y, \omega)\hat{Y}. \tag{2.39}$$

2.8 Resonant Modes Generation

The Fig. 2.2 presents how the generated modes look like. This will be able to tell us the number of resonant modes in particular direction. The transverse components of EM waves are expressed as E_x, E_y, H_x, H_y. If propagation of wave is along z-direction, E_z, H_z fields are the longitudinal components. These fields are modal solutions, solved based on Helmholtz equations using standard boundary conditions. The RDRA is basically a boundary value problem. The linear combinations of sine and cosine terms give rise to TE and TM modes. The generation of various kinds of modes in an antenna and propagation is very critical issue; it need through study. Now, rewriting Helmholtz equation for source-free medium (Fig. 2.3)

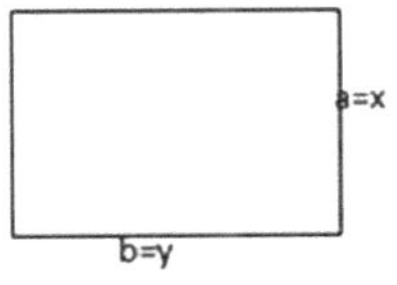

Fig. 2.2 Rectangular resonator

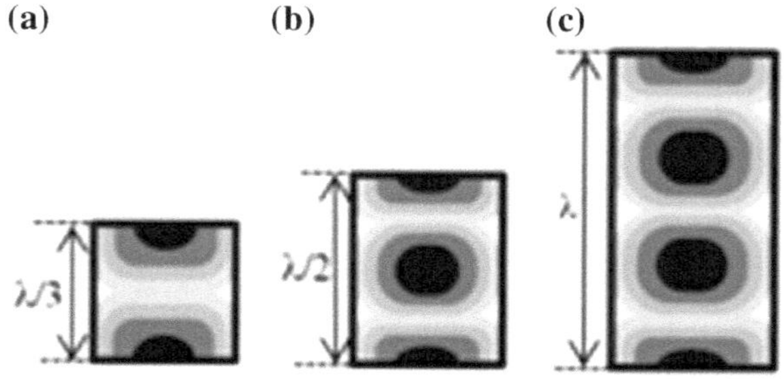

Fig. 2.3 Resonant modes generated in RDRA by HFSS. **a** $TE^x_{\delta 12}$, **b** $TE^x_{\delta 14}$, **c** $TE^x_{\delta 16}$ modes

$$\nabla^2 \Psi + k^2 \Psi = 0,$$

Here, k is the wave number and $k^2 = k_x^2 + k_y^2 + k_z^2$ and, $\Psi = \Psi_x \cdot \Psi_y \cdot \Psi_z$

$$\frac{1}{\Psi_x}\left(\frac{\partial}{\partial x}\right)^2 \Psi = -k_x^2 \tag{2.40a}$$

$$\frac{1}{\Psi_y}\left(\frac{\partial}{\partial y}\right)^2 \Psi = -k_y^2 \tag{2.40b}$$

$$\frac{1}{\Psi_z}\left(\frac{\partial}{\partial z}\right)^2 \Psi = -k_z^2 \tag{2.40c}$$

Solving above function and keeping propagation in +z-direction only, we get

$$\Psi \text{ or } H_z \text{ or } E_z = \left\{(A \sin k_x \cdot x + B \cos k_x \cdot x)\left(C \sin k_y \cdot y + D \cos k_y \cdot y\right)\right\} e^{-jk_z z}$$

From boundary conditions, we get

$$H_z = \sum_{m,n} \left\{ C_{mn} \left(\cos \frac{m\pi x}{a}\right)\left(\cos \frac{n\pi y}{b}\right)\right\} e^{-jk_z z}; \quad C_{mn} \text{ Fourier Coefficients;} \tag{2.41a}$$

$$E_z = \sum_{m,n} \left\{ D_{mn} \left(\sin \frac{m\pi x}{a}\right)\left(\sin \frac{n\pi y}{b}\right)\right\} e^{-jk_z z}; \quad D_{mn} \text{ Fourier Coefficients;} \tag{2.41b}$$

Let $\gamma = -jk_z$ and m, n are integers and a, b are dimensions;

$$\gamma^2 + \omega^2 \mu \epsilon = k_x^2 + k_y^2 = \left(\frac{m\pi}{a}\right)^2 + \left(\frac{n\pi}{b}\right)^2$$

$$k_z^2 = \omega^2 \mu \epsilon - \left(\left(\frac{m\pi}{a}\right)^2 + \left(\frac{n\pi}{b}\right)^2\right)$$

Hence, EM wave will propagate in z-direction if

$$\omega^2\mu\epsilon - \left(\left(\frac{m\pi}{a}\right)^2 + \left(\frac{n\pi}{b}\right)^2\right) > 0$$

This gives cutoff frequency as

$$\omega_c = \frac{1}{\sqrt{\mu\epsilon}}\sqrt{\left(\left(\frac{m\pi}{a}\right)^2 + \left(\frac{n\pi}{b}\right)^2\right)}$$

It means, waveguide will support all waves having ω greater than ω_c to propagate.

Now, rewriting H_z and E_z

$$H_z = \sum_{m,n}\left\{C_{mn}\left(\cos\frac{m\pi x}{a}\right)\left(\cos\frac{n\pi y}{b}\right)\right\}e^{-jk_z z} \tag{2.42}$$

$$E_z = \sum_{m,n}\left\{D_{mn}\left(\sin\frac{m\pi x}{a}\right)\left(\sin\frac{n\pi y}{b}\right)\right\}e^{-jk_z z} \tag{2.43}$$

Here, C_{mn} and D_{mn} are the coefficients of Fourier cosine and sine series.

$$\gamma_{m,n} = \sqrt{h_{m,n}^2 - \omega^2\mu\varepsilon}$$

Hence, C_{mn} and D_{mn} gives us relative amplitudes and phase. Hence, we get solution of possible amplitudes and phase of wave propagating through rectangular waveguide called as modes of propagation.

2.9 MATLAB Simulated Results

Results of resonant frequency obtained on various sizes RDRA's using HFSS have been placed in Table 2.1. The MATLAB programs are being developed for modes graphical view. Resonant modes and resonant frequencies are being obtained based on formulations. The programs and simulated results are given below:

MATLAB program no.1

```
clearall;
clc;
closeall;
c=3*10^8;
m=7;
n=10;
p=6;
E=10;
```

```
a=5*10^-3:.1*10^-3:30*10^-3;
b=10*10^-3;
d=15*10^-3;

for i=1:length(a)
f(i)=c/(2*pi)*sqrt(E)*sqrt((m*pi/a(i))^2+(n*pi/b)^2+(pp*pi/(2*d)^2));
end

a1=15*10^-3;
b1=10*10^-3:.1*10^-3:40*10^-3;
d1=20*10^-3;

for k=1:length(b1)
f1(k)=c/(2*pi)*sqrt(E)*sqrt((m*pi/a1)^2+(n*pi/b1(k))^2+(pp*pi/(2*d1)^2));
end

a2=10*10^-3;
b2=5*10^-3;
d2=10*10^-3:.1*10^-3:50*10^-3;

for t=1:length(d2)
f2(t)=c/(2*pi)*sqrt(E)*sqrt((m*pi/a2)^2+(n*pi/b2)^2+(pp*pi/(2*d2(t))^2));
end

subplot(3,1,1);plot(a,f);title('plot a vs f when a is varying');
subplot(3,1,2);plot(b1,f1);title('plot b vs f when b is varying');
subplot(3,1,3);plot(d2,f2);title('plot d vs f when d is varying');
```

Table 2.1 RDRA HFSS f_r

S. No.	Permittivity	Dimension ($a \times b \times h$) mm	Resonant frequency
1	10.0	14.3 × 25.4 × 26.1	3.5
2	10.0	14 × 8 × 8	5.5
3	10.0	15.24 × 3.1 × 7.62	6.21
4	20.0	10.2 × 10.2 × 7.89	4.635
5	20.0	10.16 × 10.2 × 7.11	4.71
6	35.0	18 × 18 × 6	2.532
7	35.0	18 × 18 × 9	2.45
8	100.0	10 × 10 × 1	7.97

The graph shown in Fig. 2.4 represents inverse relationship between height and resonant frequency as λ-wavelength is inversely proportional to resonant frequency fr. MATLAB simulation shown in Fig. 2.5 represents number of modes generated in x, y, z directions. The mathematical expression on the topic is expressed in Eqs. (2.1)–(2.31).

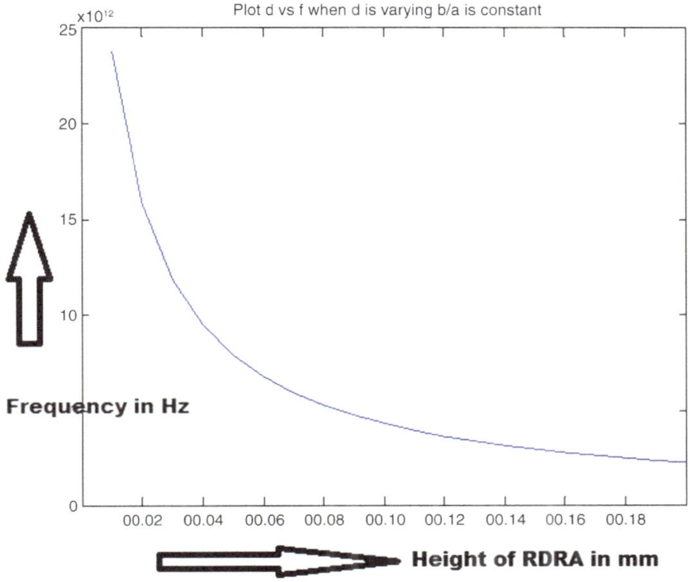

Fig. 2.4 Simulated resonant frequency plot for excited modes

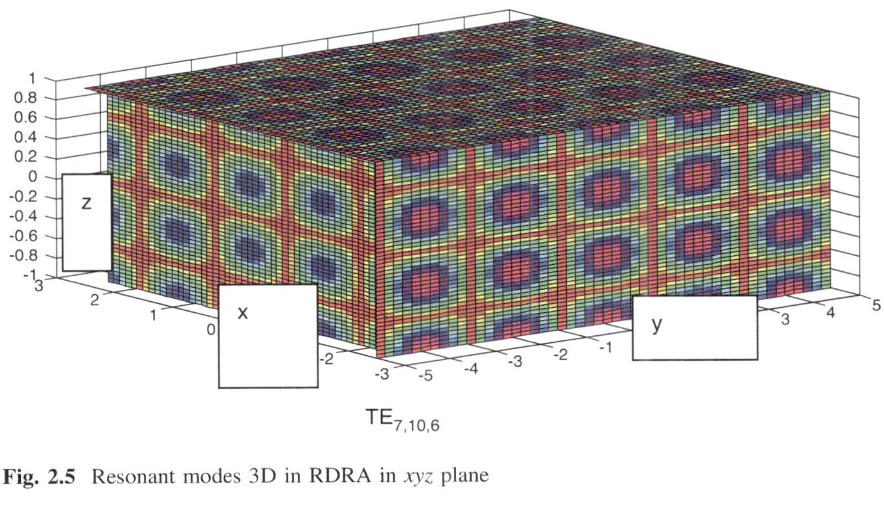

Fig. 2.5 Resonant modes 3D in RDRA in *xyz* plane

MATLAB Program2

```
m=5;
n=4;
p=3;
a=10;
b=5;
c=2;
x=linspace(-5,5,51);
y=linspace(-2.5,2.5,51);
z=linspace(-1,1,51);
[xi,yi,zi] = meshgrid(x,y,z);
Ez= (cos(m*pi*xi/a).*cos(n*pi*yi/b)).*sin(p*pi*zi/c);
Ez= Ez.^2;
Ez= sqrt(Ez);
xslice = -4.5; yslice = -2.5; zslice =1;
slice(xi,yi,zi,Ez,xslice,yslice,zslice)
colormap hsv
```

Reference

1. Okaya A, Barash LF (1962) The dielectric microwave resonator. Proc IRE 50:2081–2092

Chapter 3
Mathematical Analysis of Rectangular DRA

Abstract This chapter, mathematical analysis of electromagnetic fields in rectangular dielectric resonator antenna (RDRA) has been introduced. The investigations are based on the first applying waveguide theory, then converting it to resonator by replacing $-\gamma$ to d/dz. Initially, these fields are exploited using the Maxwell curl equations, then manipulating them to express the transverse components of the fields in terms of partial derivatives of the longitudinal components of the fields with respect to x and y axis (i.e., the transverse coordinates). Waveguide models of four rectangular DRAs with specified boundary conditions with linear permittivity have been realized.

Keywords Electromagnetic fields mathematical modeling · Resonator · Waveguide · Homogeneous medium · Boundary conditions · Surface interface

In this chapter, mathematical analysis of electromagnetic fields in rectangular dielectric resonator antenna (RDRA) has been introduced. The investigations are based on first applying waveguide theory, then converting it to resonator by replacing $-\gamma$ to d/dz. Through out this book, electromagnetic field propagation has been taken along z-axis, i.e., exp($-\gamma z$). Initially, these fields are exploited using the Maxwell curl equations, then manipulating them to express the transverse components of the fields in terms of partial derivatives of the longitudinal components of the fields with respect to x and y axis (i.e., the transverse coordinates). Waveguide models of four different rectangular DRAs with specified boundary conditions with homogeneous material having linear permittivity have been mathematically modeled. The fields are realized to determine TE and TM modes of propagating fields. These have resulted into different sine–cosine combinations. Propagation of these fields have been split as inside the RDRA and outside RDRA. The interfacing surface is having two different dielectrics. The solution is developed as transcendental equation, which purely characterized rectangular DRA frequency and propagating fields in terms of propagation constants and dominant resonant frequency. TE modes generation required H_z as longitudinal fields and E_x, E_y, H_x, and H_y as transverse fields. Excitation is applied along x-axis as partial fields, y-axis will have fixed variation, and z-axis will have desired variation in propagating

R.S. Yaduvanshi and H. Parthasarathy, *Rectangular Dielectric Resonator Antennas*, DOI 10.1007/978-81-322-2500-3_3

fields, for example, TE δ_{13} and TE δ_{43}. Similar cases can be developed for other modes, so as to propagate E_z fields as longitudinal and E_x, E_y, H_x, and H_y as transverse fields. In this case, H_z shall get vanished because of boundary conditions. Resonant modes, i.e., amplitude coefficient of these fields C_{mnp} and D_{mnp} inside the DRA can be determined by comparing magnetic energies equal to electrical energies based on principle orthonormality or law of conservation. The derivation for the quality factor and radiation pattern have been developed for deeper antenna analysis.

3.1 Rectangular DRA with Homogeneous Medium

In Rectangular DRA as shown in Fig. 3.1, top and bottom walls of RDRA are PMC and rest of the other walls are PEC. On magnetic walls (PMC), $n \cdot E = 0$, where E denotes the electric field intensity and n denotes the normal to the surface of the resonator. Similarly, $n \times H = 0$ is not necessarily satisfied at all the surfaces of the DRA by all the modes. Different resonant modes shall have different electromagnetic

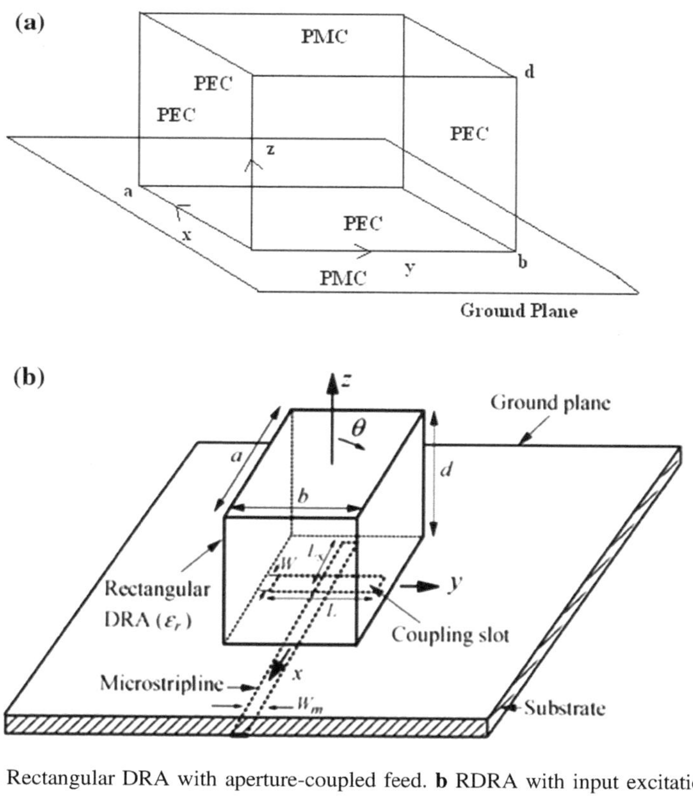

Fig. 3.1 **a** Rectangular DRA with aperture-coupled feed. **b** RDRA with input excitation

field distribution inside the RDRA, and each mode may provide a different resonant frequency and radiation pattern, i.e., eigen vector and eigen frequency. Excitation-based resonant modes can generate desired radiation pattern for different coverage requirements. By making use of this mechanism, internal as well as associated external fields distribution can be obtained.

Rectangular DRA is better choice due to flexible aspect ratio, i.e., *b/a* or *d/a* options can generate different modes. The existence of two independent aspect ratios in a rectangular DRA offers better design flexibility. Assuming the ground plane to be infinitely large, image theory is applied to replace the isolated RDRA by a grounded resonator of half-size. In this RDRA, two of the six surfaces of the resonator are assumed to be perfect magnetic walls, while the remaining four are assumed to be perfect electric walls. Electromagnetic theory is then applied to study its theoretical analysis, and later three more cases have been developed based on the different boundary conditions. For example, the fields undergo one half-wave variation along the dimension '*a*' and remains constant along dimension '*b*'. They undergo less than a half-wave variation along *z*-axis, i.e., variation along DRA height '*d*'. The resonant mode is therefore identified as $TE^z{}_{10\delta}$. The propagation direction has been assumed in *z*-direction. $TE^z{}_{310}$ resonant fields undergo three half-wave variations along length of DRA '*a*' and one half-wave variation along breadth '*b*', and no variation along height '*d*'. To adapt these formulae to an DRA, we note that the propagation constants along *z* can be $\pm\gamma$ with the linear combinations of coefficients chosen, so as to meet the boundary conditions at $z = 0$ and $z = d$, i.e., the top and bottom surfaces of the RDRA, which have been taken as PEC (permanent electrical conducting) walls. On a PEC, the tangential components $(n \times E = 0)$ of the electric field and the normal component $(n \cdot H = 0)$ of the magnetic fields get vanished. While on a PMC wall, by directly, the normal component of the electric field $(n \cdot E = 0)$ and the tangential components $(n \times H = 0)$ of the magnetic field get vanished.

To compute resonant modes, vector principle of orthonormality on half-wave Fourier analysis has been applied, i.e., radiated magnetic energies are compared with applied electrical energies in RDRA. More number of modes along z-axis in RDRA can be generated either by increasing electrical height '*d*' of RDRA or by increasing excitation resonant frequency. Given below are the two rectangular DRAs with different configurations shown in Fig. 3.1.

In Fig. 3.1, PMC and PEC walls' configuration is labeled. The mathematical solution is developed based on this configuration. The boundary conditions of interface walls shall form linear combinations of sine–cosine terms. Accordingly, they will decide whether transverse electric fields or magnetic fields will vanish. Propagation of longitudinal fields shall depend on the direction of excitation. Excitation of resonant modes in rectangular boundaries are easier as compared to cylindrical. Transcendental equation and characteristics equations have been developed for rectangular DRA. This has provided complete solution of resonant frequency and propagation constants.

3.2 Rectangular DRA Mathematical Modeling

In this chapter, four different solutions are presented, each RDRA is associated with different boundaries. The resultant field formed the resonant modes of different kinds.

Figure 3.2 described E and H fields pattern forming resonant modes, i.e., dominant or higher-order excited modes inside the RDRA.

3.2.1 *Model-1*

(a) Here, top and bottom walls are assumed as PMC and rest of the other four walls are PEC as per Fig. 3.1.

Given top and bottom surfaces of RDRA as PMC at $z = 0, d$;

$$\begin{aligned} \therefore n \times H &= 0 \\ n \cdot E &= 0; \\ H_y = H_x &= 0; \\ E_z &= 0; \end{aligned}$$

Rest of the other four walls are PEC.

$$\begin{aligned} n \times E &= 0; \\ n \cdot H &= 0; \\ x &= 0, a; \\ E_y = E_z &= 0, \\ H_x &= 0; \end{aligned}$$

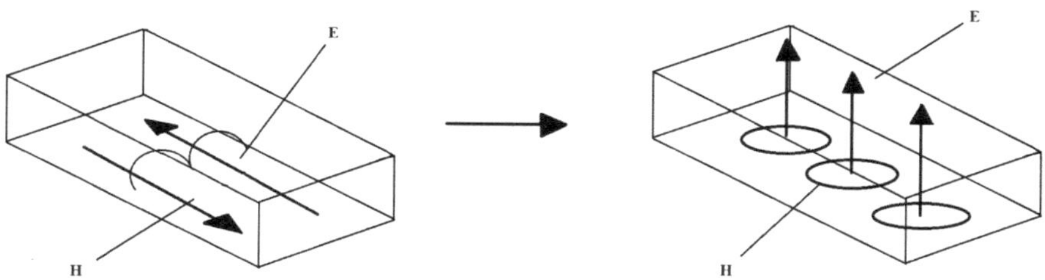

Fig. 3.2 E and H fields pattern inside RDRA

At,

$$y = 0, b;$$
$$E_x = E_z = 0,$$
$$H_y = 0;$$

From separation of variables (Refer Chap. 2),

$$E_x = \frac{1}{j\omega\epsilon\left(1 + \frac{\gamma^2}{k^2}\right)}\left[\frac{\partial H_z}{\partial y} - \frac{1}{j\omega\mu}\frac{\partial^2 E_z}{\partial z \partial x}\right] \tag{3.1}$$

$$E_y = \frac{1}{j\omega\epsilon\left(1 + \frac{\gamma^2}{k^2}\right)}\left[-\frac{1}{j\omega\mu}\frac{\partial^2 E_z}{\partial z \partial y} - \frac{\partial H_z}{\partial x}\right] \tag{3.2}$$

$$H_x = \frac{-1}{j\omega\mu\left(1 + \frac{\gamma^2}{k^2}\right)}\left[\frac{\partial E_z}{\partial y} - \frac{1}{j\omega\epsilon}\frac{\partial^2 H_z}{\partial z \partial x}\right] \tag{3.3}$$

$$H_y = \frac{-1}{j\omega\mu\left(1 + \frac{\gamma^2}{k^2}\right)}\left[\frac{1}{j\omega\epsilon}\frac{\partial^2 H_z}{\partial z \partial y} - \frac{\partial E_z}{\partial x}\right] \tag{3.4}$$

Solution of second-order differential equation is given as,

$$\psi_z = X(x)Y(y)Z(z)$$

where

$$X(x) = A_1 \sin k_x x + A_2 \cos k_x x \tag{3.5}$$

$$Y(y) = A_3 \sin k_y y + A_4 \cos k_y y \tag{3.6}$$

$$Z(z) = A_5 \sin k_z z + A_6 \cos k_z z \tag{3.7}$$

TE mode ($\boldsymbol{E_z = 0}$ and $\boldsymbol{H_z \neq 0}$)

$$\psi_{H_z} = X(x)Y(y)Z(z)$$

substituting $E_z = 0$ in Eq. (3.2) to get, E_y

$$E_y = C'\left[-\frac{\partial H_z}{\partial x}\right];$$

or,

$$E_y = C'X'(x)Y(y)Z(z);$$

Now

$$X'(x) = A_1 \cos k_x x - A_2 \sin k_x x;$$

But at

$$x = 0, a; \; E_y = 0;$$

$$\therefore \quad 0 = A_1 \cos k_x 0 - A_2 \sin k_x 0;$$

or,

$$A_1 = 0 \quad \text{and} \quad k_x = \frac{m\pi}{a};$$

Similarly from Eq. (3.1)

$$E_x = C'\left[\frac{\partial H_z}{\partial y}\right];$$

or,

$$E_x = C'X(x)Y'(y)Z(z).$$

Now

$$Y'(y) = A_3 \cos k_y y - A_4 \sin k_y y;$$

At,

$$y = 0, b; \quad E_x = 0;$$

$$\therefore \quad 0 = A_3 \cos k_y 0 - A_4 \sin k_y 0;$$

or,

$$A_3 = 0 \quad \text{and} \quad k_y = \frac{n\pi}{b};$$

From above equation,

$$H_x = C'\left[-\frac{1}{j\omega\epsilon}\frac{\partial^2 H_z}{\partial z \partial x}\right];$$

or,

$$H_x = C'X'(x)Y(y)Z'(z)$$

Now

$$Z'(z) = A_5 \cos k_z z - A_6 \sin k_z z;$$

At,

$$z = 0, d \quad H_x = 0;$$

$$\therefore \quad A_5 \cos k_z 0 - A_6 \sin k_z 0 = 0;$$

$$A_5 = 0 \quad \text{and} \quad k_z = \frac{p\pi}{d};$$

Hence,

$$H_z = A_2 A_4 A_6 \cos\left(\frac{m\pi}{a}x\right) \cos\left(\frac{n\pi}{b}y\right) \cos\left(\frac{p\pi}{d}z\right) \tag{3.8}$$

Using Eqs. (3.1)–(3.4), and (3.8), we get

$$\begin{aligned}
H_x &= C'' A_2 A_4 A_6 \left(\frac{m\pi}{a}\right)\left(\frac{p\pi}{d}\right) \sin\left(\frac{m\pi}{a}x\right) \cos\left(\frac{n\pi}{b}y\right) \sin\left(\frac{p\pi}{d}z\right);\\
H_y &= C'' A_2 A_4 A_6 \left(\frac{n\pi}{b}\right)\left(\frac{p\pi}{d}\right) \cos\left(\frac{m\pi}{a}x\right) \sin\left(\frac{n\pi}{b}y\right) \sin\left(\frac{p\pi}{d}z\right);\\
E_y &= C'' A_2 A_4 A_6 \left(\frac{m\pi}{a}\right) \sin\left(\frac{m\pi}{a}x\right) \cos\left(\frac{n\pi}{b}y\right) \cos\left(\frac{p\pi}{d}z\right);\\
E_x &= C'' A_2 A_4 A_6 \left(\frac{n\pi}{b}\right) \cos\left(\frac{m\pi}{a}x\right) \sin\left(\frac{n\pi}{b}y\right) \cos\left(\frac{p\pi}{d}z\right);
\end{aligned}$$

Similarly, for TM mode ($H_z = 0$ and $E_z \neq 0$)

$$\psi_{E_z} = X(x)Y(y)Z(z);$$

At,

$$\begin{aligned}
&x = 0, a;\\
&E_z = 0;\\
&A_1 \sin k_x 0 + A_2 \cos k_x 0 = 0;\\
&\therefore \quad A_2 = 0 \quad \text{and} \quad k_x = \frac{m\pi}{a};
\end{aligned}$$

Also, at

$$\begin{aligned}
&y = 0, b;\\
&E_z = 0;\\
&A_3 \sin k_y 0 + A_4 \cos k_y 0 = 0;\\
&A_4 = 0; \quad \text{and} \quad k_y = \frac{n\pi}{b}
\end{aligned}$$

At,

$$\begin{aligned}
&z = 0, d;\\
&E_z = 0;\\
&\therefore \quad A_5 \sin k_z 0 + A_6 \cos k_z 0 = 0;\\
&A_6 = 0 \quad \text{and} \quad k_z = \frac{p\pi}{d};
\end{aligned}$$

Hence,

$$E_z = A_1 A_3 A_5 \sin\left(\frac{m\pi}{a}x\right) \sin\left(\frac{n\pi}{b}y\right) \sin\left(\frac{p\pi}{d}z\right) \tag{3.9}$$

Using Eqs. (1.1)–(1.4), and (1.9), we get

$$\begin{aligned}
E_x &= C'' A_1 A_3 A_5 \left(\frac{m\pi}{a}\right)\left(\frac{p\pi}{d}\right) \cos\left(\frac{m\pi}{a}x\right) \sin\left(\frac{n\pi}{b}y\right) \cos\left(\frac{p\pi}{d}z\right);\\
E_y &= C'' A_1 A_3 A_5 \left(\frac{n\pi}{b}\right)\left(\frac{p\pi}{d}\right) \sin\left(\frac{m\pi}{a}x\right) \cos\left(\frac{n\pi}{b}y\right) \cos\left(\frac{p\pi}{d}z\right);\\
H_y &= C'' A_1 A_3 A_5 \left(\frac{m\pi}{a}\right) \cos\left(\frac{m\pi}{a}x\right) \sin\left(\frac{n\pi}{b}y\right) \sin\left(\frac{p\pi}{d}z\right);\\
H_x &= C'' A_1 A_3 A_5 \left(\frac{n\pi}{b}\right) \sin\left(\frac{m\pi}{a}x\right) \cos\left(\frac{n\pi}{b}y\right) \sin\left(\frac{p\pi}{d}z\right);
\end{aligned}$$

3.2.2 Model-2

(b) Top and bottom walls are PEC and rest of the other walls are PMC:
Assuming the top and bottom surface plane be at $z = 0, d$;

$$n \times E = 0;$$
$$n \cdot H = 0;$$
$$E_y = E_x = 0;$$
$$H_z = 0;$$

Rest of the other walls are PMC

$$n \times H = 0;$$
$$n \cdot E = 0;$$

At,

$$x = 0, a;$$
$$H_y = H_z = 0;$$
$$E_x = 0;$$

At,

$$y = 0, b;$$
$$H_x = H_z = 0;$$
$$E_y = 0;$$

We also know that

$$E_x = \frac{1}{j\omega\epsilon\left(1 + \frac{\gamma^2}{k^2}\right)}\left[\frac{\partial H_z}{\partial y} - \frac{1}{j\omega\mu}\frac{\partial^2 E_z}{\partial z \partial x}\right]; \tag{3.10}$$

$$E_y = \frac{1}{j\omega\epsilon\left(1 + \frac{\gamma^2}{k^2}\right)}\left[-\frac{1}{j\omega\mu}\frac{\partial^2 E_z}{\partial z \partial y} - \frac{\partial H_z}{\partial x}\right]; \tag{3.11}$$

$$H_x = \frac{-1}{j\omega\mu\left(1 + \frac{\gamma^2}{k^2}\right)}\left[\frac{\partial E_z}{\partial y} - \frac{1}{j\omega\epsilon}\frac{\partial^2 H_z}{\partial z \partial x}\right]; \tag{3.12}$$

$$H_y = \frac{-1}{j\omega\mu\left(1 + \frac{\gamma^2}{k^2}\right)}\left[\frac{1}{j\omega\epsilon}\frac{\partial^2 H_z}{\partial z \partial y} - \frac{\partial E_z}{\partial x}\right]; \tag{3.13}$$

Now, the solution of second-order differential equation is given as

$$\psi_z = X(x)Y(y)Z(z);$$

where

$$X(x) = A_1 \sin k_x x + A_2 \cos k_x x; \tag{3.14}$$

$$Y(y) = A_3 \sin k_y y + A_4 \cos k_y y; \tag{3.15}$$

$$Z(z) = A_5 \sin k_z z + A_6 \cos k_z z; \tag{3.16}$$

TE mode ($E_z = 0$ and $H_z \neq 0$)

$$\psi_{H_z} = X(x)Y(y)Z(z);$$

At,

$$\begin{aligned} &x = 0, a; \\ &H_z = 0; \\ &A_1 \sin k_x 0 + A_2 \cos k_x 0 = 0; \\ &A_2 = 0; \end{aligned}$$

and

$$k_x = \frac{m\pi}{a};$$

Also, at

$$\begin{aligned} &y = 0, b; \\ &H_z = 0; \\ &A_3 \sin k_y 0 + A_4 \cos k_y 0 = 0; \\ &A_4 = 0; \end{aligned}$$

and

$$k_y = \frac{n\pi}{b};$$

At,

$$
\begin{aligned}
&z = 0, d;\\
&H_z = 0;\\
&A_5 \sin k_z 0 + A_6 \cos k_z 0 = 0;\\
&A_6 = 0;
\end{aligned}
$$

and

$$k_z = \frac{p\pi}{d};$$

Hence,

$$H_z = A_1 A_3 A_5 \sin\left(\frac{m\pi}{a}x\right) \sin\left(\frac{n\pi}{b}y\right) \sin\left(\frac{p\pi}{d}z\right); \tag{3.17}$$

Using Eqs. (1.1)–(1.4), and (1.8), we get

$$
\begin{aligned}
H_x &= C'' A_1 A_3 A_5 \left(\frac{m\pi}{a}\right)\left(\frac{p\pi}{d}\right) \cos\left(\frac{m\pi}{a}x\right) \sin\left(\frac{n\pi}{b}y\right) \cos\left(\frac{p\pi}{d}z\right),\\
H_y &= C'' A_1 A_3 A_5 \left(\frac{n\pi}{b}\right)\left(\frac{p\pi}{d}\right) \sin\left(\frac{m\pi}{a}x\right) \cos\left(\frac{n\pi}{b}y\right) \cos\left(\frac{p\pi}{d}z\right),\\
E_y &= C'' A_1 A_3 A_5 \left(\frac{m\pi}{a}\right) \cos\left(\frac{m\pi}{a}x\right) \sin\left(\frac{n\pi}{b}y\right) \sin\left(\frac{p\pi}{d}z\right);\\
E_x &= C'' A_1 A_3 A_5 \left(\frac{n\pi}{b}\right) \sin\left(\frac{m\pi}{a}x\right) \cos\left(\frac{n\pi}{b}y\right) \sin\left(\frac{p\pi}{d}z\right);
\end{aligned}
$$

TM mode ($H_z = 0$ and $E_z \neq 0$)

$$\psi_{E_z} = X(x)Y(y)Z(z);$$

From Eq. (3.2) after substituting $H_z = 0$, we get

$$E_y = C'\left[-\frac{1}{j\omega\mu}\frac{\partial^2 E_z}{\partial z \partial y}\right];$$

$$E_y = C' X(x) Y'(y) Z'(z);$$

Now

$$Y'(y) = A_3 \cos k_y y - A_4 \sin k_y y;$$

At,

$$y = 0, b; \quad E_y = 0;$$
$$0 = A_3 \cos k_y 0 - A_4 \sin k_y 0;$$
$$A_3 = 0 \quad \text{and} \quad k_y = \frac{n\pi}{b};$$

Similarly, from the above equations,

$$E_x = C'\left[-\frac{1}{j\omega\mu}\frac{\partial^2 E_z}{\partial z \partial x}\right];$$
$$E_x = C'X'(x)Y(y)Z'(z);$$
$$X'(x) = A_1 \cos k_x x - A_2 \sin k_x x;$$
$$x = 0, a;$$
$$E_x = 0;$$
$$0 = A_1 \cos k_x 0 - A_2 \sin k_x 0;$$
$$A_1 = 0;$$

and

$$k_x = \frac{m\pi}{a};$$

Also, from above equations,

$$Z'(z) = A_5 \cos k_z z - A_6 \sin k_z z;$$

At,

$$z = 0, d; \quad E_x = 0;$$
$$\therefore \quad 0 = A_5 \cos k_z 0 - A_6 \sin k_z 0;$$

or,

$$A_5 = 0 \quad \text{and} \quad k_z = \frac{p\pi}{d};$$

Hence,

$$E_z = A_2 A_4 A_6 \cos\left(\frac{m\pi}{a}x\right) \cos\left(\frac{n\pi}{b}y\right) \cos\left(\frac{p\pi}{d}z\right); \tag{3.18}$$

Using Eqs. (3.11–3.14) and (3.18), we get

$$E_x = A_2A_4A_6\left(\frac{m\pi}{a}\right)\left(\frac{p\pi}{d}\right)\sin\left(\frac{m\pi}{a}x\right)\cos\left(\frac{n\pi}{b}y\right)\sin\left(\frac{p\pi}{d}z\right);$$
$$E_y = A_2A_4A_6\left(\frac{n\pi}{b}\right)\left(\frac{p\pi}{d}\right)\cos\left(\frac{m\pi}{a}x\right)\sin\left(\frac{n\pi}{b}y\right)\sin\left(\frac{p\pi}{d}z\right);$$
$$H_x = A_2A_4A_6\left(\frac{n\pi}{b}\right)\cos\left(\frac{m\pi}{a}x\right)\sin\left(\frac{n\pi}{b}y\right)\cos\left(\frac{p\pi}{d}z\right);$$
$$H_y = A_2A_4A_6\left(\frac{m\pi}{a}\right)\sin\left(\frac{m\pi}{a}x\right)\cos\left(\frac{n\pi}{b}y\right)\cos\left(\frac{p\pi}{d}z\right);$$

3.2.3 Model-3

(c) Solution of RDRA, when all six walls are PEC (perfect electrical walls):
Using Maxwell equations:

$$\nabla \times E = -j\omega B = -j\omega\mu H;$$
$$\nabla \times H = j\omega\epsilon E;$$
$$\nabla \times E = -j\omega\mu H;$$
$$\begin{vmatrix} \hat{x} & \hat{y} & \hat{z} \\ \frac{\partial}{\partial x} & \frac{\partial}{\partial y} & \frac{\partial}{\partial z} \\ E_x & E_y & E_z \end{vmatrix} = -j\omega\mu H;$$
$$\hat{x}\left(\frac{\partial E_z}{\partial y} - \frac{\partial E_y}{\partial z}\right) + \hat{y}\left(\frac{\partial E_x}{\partial z} - \frac{\partial E_z}{\partial x}\right) + \hat{z}\left(\frac{\partial E_y}{\partial x} - \frac{\partial E_x}{\partial y}\right) = -j\omega\mu H;$$

On comparing (x, y, z) components both the sides

$$\frac{\partial E_z}{\partial y} - \frac{\partial E_y}{\partial z} = -j\omega\mu H_x; \quad (3.19)$$

$$\frac{\partial E_x}{\partial z} - \frac{\partial E_z}{\partial x} = -j\omega\mu H_y; \quad (3.20)$$

$$\frac{\partial E_y}{\partial x} - \frac{\partial E_x}{\partial y} = -j\omega\mu H_z; \quad (3.21)$$

Similarly, using $\nabla \times H = j\omega\epsilon E$; We get

$$\frac{\partial H_z}{\partial y} - \frac{\partial H_y}{\partial z} = j\omega\epsilon E_x; \quad (3.22)$$

$$\frac{\partial H_x}{\partial z} - \frac{\partial H_z}{\partial x} = j\omega\epsilon E_y; \tag{3.23}$$

$$\frac{\partial H_y}{\partial x} - \frac{\partial H_x}{\partial y} = j\omega\epsilon E_z; \tag{3.24}$$

Comparing above equations,

$$E_x = \frac{1}{j\omega\epsilon}\left[\frac{\partial H_z}{\partial y} + \frac{1}{j\omega\mu}\left(\frac{\partial^2 E_x}{\partial z^2} - \frac{\partial^2 E_z}{\partial x \partial z}\right)\right]; \tag{3.25}$$

$$E_x + \frac{1}{k^2}\frac{\partial^2 E_x}{\partial z^2} = \frac{1}{j\omega\epsilon}\left[\frac{\partial H_z}{\partial y} - \frac{1}{j\omega\mu}\frac{\partial^2 E_z}{\partial x \partial z}\right]; \tag{3.26}$$

$$E_x\left(1 + \frac{\gamma^2}{k^2}\right) = \frac{1}{j\omega\epsilon}\left[\frac{\partial H_z}{\partial y} - \frac{\gamma}{j\omega\mu}\frac{\partial E_z}{\partial x}\right];$$

E_y, H_x, and H_y are expressed in E_z and H_z fields:

$$E_y\left(1 + \frac{\gamma^2}{k^2}\right) = \frac{1}{j\omega\epsilon}\left[\frac{-\gamma}{j\omega\mu}\frac{\partial E_z}{\partial y} - \frac{\partial H_z}{\partial x}\right];$$

$$H_x\left(1 + \frac{\gamma^2}{k^2}\right) = -\frac{1}{j\omega\mu}\left[\frac{\partial E_z}{\partial y} - \frac{\gamma}{j\omega\epsilon}\frac{\partial H_z}{\partial x}\right];$$

$$H_y\left(1 + \frac{\gamma^2}{k^2}\right) = -\frac{1}{j\omega\mu}\left[\frac{\gamma}{j\omega\mu}\frac{\partial H_z}{\partial y} - \frac{\partial E_z}{\partial x}\right]$$

Separation of variables with given boundary conditions, solution is obtained.

$$\begin{aligned}\psi &= X(x)Y(y)Z(z);\\ &= (A_1 \sin k_x x + A_2 \cos k_x x)\big(A_3 \sin k_y y + A_y \sin k_y y\big)(A_5 \sin k_z z + A_6 \cos k_z z);\end{aligned}$$

TM mode of propagation, $H_z = 0$;

Boundary conditions

Electrical walls $\rightarrow E_{\tan} = 0 = n \times E;$

$\rightarrow H_n = 0 = n \cdot H;$

At, $x = 0$;

$$E_z = X(x) = A_z \cos(0); \quad \text{so } A_2 \text{ must be zero.}$$
$$y = 0, Y(y) = A_y \cos(0); \quad A_4 \text{ must be zero.}$$

For standing wave in direction of z,
Therefore,

$$\frac{\partial}{\partial z} Z(z) = 0;$$

$$A_5 \cos(k_z z) - A_6 \sin(k_z z) = 0;$$

Therefore, at

$$z = 0, d;$$
$$A_5 \text{ must be zero;}$$

Hence, we are left with

$$E_z = A_1, A_3, A_5, \sin k_x x \sin k_y y \cos k_z z;$$

Next, boundary conditions are
At,

$$x = a,$$
$$X(x) = A_1 \sin k_x a = 0;$$

$$k_x = \frac{m\pi}{a};$$

At,

$$Y = b; \quad Y(y) = A_2 \sin k_y d = 0;$$

$$k_y = \frac{n\pi}{b};$$

At,

$$z = 0\, z(z)\, A_4 \sin k_z d = 0;$$

$$k_z = \frac{p\pi}{d};$$

As, we know that

$$k_0^2 = k_x^2 + k_y^2 + k_z^2;$$

$$E_x = \frac{1}{j\omega\epsilon\left(1 + \frac{y^2}{k^2}\right)}\left(\frac{\partial H_z}{\partial y} - \frac{1}{j\omega\mu}\frac{\partial^2 E_z}{\partial x \partial z}\right);$$

$$E_z = 0;$$

$$E_x = \frac{1}{k^2\left(1 + \frac{y^2}{k^2}\right)}\left(A_1 A_3 A_5 k_x k_z \cos k_x x \sin k_y y \sin k_z z\right);$$

$$E_y = \frac{1}{j\omega\epsilon\left(1 + \frac{y^2}{k^2}\right)}\left(-\frac{\partial H_z}{\partial x} - \frac{1}{j\omega\mu}\frac{\partial^2 E_z}{\partial y \partial z}\right);$$

$$E_y = \frac{-A_1 A_3 A_5 k_y}{k^2\left(1 + \frac{y^2}{k^2}\right)} k_z\left(\sin k_x x \cos k_y y \sin k_z z\right);$$

$$\begin{aligned} H_x &= \frac{-1}{j\omega\mu\left(1 + \frac{y^2}{k^2}\right)}\left(\frac{\partial E_z}{\partial y} - \frac{1}{j\omega\epsilon}\frac{\partial^2 H_z}{\partial x \partial z}\right); \\ &= \frac{-k_y A_1 A_3 A_5}{\omega^2\mu\epsilon\left(1 + \frac{y^2}{k^2}\right)}\left(\sin k_x x \cos k_y y \cos k_z z\right); \end{aligned}$$

$$\begin{aligned} H_y &= \frac{-1}{j\omega\mu\left(1 + \frac{y^2}{k^2}\right)}\left(\frac{1}{j\omega\epsilon}\frac{\partial^2 H_z}{\partial y \partial z} - \frac{\partial E_z}{\partial x}\right); \\ &= \frac{-k_z k_x A_1 A_3 A_5}{\omega^2\mu\epsilon\left(1 + \frac{y^2}{k^2}\right)}\left(\cos k_x x \sin k_y y \cos k_z z\right); \\ H_y &= \frac{k_x A_1 A_3 A_5}{j\omega\mu\left(1 + \frac{y^2}{k^2}\right)}\left(\cos k_x x \sin k_y y \cos k_z z\right); \end{aligned}$$

For, TE mode

$$\psi = A_1\left(\sin k_x x + A_2 \cos kx\right)\left(A_3 \sin k_y y + A_y \sin ky\right)\left(A_5 \sin k_z z + A_6 \cos k_z z\right);$$

For PEC walls, electric field components are assumed to be varying with H_z *in* direction of (x, y, z)

$$
\begin{aligned}
E_x &= C' \frac{\partial}{\partial y} H_z \\
&= C' X(x) Y'(y) Z(z); \\
y &= 0, b; \\
Y'(y) &= A_3 \cos k_y y - A_4 \sin k_y y = 0; \\
A_3 &= 0; \\
k_y &= \frac{n\pi}{b};
\end{aligned}
$$

Similarly, $E_y = C'' \frac{\partial}{\partial x} H_z$;

$$
\begin{aligned}
A_1 &= 0; \\
k_x &= \frac{m\pi}{a}; \\
Z(z) &= A_5 \sin k_z z + A_6 \cos k_z z;
\end{aligned}
$$

At,

$$
\begin{aligned}
z &= 0, d; \\
A_6 &= 0; \\
k_z &= \frac{p\pi}{d}; \\
H_z &= A_2 A_4 A_5 \cos k_x x \cos k_y y \sin k_z z;
\end{aligned}
$$

Therefore,

$$
\begin{aligned}
E_x &= \frac{1}{j\epsilon\omega\left(1 + \frac{y^2}{k^2}\right)} (A_2 A_4 A_5 \cos k_x x \sin ky \sin kz); \\
E_y &= \frac{-A_2 A_4 A_5 k_x}{j\epsilon\omega\left(1 + \frac{y^2}{k^2}\right)} \left(\sin k_x x \cos k_y y \sin k_z z\right); \\
H_x &= \frac{k_x k_z A_1 A_3 A_5}{k^2 \left(1 + \frac{y^2}{k^2}\right)} \left(\sin k_x x \cos k_y y \cos k_z z\right); \\
H_y &= \frac{-k_z k_y A_1 A_3 A_5}{\omega^2 \mu\epsilon\left(1 + \frac{y^2}{k^2}\right)} \left(\cos k_x x \sin k_y y \cos k_z z\right);
\end{aligned}
$$

3.2.4 Model-4

(d) When all the six walls of RDRA are assumed to be PMC (permanent magnetic walls),

$\psi_z = X(x)Y(y)Z(z)$ where ψ_z is wave function in x, y, and z direction as space.

$$\text{Or} = (A_1 \sin k_x x + A_2 \cos k_x x)(A_3 \sin k_y y + A_4 \cos k_y y)(A_5 \sin k_z z + A_6 \cos k_z z) \tag{3.27}$$

where $A1$–$A6$ are constants and $(A_1 \sin k_x x + A_2 \cos k_x x)$ is solution of second-order differential equation in x direction, i.e., $X(x)$.

When all six walls are PMC, the rectangular DRA solution is

$$H_{\text{tan}} = n \times H = 0;$$
$$H_{\text{nor}} = n \cdot E = 0;$$

Applying boundaries,
At,

$$x = 0, a \Rightarrow H_y \text{ and } H_z = 0; E_x = 0;$$

At,

$$y = 0, b \Rightarrow H_x \text{ and } H_z = 0; E_y = 0;$$

At,

$$z = 0, d \Rightarrow H_x \text{ and } H_y = 0; E_z = 0;$$

TE mode of propagation $(\boldsymbol{E_z} = 0; \boldsymbol{H_z} \neq 0)$
Using boundary conditions
At,

$$x = 0, a; \quad H_z = 0 \Rightarrow A_2 = 0 \quad \text{and} \quad k_x = \frac{m\pi}{a};$$

At,

$$y = 0, b; \quad H_z = 0 \Rightarrow A_4 = 0 \quad \text{and} \quad k_y = \frac{n\pi}{b};$$

Now,

$$H_x = C'' \frac{\partial^2 H_z}{\partial x \partial z} = C'' X'(x) Y(y) z'(z)$$

$$z'(z) = A_5 \cos k_z z - A_6 \sin k_z z$$

At,

$$z = 0, d \Rightarrow d \Rightarrow H_x = 0;$$
$$\Rightarrow H_x = 0 \Rightarrow A_5 = 0; k_z = \frac{p\pi}{d};$$

Hence,

$$H_z = A_1 A_3 A_6 \sin\left(\frac{m\pi x}{a}\right) \sin\left(\frac{n\pi y}{b}\right) \cos\left(\frac{p\pi z}{d}\right) \tag{3.28}$$

TM mode of propagation $(\boldsymbol{E}_z \neq 0, \boldsymbol{H}_z = 0)$
We again look for the conditions, when H_z = 0, i.e., to get the value of E_z

$$\begin{aligned}
H_z &= \frac{C' \partial E_z}{\partial y} \\
&= C' X(x) Y(y') Z(z); \\
Y'(y) &= A_3 \cos k_y y - A_4 \sin k_y y; \\
H_x &= 0 \text{ at } Y = 0, b; \\
&\Rightarrow A_3 = 0 \text{ at } k_y = \frac{n\pi}{b};
\end{aligned} \tag{3.29}$$

Similarly,

$$H_y = C'' \frac{\partial E_z}{\partial_x};$$

$$\begin{aligned}
&C'' X'(x) Y(y') Z(z); \\
&X'(x) = A_1 \cos k_x x - A_2 \sin k_x x; \\
&\Rightarrow H_y = 0 \text{ at } x = 0, a, \\
&\Rightarrow A_1 = 0, \\
&k_x = \frac{m\pi}{a};
\end{aligned}$$

At,

$$z = 0,\ d \Rightarrow E_z = 0;$$
$$\Rightarrow A_5 = 0 \text{ and } k_z = \frac{p\pi}{d};$$

At,

$$z = 0,\ d \Rightarrow E_z = 0;$$
$$\Rightarrow A_5 = 0 \text{ and } k_z = \frac{p\pi}{d};$$

Hence,

$$E_z = A_2 A_4 A_5 \cos\left(\frac{m\pi x}{a}\right) \cos\left(\frac{n\pi y}{b}\right) \sin\left(\frac{p\pi z}{d}\right). \tag{3.30}$$

3.2.5 Basic Theory

Depending on the nature of the surfaces, different linear combinations of the $\pm\gamma$ modes are formed. The propagation constant (γ) itself is taking discrete values. This forces the natural frequencies of the field oscillations to take discrete values (mnp), indexed by three positive integers m, n, and p. The solutions of the waveguide problem yield discrete values of γ, i.e., $\gamma(m, n, \omega)$ for a given frequency ω by applying boundary conditions to the electromagnetic fields on the side walls. The corresponding field amplitudes are solutions to the 2-D Helmholtz equations corresponding to the transverse Laplacian $\nabla_\perp^2$. These amplitudes are called "the waveguide modes" and are of the form

$$\mathcal{L} \oint \left\{\cos\left(\frac{n\pi x}{a}\right), \sin\left\{\frac{n\pi x}{a}\right\}\right\} \otimes \mathcal{L} \oint \left\{\cos\left(\frac{m\pi y}{b}\right), \sin\left\{\frac{m\pi y}{b}\right\}\right\}$$

where $\mathcal{L}$ denotes linear components. It turns out that, depending on the nature of wall surfaces (PEC or PMC), four possible linear combinations can appear $(\cos \otimes \sin, \sin \otimes \cos, \sin \otimes \sin,\ \text{and}\ \cos \otimes \cos)$.

In rectangular DRA, we have got to applying in additional boundary conditions on top and bottom surfaces to be the linear combinations as compared to the waveguide.

$$C_1 \exp\{(-\gamma(m, n, \omega)z\} + C_2 \exp\{+\gamma(m, n, \omega)z\}$$

and these cases are $\gamma(m, n, \omega) = \frac{\pi p}{d}$, when $p = 1, 2, 3\ldots$ and have two possible linear combinations of $\sin\left(\frac{\pi p z}{d}\right)$ and $\cos\left(\frac{\pi p z}{d}\right)$.

Thus, the possible frequencies ω obtained by solving $\gamma(m, n, \omega) = \frac{\pi p}{d}$ and then comes out to be

$$\omega(m, n, p) = \pi \left[\frac{m^2}{a^2} + \frac{n^2}{b^2} + \frac{p^2}{d^2} \right]^{1/2}$$

An equivalent but computationally simpler way to pass on from waveguide physics to resonator physics is to just replace γ by $-\frac{\partial}{\partial z}$ in all the waveguide formulae that express the tangential field components in terms of the longitudinal components. This is done after solving the full 3-D Helmholtz equations using separation of variable in x, y, and z.

$$\left(\nabla^2 + \frac{\omega^2}{c^2} \right) \begin{pmatrix} E_z \\ H_z \end{pmatrix} = 0$$

The discrete modes $\omega(mnp)$ enable us to visualize the resonator as collection of L, C oscillators with different L, C values. The outcome of all this analysis enables us to write down the $\underline{E}$ and $\underline{H}$ fields inside the resonator, as superposition of four or three vector-valued basis functions.

$$\begin{aligned} \underline{E}(x, y, z, t) = & \sum_{m,n,p=1}^{\infty} \mathrm{Re}\left\{ C(mnp) e^{j\omega(mnp)t} \underline{\psi}_{mnp}^{E}(x, y, z) \right\} \\ & + \sum_{m,n,p=1}^{\infty} \mathrm{Re}\left\{ D(mnp) e^{j\omega(mnp)t} \underline{\bar{\phi}}_{mnp}^{E}(x, y, z) \right\} \end{aligned} \tag{3.31}$$

and

$$\begin{aligned} \underline{H}(x, y, z, t) = & \sum_{m,n,p=1}^{\infty} \mathrm{Re}\left\{ C(mnp) e^{j\omega(mnp)t} \underline{\psi}_{mnp}^{H}(x, y, z) \right\} \\ & + \sum_{m,n,p=1}^{\infty} \mathrm{Re}\left\{ D(mnp) e^{j\omega(mnp)t} \underline{\bar{\phi}}_{mnp}^{H}(x, y, z) \right\} \end{aligned} \tag{3.32}$$

We note that there are only two sets of amplitude coefficients $\{C(mnp)\}$ and $\{D(mnp)\}$ of linear combination of coefficients using from the E_z and H_z expansions. The vector-valued complex functions are $\underline{\psi}_{mnp}^{E}, \underline{\bar{\phi}}_{mnp}^{E}, \underline{\psi}_{mnp}^{H}, \underline{\bar{\phi}}_{mnp}^{H} \epsilon R^3$ (where R is autocorrelation) and contains components $\{\cos, \sin\} \otimes \{\cos, \sin\} \otimes \{\cos, \sin\}$, functions and hence for $(m'n'p') \neq (mnp)$, each function of the set

$$\left\{ \underline{\psi}_{mnp}^{E}, \underline{\bar{\phi}}_{mnp}^{E}, \underline{\psi}_{mnp}^{H}, \underline{\bar{\phi}}_{mnp}^{H} \right\};$$

is orthogonal to each functions of the set

$$\left\{ \underline{\psi}^{E}_{m'n'p'}, \bar{\underline{\phi}}^{E}_{m'n'p'}, \underline{\psi}^{H}_{mnp}, \bar{\underline{\phi}}^{H}_{m'n'p'} \right\};$$

w.r.t. The measure of dx dy dz over $[0, a] \times [0, b] \times [0, d]$;

The exact form of the function $\bar{\underline{\phi}}^{E}, \bar{\underline{\phi}}^{H}, \underline{\psi}^{E}, \underline{\psi}^{H}$ depends on the nature of the boundaries. The next problem addressed can be on excitations of RDRA. To calculate the amplitude coefficients $\{C(mnp)\}$ and $\{D(mnp)\}$, we assume that at $z = 0$, an excitation $E_x^{(e)}(x, y, t)$ or $E_y^{(e)}(x, y, t)$ is applied for some time say $t \in [0, T]$ and then removed. Then, the Fourier components in this excitation corresponding to the frequencies $\{\omega(mnp)\}$ are excited and their solutions are the oscillations, while the waveguide for $t > T$. The other Fourier components decay within the resonator.

$\{C(mnp),\ D(mnp)\}$ are components of the form,

$$\begin{aligned} E_x^{(e)}(x, y, t) = \sum_{m,n,p} & \operatorname{Re}(C(mnp))e^{j\omega(mnp)t}\psi^{E}_{mnp\,x}(x, y, 0) \\ & + \operatorname{Re}\left\{ D(mnp)e^{j\omega(mnp)t}\bar{\underline{\phi}}^{E}_{mnp\,x} x, y, 0 \right\} \end{aligned} \tag{3.33}$$

and

$$\begin{aligned} E_y^{(e)}(x, y, t) = \sum_{m,n,p} & \operatorname{Re}\ (C(mnp))e^{j\omega(mnp)t}\psi^{E}_{mnp\,y}(x, y, 0) \\ & + \operatorname{Re}\left\{ D(mnp)e^{j\omega(mnp)t}\bar{\underline{\phi}}^{E}_{mnp\,y}(x, y, 0) \right\} \end{aligned} \tag{3.34}$$

By using orthogonality of $\left\{ \psi^{E}_{mnp\,x}(x, y, 0), \bar{\underline{\phi}}^{E}_{mnp\,x}(x, y, 0) \right\}$, for different (m, n), we can write p to be fixed and likewise of $\left\{ \psi^{E}_{mnp\,y}(x, y, 0), \bar{\underline{\phi}}^{E}_{mnp\,y}(x, y, 0) \right\}$;

In addition, we need to use KAM (Kolmogorov–Arnold–Moser) type of time averaging to yield

$$\begin{aligned} & C(mnp)\psi^{E}_{mnp\,x}(x, y, 0) + D(mnp)\bar{\underline{\phi}}^{E}_{mnp\,x}(x, y, 0) \\ & = \lim_{T \to \infty} \frac{1}{2T} \int_{-T}^{T} E_x^{(e)}(x, y, t)e^{-j\omega(mnp)t}\mathrm{d}t \end{aligned}$$

and likewise

$$\begin{aligned}
&C(mnp)\psi^{E}_{mnp\ y}(x,y,0) + D(mnp)\underline{\bar{\phi}}^{E}_{mnp\ y}(x,y,0) \\
&= \lim_{T \to \infty} \frac{1}{2T} \int_{-T}^{T} E_y^{(e)}(x,y,t) e^{j\omega(mnp)t} \mathrm{d}t
\end{aligned}$$

Chapter 4
Mathematical Analysis of Transcendental Equation in Rectangular DRA

Abstract Mathematical analysis of transcendental equation in rectangular DRA has been derived. Transcendental equation of rectangular DRA provides complete solution of propagation constants, i.e., k_x, k_y, and k_z. The propagation constant gives rise to resonant frequency with the help of characteristic equation. The wave numbers k_x, k_y, and k_z are in x, y, and z direction, respectively. The free space wave number is k_0. The exact solution of RDRA resonant frequency can be determined from combined solution of transcendental equation and characteristic equation. These equations have unique solution. RDRA depends upon boundary conditions. MATLAB-based simulation has been worked for RDRAs. They have been depicted with examples. This chapter has given a complete design solution of rectangular DRAs.

Keywords Mathematical analysis · Transcendental equation · Rectangular DRA · Propagation constant · Eigen vectors · Effective electrical length · Characteristic equation

Transcendental equation of rectangular DRA provides complete solution of propagation constants, i.e., k_x, k_y, and k_z. The propagation constant gives rise to resonant frequency with the help of characteristic equation. The wave numbers k_x, k_y, and k_z are in x, y, and z direction, respectively. The free space wave number is k_0. The exact solution of RDRA resonant frequency can be determined from combined solution of transcendental equation and characteristic equation. These equations have unique solution if RDRA boundary conditions are fixed. For example, top and bottom walls are PMC and rest of the four walls is PEC and vice versa, only two different transcendental equations will be developed.

To get this solution, H_z fields and derivative of H_z fields need to be solved. They are solved for continuous propagating fields conditions. The fields are assumed continuous at interface of RDRA. The RDRA along with eigen vectors is shown in Fig. 4.1a, b.

CASE#1 RDRA solution:
See Fig. 4.2.

R.S. Yaduvanshi and H. Parthasarathy, *Rectangular Dielectric Resonator Antennas*, DOI 10.1007/978-81-322-2500-3_4

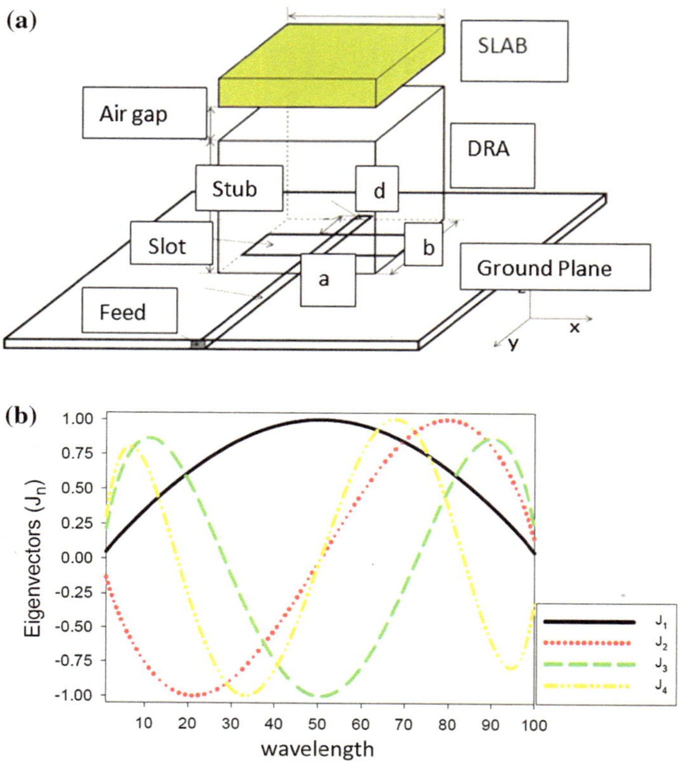

Fig. 4.1 **a** Rectangular DRA. **b** Eigen currents (current vectors) versus wavelength

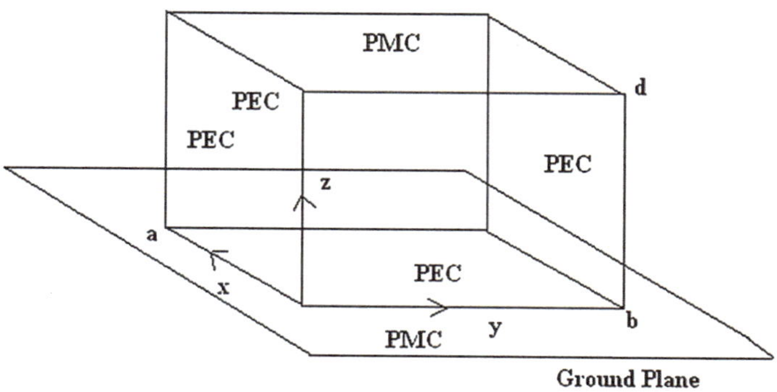

Fig. 4.2 RDRA under defined boundaries

To derive transcendental equation, the fields inside the resonator and outside the resonator are required.

$$\tan(k_z d) = \frac{k_z}{\sqrt{(\epsilon_r - 1){k_0}^2 - {k_z}^2}} \text{(transcendental equation)} \quad (4.1)$$

$$\epsilon_r {k_0}^2 = {k_x}^2 + {k_y}^2 + {k_z}^2 \text{(wave equation)} \quad (4.2)$$

$$k_x = n\pi/a \quad (4.3a)$$

$$k_y = n\pi/b \quad (4.3b)$$

$$k_z = p\pi/d \quad (4.3c)$$

where a, b, and d are dimensions; m, n, and p are the indices.

$TE_{\delta 11}$, $TE_{1\delta 1}$, and $TE_{11\delta}$ are dominant modes.

The solution of resonant frequency can be had if solution of k_y propagation constant is obtained from characteristic equation, $\epsilon_r {k_0}^2 = {k_x}^2 + {k_y}^2 + {k_z}^2$, and then substituted in transcendental equation to compute resonant frequency f_0.

Boundary condition

Propagation constant, $\gamma_{mn}^2 = {k_0}^2 + {h_{mn}}^2$

$$k = 2\pi/\lambda = \omega\sqrt{\mu\epsilon} = \omega/\text{c};$$

$$\int E^2 \text{d}V = \int H^2 \text{d}V$$

Time average electric energy = time average magnetic energy

$$\epsilon_r {k_0}^2 = {k_x}^2 + {k_y}^2 + {k_z}^2 \quad (4.4)$$

$$\epsilon_0 {k_0}^2 = {k_x}^2 + {k_y}^2 + {k_z'}^2 \quad (4.5)$$

$$k_z = p\pi/d$$

Subtracting Eq. (4.1) from Eq. (4.2), we get

$$k_z'^2 - k_z^2 = \epsilon_0 {k_0'}^2 - \epsilon_r {k_0}^2$$

$$k_z'^2 - k_z^2 = \epsilon_0 \mu_0 \omega^2 - \epsilon_r \mu_0 \omega^2$$

Taking value of $\epsilon_0 = 1$ and $\mu_0 = \mu$, we get

$$k_z'^2 - k_z^2 = \omega^2 \mu (1 - \epsilon_r) \quad (4.6)$$

4.1 Case-1: Top and Bottom Walls as PMC and Rest of the Four Walls are PEC

See Fig. 4.3.

Assuming that the top and bottom surface plane be at $z = 0, d$ to be PMC

$$\therefore \quad n \times H = 0$$

And

$$n \cdot E = 0$$

or,

$$\begin{aligned} H_y = H_x &= 0 \\ E_z &= 0 \end{aligned}$$

Rest of the other walls is PEC

$$\therefore \quad n \times E = 0$$

And

$$n \cdot H = 0$$

At

$$\begin{aligned} x = 0, a \quad E_y = E_z &= 0 \\ H_x &= 0 \end{aligned}$$

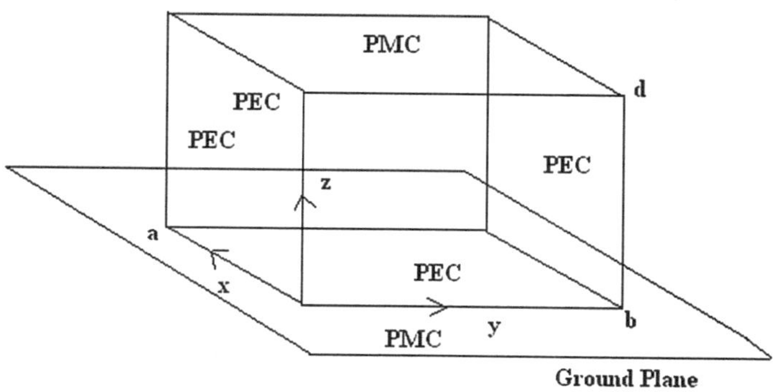

Fig. 4.3 RDRA with boundaries

At

$$y = 0, b \quad E_x = E_z = 0$$
$$H_y = 0$$

We also know

$$E_x = \frac{1}{j\omega\epsilon\left(1+\frac{\gamma^2}{k^2}\right)}\left[\frac{\partial H_z}{\partial y} - \frac{1}{j\omega\mu}\frac{\partial^2 E_z}{\partial z \partial x}\right] \tag{4.7a}$$

$$E_y = \frac{1}{j\omega\epsilon\left(1+\frac{\gamma^2}{k^2}\right)}\left[-\frac{1}{j\omega\mu}\frac{\partial^2 E_z}{\partial z \partial y} - \frac{\partial H_z}{\partial x}\right] \tag{4.7b}$$

$$H_x = \frac{-1}{j\omega\mu\left(1+\frac{\gamma^2}{k^2}\right)}\left[\frac{\partial E_z}{\partial y} - \frac{1}{j\omega\epsilon}\frac{\partial^2 H_z}{\partial z \partial x}\right] \tag{4.7c}$$

$$H_y = \frac{-1}{j\omega\mu\left(1+\frac{\gamma^2}{k^2}\right)}\left[\frac{1}{j\omega\epsilon}\frac{\partial^2 H_z}{\partial z \partial y} - \frac{\partial E_z}{\partial x}\right] \tag{4.7d}$$

Now, the solution of second-order differential equation is given as follows:

$$\psi_z = X(x)Y(y)Z(z)$$

where

$$X(x) = A_1 \sin k_x x + A_2 \cos k_x x \tag{4.8a}$$

$$Y(y) = A_3 \sin k_y y + A_4 \cos k_y y \tag{4.8b}$$

$$Z(z) = A_5 \sin k_z z + A_6 \cos k_z z \tag{4.8c}$$

For TE mode ($E_z = 0$ and $H_z \neq 0$)

$$\psi_{H_z} = X(x)Y(y)Z(z)$$

After putting $E_z = 0$, we get

$$E_y = C'\left[-\frac{\partial H_z}{\partial x}\right]$$

or,

$$E_y = C'X'(x)Y(y)Z(z)$$

Now

$$X'(x) = A_1 \cos k_x x - A_2 \sin k_x x$$

But at

$$\begin{aligned} &x = 0, a \quad E_y = 0 \\ \therefore \quad &0 = A_1 \cos k_x 0 - A_2 \sin k_x 0 \end{aligned}$$

or,

$$A_1 = 0 \;\text{ and }\; k_x = \frac{m\pi}{a}$$

Similarly,

$$E_x = C'\left[\frac{\partial H_z}{\partial y}\right]$$

or,

$$E_x = C'X(x)Y(y)Z(z)$$

Now

$$Y'(y) = A_3 \cos k_y y - A_4 \sin k_y y$$

But at

$$\begin{aligned} &y = 0, b \quad E_x = 0 \\ \therefore \quad &0 = A_3 \cos k_y 0 - A_4 \sin k_y 0 \end{aligned}$$

or,

$$A_3 = 0 \;\text{ and }\; k_y = \frac{n\pi}{b}$$

from

$$H_x = C'\left[-\frac{1}{j\omega\epsilon}\frac{\partial^2 H_z}{\partial z \partial x}\right]$$

or,

$$H_x = C'X'(x)Y(y)Z'(z)$$

Now

$$Z'(z) = A_5 \cos k_z z - A_6 \sin k_z z$$

At

$$z = 0, d \quad H_x = 0$$

$$\therefore \quad A_5 \cos k_z 0 - A_6 \sin k_z 0 = 0$$
$$A_5 = 0 \text{ and } k_z = \frac{p\pi}{d}$$

Hence,

$$H_z = A_2 A_4 A_6 \cos\left(\frac{m\pi}{a}x\right) \cos\left(\frac{n\pi}{b}y\right) \cos\left(\frac{p\pi}{d}z\right) \tag{4.9}$$

Using Eqs. (4.1)–(4.4), and (4.8a)–(4.8c), we get

$$H_x = C'' A_2 A_4 A_6 \left(\frac{m\pi}{a}\right)\left(\frac{p\pi}{d}\right) \sin\left(\frac{m\pi}{a}x\right) \cos\left(\frac{n\pi}{b}y\right) \sin\left(\frac{p\pi}{d}z\right) \tag{4.10a}$$

$$H_y = C'' A_2 A_4 A_6 \left(\frac{n\pi}{b}\right)\left(\frac{p\pi}{d}\right) \cos\left(\frac{m\pi}{a}x\right) \sin\left(\frac{n\pi}{b}y\right) \sin\left(\frac{p\pi}{d}z\right) \tag{4.10b}$$

$$E_y = C'' A_2 A_4 A_6 \left(\frac{m\pi}{a}\right) \sin\left(\frac{m\pi}{a}x\right) \cos\left(\frac{n\pi}{b}y\right) \cos\left(\frac{p\pi}{d}z\right) \tag{4.10c}$$

$$E_x = C'' A_2 A_4 A_6 \left(\frac{n\pi}{b}\right) \cos\left(\frac{m\pi}{a}x\right) \sin\left(\frac{n\pi}{b}y\right) \cos\left(\frac{p\pi}{d}z\right) \tag{4.10d}$$

Now, evaluate E_x and H_z at the boundary walls of the dielectric waveguide.

As we know that at the PMC wall, the tangential component of magnetic field and normal component of electric field are equal to "zero" at the interface $z = 0, d$.

Hence,

$$H_x, H_y = 0$$

and

$$E_z = 0$$

Also, for propagation to be possible, we need two normal components of E and H. Thus, we take E_x and H_y.

Now, the propagating wave is continuous at the interface, i.e., $E_x = E'_x$.

Therefore,

$$A\cos\left(\frac{m\pi}{a}x\right)\sin\left(\frac{n\pi}{b}y\right)\left(C_1e^{jk_zz}+C_2e^{-jk_zz}\right)=A\cos\left(\frac{m\pi}{a}x\right)\sin\left(\frac{n\pi}{b}y\right)C_2'e^{-jk_z'z} \tag{4.11}$$

or,

$$C_1e^{jk_zz}+C_2e^{-jk_zz}=C_2'e^{-jk_z'z} \tag{4.12}$$

But at $z = 0$, only the inside waveform exists.

Therefore,

$$C_1e^{jk_zz}+C_2e^{-jk_zz}=0$$

Now substituting the value of $z = 0$, we get

$$\begin{aligned} &C_1+C_2=0\\ &\text{or,}\quad C_1=-C_2 \end{aligned} \tag{4.13}$$

As H_z is continuous at the interface $z = d$,

$$H_z=H_z'\quad\text{and}\quad\frac{\partial H_z}{\partial x}=\frac{\partial H_z'}{\partial x}$$

From Eq. (4.9),

$$H_z=B\cos\left(\frac{m\pi}{a}x\right)\cos\left(\frac{n\pi}{b}y\right)\cos(k_zz) \tag{4.14a}$$

and

$$H_z'=B\cos\left(\frac{m\pi}{a}x\right)\cos\left(\frac{n\pi}{b}y\right)\cos\left(k_z'z\right) \tag{4.14b}$$

Equating Eqs. (4.14a) and (4.14b), we get

$$B\cos\left(\frac{m\pi}{a}x\right)\cos\left(\frac{n\pi}{b}y\right)\left(C_1e^{jk_zz}+C_2e^{-jk_zz}\right)=B\cos\left(\frac{m\pi}{a}x\right)\cos\left(\frac{n\pi}{b}y\right)\left(C_2'e^{-jk_z'z}\right)$$

or,

$$C_1e^{jk_zz}+C_2e^{-jk_zz}=C_2'e^{-jk_z'z} \tag{4.15}$$

From Eq. (4.15), i.e., $C_1 = -C_2$, we get, at $z = d$,

$$2jC_1 \sin(k_z d) = C_2' e^{-jk_z d} \tag{4.16}$$

Now, equating the derivative of H_z, we get

$$jk_z\left(C_1 e^{jk_z z} - C_2 e^{-jk_z z}\right) = -jk_z' C_2' e^{-jk_z' z} \tag{4.17}$$

or,

$$2k_z C_1 \cos(k_z d) = -k_z' C_2' e^{-jk_z' z}.$$

Dividing equation (4.16) by (4.17)

$$\frac{j \tan k_z d}{k_z} = \frac{-1}{k_z'} \tag{4.18a}$$

Squaring both sides and substituting the value of k_z^2 from Eq. (4.3c),

$$k_z'^2 = k_z^2 - \omega^2 \mu(\epsilon_r - 1)$$

and substituting $\mu = 1$. We get,

$$\tan(k_z d) = \frac{k_z}{\sqrt{k_0^2(\epsilon_r - 1) - k_z^2}} \tag{4.18b}$$

The above equation is the required transcendental equation.

4.2 Case-2

For transcendental equation, we need to compute the fields inside the resonator and outside it.

$$\tan(k_z d) = \frac{k_z}{\sqrt{(\epsilon_r - 1)k_0^2 - k_z^2}} \tag{4.19}$$

where $\epsilon_r k_0^2 = k_x^2 + k_y^2 + k_z^2$ (characteristic wave equation)

$$k_x = \frac{m\pi}{a} \tag{4.20a}$$

$$k_y = \frac{n\pi}{b} \tag{4.20b}$$

$$k_z = \frac{p\pi}{d} \tag{4.20c}$$

where a, b, and d are dimensions; m, n, and p are modes.

$TE_{\delta 11}$, $TE_{1\delta 1}$, and $TE_{11\delta}$ are dominant modes.

Boundary condition

Propagation constant, $\gamma_{mn}^2 = k_0^2 + hmn^2$ where $k = \frac{2\pi}{\lambda} = \omega\sqrt{\mu\epsilon} = \frac{\omega}{c}$.

From the energy conservation principle,

$$\int E^2 \mathrm{d}V = \int H^2 \mathrm{d}V.$$

i.e., time average electric energy = time average magnetic energy.

When top and bottom walls are PMC, rest of the other walls is PEC

Assuming that the top and bottom surface plane be at $z = 0, d$

$$\therefore \quad n \times H = 0$$

And

$$n \cdot E = 0$$

or,

$$\begin{aligned} H_y &= H_x = 0 \\ E_z &= 0 \end{aligned}$$

Rest of the other walls is PEC

$$\therefore \quad n \times E = 0$$

And

$$n \cdot H = 0$$

At

$$\begin{aligned} x = 0, a \quad E_y &= E_z = 0 \\ H_x &= 0 \end{aligned}$$

At

$$y = 0, b \quad E_x = E_z = 0$$
$$H_y = 0$$

We also know

$$E_x = \frac{1}{j\omega\epsilon\left(1 + \frac{\gamma^2}{k^2}\right)} \left[\frac{\partial H_z}{\partial y} - \frac{1}{j\omega\mu} \frac{\partial^2 E_z}{\partial z \partial x}\right] \tag{4.21a}$$

$$E_y = \frac{1}{j\omega\epsilon\left(1 + \frac{\gamma^2}{k^2}\right)} \left[-\frac{1}{j\omega\mu} \frac{\partial^2 E_z}{\partial z \partial y} - \frac{\partial H_z}{\partial x}\right] \tag{4.21b}$$

$$H_x = \frac{-1}{j\omega\mu\left(1 + \frac{\gamma^2}{k^2}\right)} \left[\frac{\partial E_z}{\partial y} - \frac{1}{j\omega\epsilon} \frac{\partial^2 H_z}{\partial z \partial x}\right] \tag{4.21c}$$

$$H_y = \frac{-1}{j\omega\mu\left(1 + \frac{\gamma^2}{k^2}\right)} \left[\frac{1}{j\omega\epsilon} \frac{\partial^2 H_z}{\partial z \partial y} - \frac{\partial E_z}{\partial x}\right] \tag{4.21d}$$

Now, the solution of second-order differential equation is given as follows:

$$\psi_z = X(x)Y(y)Z(z) \tag{4.22}$$

where

$$X(x) = A_1 \sin k_x x + A_2 \cos k_x x$$
$$Y(y) = A_3 \sin k_y y + A_4 \cos k_y y$$
$$Z(z) = A_5 \sin k_z z + A_6 \cos k_z z$$

For TE mode ($E_z = 0$ and $H_z \neq 0$)

$$\psi_{H_z} = X(x)Y(y)Z(z)$$
$$E_z = 0$$

we get

$$E_y = C' \left[-\frac{\partial H_z}{\partial x}\right]$$

or,

$$E_y = C'X'(x)Y(y)Z(z)$$

Now

$$X'(x) = A_1 \cos k_x x - A_2 \sin k_x x$$

But at

$$x = 0, a \;\; E_y = 0$$
$$\therefore \quad 0 = A_1 \cos k_x 0 - A_2 \sin k_x 0$$

or,

$$A_1 = 0 \;\text{ and }\; k_x = \frac{m\pi}{a}$$

Similarly,

$$E_x = C'\left[\frac{\partial H_z}{\partial y}\right]$$

or,

$$E_x = C'X(x)Y'(y)Z(z)$$

Now

$$Y'(y) = A_3 \cos k_y y - A_4 \sin k_y y$$

But at

$$y = 0, b \;\; E_x = 0$$
$$\therefore \quad 0 = A_3 \cos k_y 0 - A_4 \sin k_y 0$$

or,

$$A_3 = 0 \;\text{ and }\; k_y = \frac{n\pi}{b}$$

$$H_x = C'\left[-\frac{1}{j\omega\epsilon}\frac{\partial^2 H_z}{\partial z \partial x}\right]$$

or,

$$H_x = C'X'(x)Y(y)Z'(z)$$

Now

$$Z'(z) = A_5 \cos k_z z - A_6 \sin k_z z$$

At

$$\begin{aligned} z &= 0, d \;\; H_x = 0 \\ \therefore \;\; & A_5 \cos k_z 0 - A_6 \sin k_z 0 = 0 \\ & A_5 = 0 \;\; \text{and} \;\; k_z = \frac{p\pi}{d} \end{aligned}$$

Hence,

$$H_z = A_2 A_4 A_6 \cos\left(\frac{m\pi}{a}x\right) \cos\left(\frac{n\pi}{b}y\right) \cos\left(\frac{p\pi}{d}z\right) \tag{4.23}$$

Using Eqs. (4.1)–(4.4), and (4.8a)–(4.8c), we get

$$H_x = C'' A_2 A_4 A_6 \left(\frac{m\pi}{a}\right)\left(\frac{p\pi}{d}\right) \sin\left(\frac{m\pi}{a}x\right) \cos\left(\frac{n\pi}{b}y\right) \sin\left(\frac{p\pi}{d}z\right) \tag{4.24a}$$

$$H_y = C'' A_2 A_4 A_6 \left(\frac{n\pi}{b}\right)\left(\frac{p\pi}{d}\right) \cos\left(\frac{m\pi}{a}x\right) \sin\left(\frac{n\pi}{b}y\right) \sin\left(\frac{p\pi}{d}z\right) \tag{4.24b}$$

$$E_y = C'' A_2 A_4 A_6 \left(\frac{m\pi}{a}\right) \sin\left(\frac{m\pi}{a}x\right) \cos\left(\frac{n\pi}{b}y\right) \cos\left(\frac{p\pi}{d}z\right) \tag{4.24c}$$

$$E_x = C'' A_2 A_4 A_6 \left(\frac{n\pi}{b}\right) \cos\left(\frac{m\pi}{a}x\right) \sin\left(\frac{n\pi}{b}y\right) \cos\left(\frac{p\pi}{d}z\right) \tag{4.24d}$$

Above equations can also be written as follows:

$$\begin{aligned} H_x &= \frac{k_x k_z}{j\omega\mu_0} \sin(k_x x) \cos(k_y y) \sin(k_z z) \\ H_y &= \frac{k_y k_z}{j\omega\mu_0} \cos(k_x x) \sin(k_y y) \sin(k_z z) \\ E_y &= -k_x \sin(k_x x) \cos(k_y y) \cos(k_z z) \\ E_x &= k_y \cos(k_x x) \sin(k_y y) \cos(k_z z) \\ H_z &= \frac{k_x^2 + k_y^2}{j\omega\mu_0} \cos(k_x x) \cos(k_y y) \cos(k_z z) \end{aligned}$$

Since H_z is continuous, i.e., $\frac{dH_z}{dz} \neq 0$,

$$H'_z = \frac{k_x^2 + k_y^2}{j\omega\mu_0} \cos(k_x x)\cos(k_y y)\cos(k'_z z)$$

Now, H_y can be written as follows:

$$H_y = \frac{k_y k_z}{j\omega\mu_0} \cos(k_x x)\sin(k_y y)(C_1 e^{jk_z d} - C_2 e^{-jk_z d})$$

But

$$H_y = 0 \text{ at } d = 0$$

$$C_1 - C_2 = 0$$

or,

$$C_1 = C_2$$

$$\frac{\mathrm{d}H_y}{\mathrm{d}z} = A' jk_z \cos(k_x x)\sin(k_y y)(C_1 e^{jk_z d} + C_2 e^{-jk_z d})$$

or,

$$\frac{\mathrm{d}H_y}{\mathrm{d}z} = C_1 jk_z \cos(k_x x)\sin(k_y y)(e^{jk_z d} + e^{-jk_z d})$$

or,

$$\frac{\mathrm{d}H_y}{\mathrm{d}z} = C_1 2jk_z \cos(k_x x)\sin(k_y y)\cos(k_z d)$$

$$H'_y = C'_1 \cos(k_x x)\sin(k_y y)e^{-jk'_z d} \text{ outside the cavity}$$

For H_z to be continuous,

$$\frac{\mathrm{d}H_y}{\mathrm{d}z} = \frac{\mathrm{d}H'_y}{\mathrm{d}z}$$

or,

$$C_1 2jk_z \cos(k_x x)\sin(k_y y)\cos(k_z d) = -jk'_z C'_1 \cos(k_x x)\sin(k_y y)e^{-jk'_z d}$$

or,

$$2C_1 k_z \cos(k_z d) = -k'_z C'_1 e^{-jk'_z d} \tag{4.25}$$

From above equations, we have

$$E_x = k_y \cos(k_x x) \sin(k_y y) \left(C_1 e^{jk_z d} + C_2 e^{-jk_z d}\right)$$

At

$$d = 0,\ E_x = 0,$$

so,

$$C_1 + C_2 = 0$$

or,

$$C_1 = -C_2$$
$$\therefore \quad E_x = k_y \cos(k_x x) \sin(k_y y) C_1 \left(e^{jk_z d} - e^{-jk_z d}\right)$$

or,

$$E_x = 2jC_1 k_y \cos(k_x x) \sin(k_y y) \sin(k_z d)$$

Also

$$E'_x = k_y \cos(k_x x) \sin(k_y y) \cos(k'_z z)$$

or,

$$E'_x = C'_1 k_y \cos(k_x x) \sin(k_y y) e^{-jk_z d}$$

For H_z to be continuous,

$$E_x = E'_x$$

or,

$$jC_1 k_y \cos(k_x x) \sin(k_y y) \sin(k_z d) = C'_1 k_y \cos(k_x x) \sin(k_y y) e^{-jk_z d}$$

or,

$$2jC_1 \sin(k_z d) = C'_1 e^{-jk_z d} \tag{4.26}$$

Dividing Eq. (4.16) by Eq. (4.25), we get

$$\frac{2C_1 j\sin(k_z d)}{2C_1 k_z\cos(k_z d)} = \frac{-C_1' e^{-jk_z d}}{k_z' C_1' e^{-jk_z' d}}$$

or,

$$\frac{j\tan k_z d}{k_z} = \frac{-1}{k_z'}$$

or,

$$j\tan k_z d = -\frac{k_z}{k_z'}$$

On squaring and putting $k_z'^2 = k_z^2 + \omega^2\mu_0(1-\epsilon_r)$

$$\tan^2 k_z d = -\frac{k_z^2}{k_z^2 + \omega^2\mu_0(1-\epsilon_r)}$$

or,

$$\tan^2 k_z d = \frac{k_z^2}{\omega^2\mu_0(\epsilon_r - 1) - k_z^2}$$

$$\tan k_z d = \frac{k_z}{\sqrt{(\epsilon_r - 1)k_0^2 - k_z^2}} \tag{4.27}$$

With the help of transcendental equation, we can find the propagation factor. Also with the help of this equation, we can obtain resonant frequency.

CASE#3

For transcendental equation, we need to compute the fields inside the resonator and outside it.

$$k_z\tan(k_z d) = \sqrt{(\epsilon_r - 1)k_0^2 - k_z^2};$$
$$\epsilon_r k_0^2 = k_x^2 + k_y^2 + k_z^2;$$

and

$$k_x = m\pi/a \tag{4.28a}$$

$$k_y = n\pi/b \tag{4.28b}$$

$$k_z = p\pi/d \tag{4.28c}$$

where a, b, and d are dimensions; m, n, and p are the indices.

$TE_{\delta 11}$, $TE_{1\delta 1}$, $TE_{11\delta}$ are dominant modes.

Boundary conditions
Propagation constant, $\gamma_{mn}^2 = k_0^2 + h_{mn}^2$

$k = 2\pi/\lambda = \omega\sqrt[2]{\mu\epsilon} = \omega/c;$

$$\int E^2 \mathrm{d}V = \int H^2 \mathrm{d}V$$

Time average electric energy = time average magnetic energy

$$\epsilon_0 \epsilon_r k_0^2 = k_x^2 + k_y^2 + k_z^2 \tag{4.29a}$$

$$\epsilon_0 k_0^2 = k_x^2 + k_y^2 + k_z'^2 \tag{4.29b}$$

$$k_z' \neq p\pi/d$$

Subtracting Eq. (4.1) from Eq. (4.2), we get

$$\begin{aligned} k_z'^2 - k_z^2 &= \epsilon_0 k_0'^2 - \epsilon_0 \epsilon_r k_0^2 \\ k_z'^2 - k_z^2 &= \epsilon_0 \mu_0 \omega^2 - \epsilon_0 \epsilon_r \mu_0 \omega^2 \end{aligned}$$

Taking the value of $\epsilon_0 = 1$ and $\mu_0 = \mu$, we get

$$k_z'^2 - k_z^2 = \omega^2 \mu \epsilon_0 \, (1 - \epsilon_r) \tag{4.30}$$

When top and bottom walls are PEC, rest of the other walls is PMC.
Now,
Assuming that the top and bottom surface plane be at $z = 0, d$

$$\therefore \quad n \times E = 0$$

and

$$n \cdot H = 0$$

or,

$$E_y = E_x = 0$$

$$H_z = 0$$

Rest of the other walls is PMC

$$\therefore \quad n \times H = 0$$

And

$$n \cdot E = 0$$

At

$$\begin{aligned} x = 0, a \quad & H_y = H_z = 0 \\ E_x = 0 & \end{aligned}$$

At

$$\begin{aligned} y = 0, b \quad & H_x = H_z = 0 \\ E_y = 0 & \end{aligned}$$

We also know

$$E_x = \frac{1}{j\omega\epsilon\left(1 + \frac{\gamma^2}{k^2}\right)} \left[\frac{\partial H_z}{\partial y} - \frac{1}{j\omega\mu} \frac{\partial^2 E_z}{\partial z \partial x}\right] \tag{4.31a}$$

$$E_y = \frac{1}{j\omega\epsilon\left(1 + \frac{\gamma^2}{k^2}\right)} \left[-\frac{1}{j\omega\mu} \frac{\partial^2 E_z}{\partial z \partial y} - \frac{\partial H_z}{\partial x}\right] \tag{4.31b}$$

$$H_x = \frac{-1}{j\omega\mu\left(1 + \frac{\gamma^2}{k^2}\right)} \left[\frac{\partial E_z}{\partial y} - \frac{1}{j\omega\epsilon} \frac{\partial^2 H_z}{\partial z \partial x}\right] \tag{4.31c}$$

$$H_y = \frac{-1}{j\omega\mu\left(1 + \frac{\gamma^2}{k^2}\right)} \left[\frac{1}{j\omega\epsilon} \frac{\partial^2 H_z}{\partial z \partial y} - \frac{\partial E_z}{\partial x}\right] \tag{4.31d}$$

Now, the solution of second-order differential equation is given as follows:

$$\psi_z = X(x)Y(y)Z(z)$$

where

$$
\begin{aligned}
X(x) &= A_1 \sin k_x x + A_2 \cos k_x x \\
Y(y) &= A_3 \sin k_y y + A_4 \cos k_y y \\
Z(z) &= A_5 \sin k_z z + A_6 \cos k_z z
\end{aligned}
$$

(i) For TE mode ($E_z = 0$ and $H_z \neq 0$)

$$\psi_{H_z} = X(x)Y(y)Z(z)$$

At

$$x = 0, a \quad H_z = 0,$$

or,

$$
\begin{aligned}
&A_1 \sin k_x 0 + A_2 \cos k_x 0 = 0 \\
&\therefore \quad A_2 = 0 \text{ and } k_x = \frac{m\pi}{a}
\end{aligned}
$$

Also at

$$y = 0, b \quad H_z = 0$$

or,

$$
\begin{aligned}
&A_3 \sin k_y 0 + A_4 \cos k_y 0 = 0 \\
&\therefore \quad A_4 = 0 \text{ and } k_y = \frac{n\pi}{b}
\end{aligned}
$$

At

$$z = 0, d \quad H_z = 0$$

$$
\begin{aligned}
&\therefore \quad A_5 \sin k_z 0 + A_6 \cos k_z 0 = 0 \\
&A_6 = 0 \quad \text{and } k_z = \frac{p\pi}{d}
\end{aligned}
$$

Hence,

$$H_z = A_1 A_3 A_5 \sin\left(\frac{m\pi}{a}x\right) \sin\left(\frac{n\pi}{b}y\right) \sin\left(\frac{p\pi}{d}z\right) \tag{4.32}$$

Using Eqs. (4.1)–(4.4), and (4.8a)–(4.8c), we get

$$H_x = C''A_1A_3A_5\left(\frac{m\pi}{a}\right)\left(\frac{p\pi}{d}\right)\cos\left(\frac{m\pi}{a}x\right)\sin\left(\frac{n\pi}{b}y\right)\cos\left(\frac{p\pi}{d}z\right) \tag{4.33a}$$

$$H_y = C''A_1A_3A_5\left(\frac{n\pi}{b}\right)\left(\frac{p\pi}{d}\right)\sin\left(\frac{m\pi}{a}x\right)\cos\left(\frac{n\pi}{b}y\right)\cos\left(\frac{p\pi}{d}z\right) \tag{4.33b}$$

$$E_y = C''A_1A_3A_5\left(\frac{m\pi}{a}\right)\cos\left(\frac{m\pi}{a}x\right)\sin\left(\frac{n\pi}{b}y\right)\sin\left(\frac{p\pi}{d}z\right) \tag{4.33c}$$

$$E_x = C''A_1A_3A_5\left(\frac{n\pi}{b}\right)\sin\left(\frac{m\pi}{a}x\right)\cos\left(\frac{n\pi}{b}y\right)\sin\left(\frac{p\pi}{d}z\right) \tag{4.33d}$$

Now, evaluate H_x and H_z at the boundary walls of the dielectric waveguide.

As we know that at the PEC wall, the tangential component of electric field and normal component of magnetic field is equal to "zero" at the interface $z = 0, d$.

Hence,

$$E_x, E_y = 0$$

and

$$H_z = 0$$

Also, for propagation to be possible, we need two normal components of E and H.

Thus we take E_y and H_x.

Now, the propagating wave is continuous at the interface, i.e., $H_x = H'_x$.

Therefore,

$$A\cos\left(\frac{m\pi}{a}x\right)\sin\left(\frac{n\pi}{b}y\right)\left(C_1e^{jk_zz} + C_2e^{-jk_zz}\right) = A\cos\left(\frac{m\pi}{a}x\right)\sin\left(\frac{n\pi}{b}y\right)C'_2e^{-jk'_zz}$$

or,

$$C_1e^{jk_zz} + C_2e^{-jk_zz} = C'_2e^{-jk'_zz} \tag{4.34}$$

But at $z = 0$, only the inside waveform exists.

Therefore,

$$C_1e^{jk_zz} + C_2e^{-jk_zz} = 0$$

Now, substituting the value of $z = 0$, we get

$$C_1 + C_2 = 0$$

or,

$$C_1 = -C_2 \tag{4.35}$$

As H_z is continuous at the interface $z = d$.
Therefore,

$$H_z = H'_z \text{ and } \frac{\partial H_z}{\partial x} = \frac{\partial H'_z}{\partial x}$$

$$H_z = B \sin\left(\frac{m\pi}{a}x\right)\sin\left(\frac{n\pi}{b}y\right)\sin(k_z z) \tag{4.36a}$$

and

$$H'_z = B \sin\left(\frac{m\pi}{a}x\right) \sin\left(\frac{n\pi}{b}y\right) \sin(k'_z z) \tag{4.36b}$$

Equating Eqs. (i) and (ii), we get

$$B \sin\left(\frac{m\pi}{a}x\right) \sin\left(\frac{n\pi}{b}y\right)(C_1 e^{jk_z z} - C_2 e^{-jk_z z}) = B \sin\left(\frac{m\pi}{a}x\right) \sin\left(\frac{n\pi}{b}y\right) C'_2 e^{-jk'_z z}$$

or,

$$C_1 e^{jk_z z} - C_2 e^{-jk_z z} = C'_2 e^{-jk'_z z}$$

From Eq. (1b), i.e., $C_1 = -C_2$, we get, at $z = d$,

$$2C_1 \cos(k_z d) = C'_2 e^{-jk_z d} \tag{4.37}$$

Now, equating the derivative of H_z, we get

$$jk_z(C_1 e^{jk_z z} + C_2 e^{-jk_z z}) = -jk'_z C'_2 e^{-jk'_z z}$$

or,

$$2jk_z C_1 \sin(k_z d) = -k'_z C'_2 e^{-jk'_z z} \tag{4.38}$$

Dividing equation (iv) by (iii), we get

$$jk_z \tan k_z d = -k'_z$$

Squaring both sides and substituting the value of $k_z'^2$ from Eq. (4.3c), we get

$$k_z' = k_z^2 - \omega^2\mu(\epsilon_r - 1)$$

and substituting $\mu = 1$, we get isolated DRA case as:

$$k_z \tan(k_z d) = \sqrt{(\epsilon_r - 1)k_0^2 - k_z^2} \tag{4.39a}$$

DRA with ground plane case as:

$$k_z \tan(k_z d/2) = \sqrt{(\epsilon_r - 1)k_0^2 - k_z^2} \tag{4.39b}$$

Hence, the solution of transcendental equation is completely obtained.

4.3 MATLAB Simulation Results

The same can be seen if MATLAB simulation is obtained as given below:

```
clear
clear all
er=9.8;
c=3*10^8;
d=10*10^-3;
for p=1:1:10
  f=c*p*(sqrt(1+tan(p*pi/2).^2))/2*d*(sqrt(er-1));
 end
plot(p,f);
title('pvsf')
xlabel('p-------->>>');
ylabel('f-------->>');
grid on;
```

Relationship between delta distance and its impact on resonant frequency is shown in Fig. 4.4.

The resonant frequency is increasing as the delta length is increasing as shown in Fig. 4.4. Also, radiation lobe is increasing as the number of resonant mode is increasing as shown in Fig. 4.5.

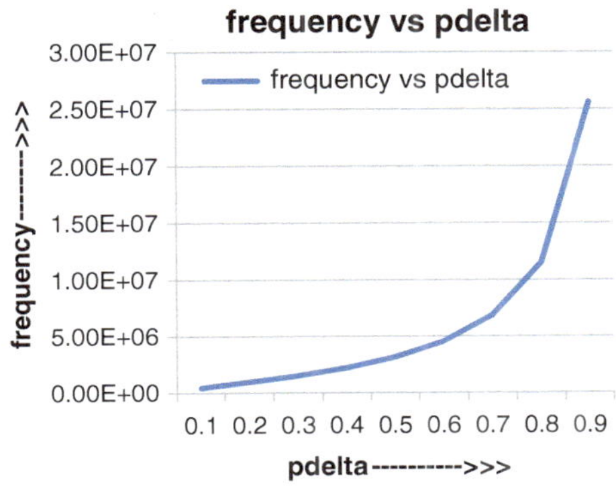

Fig. 4.4 Frequency versus delta distance

Radiation Lobes: RDRA dimensions are given to compute resonant modes using MATLAB.

Program for RDRA (a=length=10mm, b=width=5mm ,d=height=2mm)

```
m=3;
n=3;
a=10;
b=5;
x=linspace(-5,5,51);
y=linspace(-2.5,2.5,51);
[xi,yi] = meshgrid(x,y);
Ez= cos(m*pi*xi/a).*cos(n*pi*yi/b);
Ez= Ez.^2;
Ez= sqrt(Ez);
surf(xi,yi,Ez)
view([-45,60])
%%view([180,0])
drawnow
```

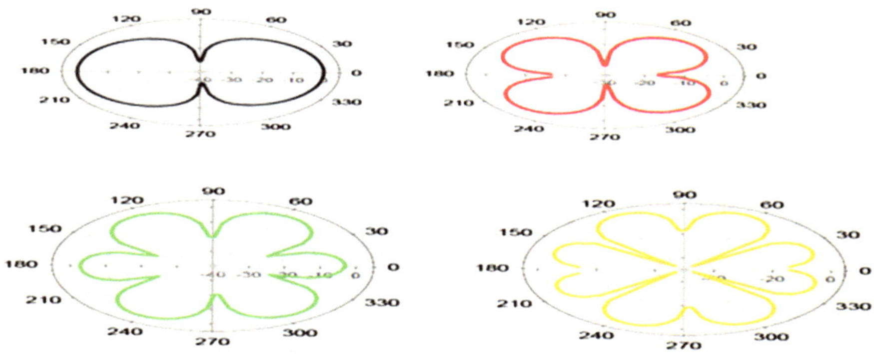

Fig. 4.5 Radiation lobes of radiation pattern in RDRA

MATLAB Program for Ez field

```
m=5;

n=4;

p=3;

a=10;

b=5;

c=2;

x=linspace(-5,5,51);

y=linspace(-2.5,2.5,51);

z=linspace(-1,1,51);

[xi,yi,zi] = meshgrid(x,y,z);

Ez= (cos(m*pi*xi/a).*cos(n*pi*yi/b)).*sin(p*pi*zi/c);

Ez= Ez.^2;

Ez= sqrt(Ez);

xslice = -4.5; yslice = -2.5; zslice =1;

slice(xi,yi,zi,Ez,xslice,yslice,zslice)

colormap hsv
```

MATLAB program for transcendental equation and resonant frequency of RDRA:

```
d=9;

w=6;

h=7.6;

c=3e8;

cons=9.8;

syms y real

kx=pi/d;

kz=pi/2/h;

ko=sqrt((kx^2+y^2+kz^2)/cons);

f=real(y*tan(y*w/2)-sqrt((cons-1)*ko^2-y^2));

ky=fzero(inline(f),[0,(pi/w)-0.01]);

fresonance = c/2/pi*sqrt((kx^2+ky^2+kz^2)/cons)/1e7;
```

The MATLAB-simulated resonant modes in Figs. 4.6, 4.7, 4.8, 4.9, 4.10, 4.11 and 4.12 have been drawn, and resonant frequency using transcendental equation is placed in table form.

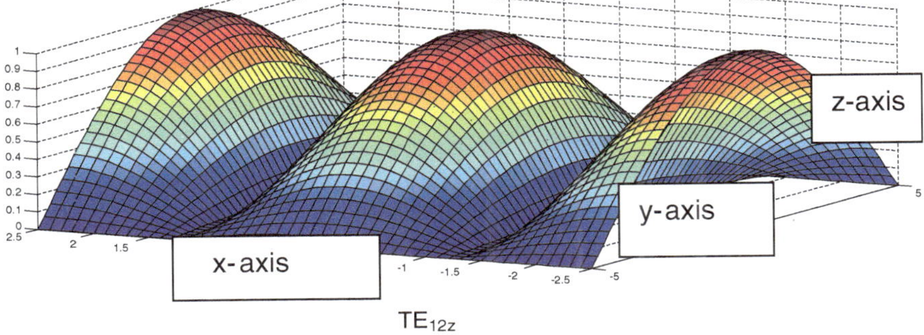

Fig. 4.6 Resonant modes in *xy* plane

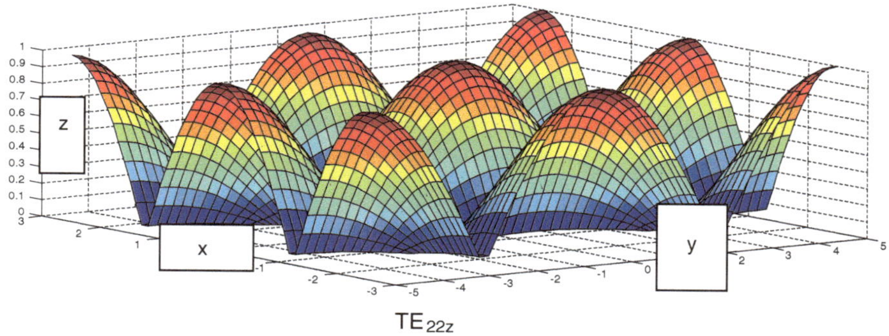

Fig. 4.7 Resonant modes in *xy* plane

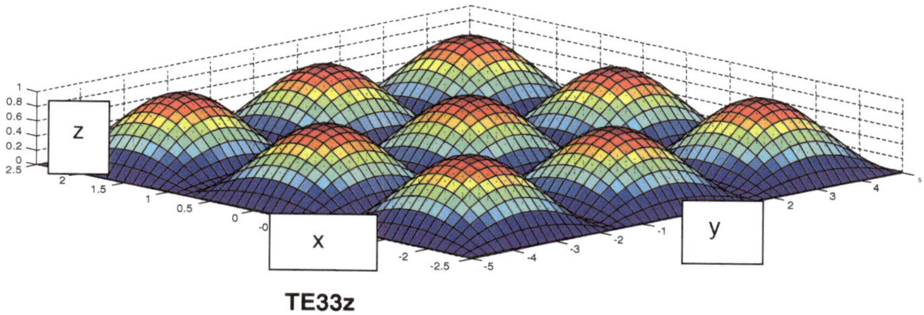

Fig. 4.8 Resonant modes in RDRA in *xy* plane

Solved examples of RDRA resonant frequency:

Example 1 Calculate the dimension of "*d*" in RDRA:
For TE_{111} mode when

$$\epsilon_r = 100$$
$$a = 10\,\text{mm}$$
$$b = 10\,\text{mm}$$
$$f_r = 7.97\,\text{GHz}$$

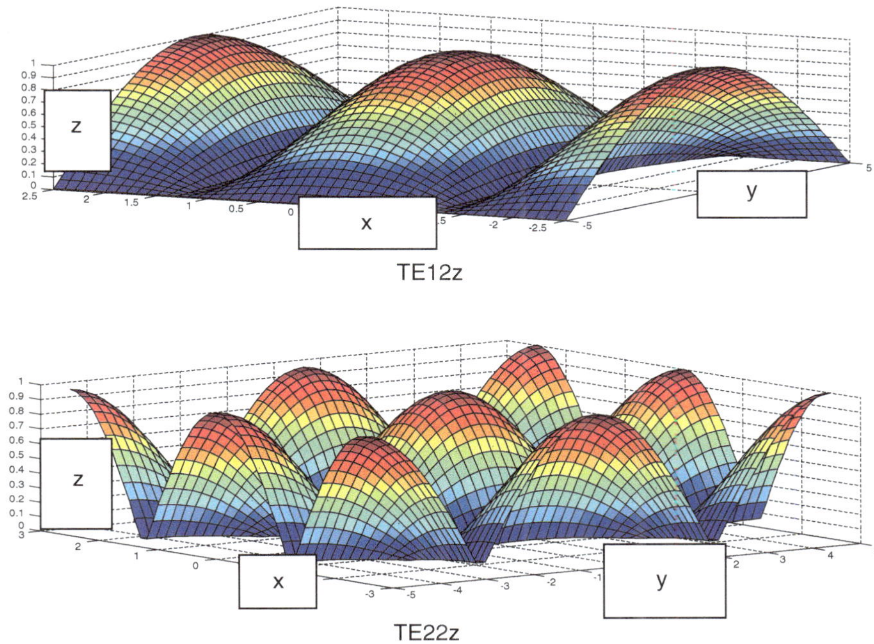

Fig. 4.9 Resonant modes in RDRA in *xy* plane

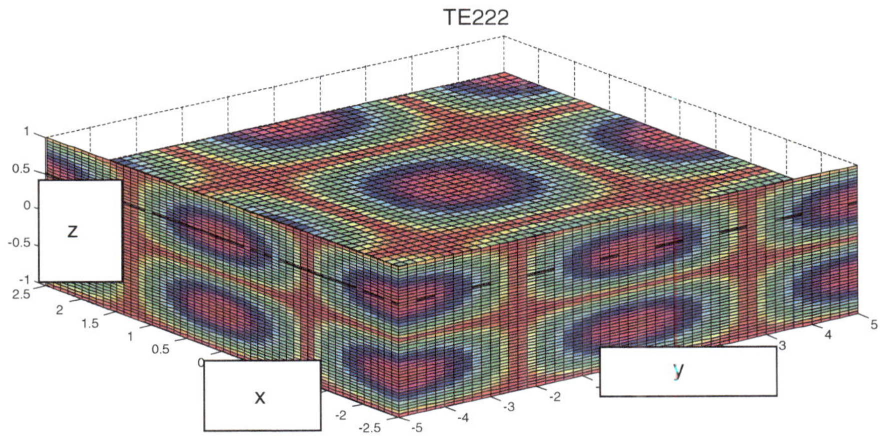

Fig. 4.10 Resonant modes 3D in RDRA in *xyz* plane

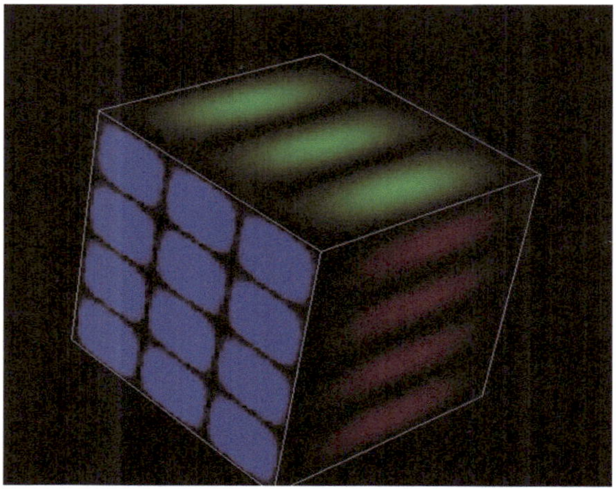

Fig. 4.11 TE 341 resonant modes

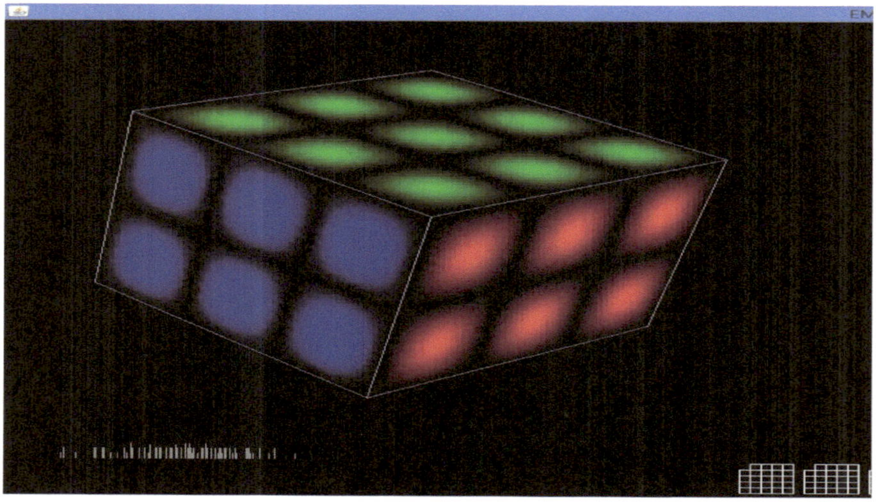

Fig. 4.12 TE 323 resonant modes

Solution Resonant frequency:

$$
\begin{aligned}
f_r &= \frac{c}{2\sqrt{\epsilon_r}}\sqrt{\frac{m2}{a}+\frac{n2}{b}+\frac{p2}{d}} \\
7.97\times 10^9 &= \frac{3\times 10^8}{2\sqrt{100}}\sqrt{100^2+100^2+\frac{1^2}{d}} \\
531.33 &= \sqrt{20000+\frac{1^2}{d}} \\
\frac{1}{d} &= 512.167 \\
d &= 1.95\,\text{mm}
\end{aligned}
$$

Example 2 RDRA with following data:

$$
\begin{aligned}
\epsilon_r &= 35 \\
a &= 18\,\text{mm} \\
b &= 18\,\text{mm} \\
f_r &= 2.45\,\text{GHz}
\end{aligned}
$$

Solution

$$
\begin{aligned}
f_r &= \frac{c}{2\sqrt{\epsilon_r}}\sqrt{\frac{m2}{a}+\frac{n2}{b}+\frac{p2}{d}} \\
2.45\times 10^9 &= \frac{3\times 10^8}{2\sqrt{35}}\sqrt{\frac{1000^2}{18}+\frac{1000^2}{18}+\frac{1^2}{d}} \\
9337.222 &= \sqrt{2\left(\frac{1000^2}{18}\right)+\frac{1^2}{d}} \\
\frac{1}{d} &= 56.252 \\
d &= 17.77\,\text{mm}
\end{aligned}
$$

Example 3 Calculate the resonant frequency for TE_{111} mode using the given data of RDRA:

$$
\begin{aligned}
\epsilon_r &= 10 \\
a &= 14\,\text{mm} \\
b &= 8\,\text{mm} \\
d &= 8\,\text{mm}
\end{aligned}
$$

Solution

$$f_r = \frac{c}{2\sqrt{\epsilon_r}}\sqrt{\frac{m2}{a}+\frac{n2}{b}+\frac{p2}{d}}$$

$$f_r = \frac{3\times 10^8}{2\sqrt{10}}\sqrt{\frac{1000^2}{14}+\frac{1000^2}{8}+\frac{1000^2}{8}}$$

$$f_r = 9.04\,\text{GHz}$$

Example 4

$$\epsilon_r = 10$$
$$a = 14\,\text{mm}$$
$$b = 8\,\text{mm}$$
$$d = 16\,\text{mm}$$

Solution

$$f_r = \frac{c}{2\sqrt{\epsilon_r}}\sqrt{\frac{m2}{a}+\frac{n2}{b}+\frac{p2}{d}}$$

$$f_r = \frac{3\times 10^8}{2\sqrt{10}}\sqrt{\frac{1000^2}{14}+\frac{1000^2}{8}+\frac{1000^2}{16}}$$

$$f_r = 7.44\,\text{GHz}$$

Example 5 Calculate the resonant frequency for the $TE_{11\delta}$ mode using the given data:

$$\epsilon_r = 10$$
$$a = 14\,\text{mm}$$
$$b = 8\,\text{mm}$$
$$d = 8\,\text{mm}$$

Solution

$$f_r = \frac{c}{2\sqrt{\epsilon_r}}\sqrt{\frac{m2}{a}+\frac{n2}{b}+\frac{\delta^2}{d}}$$

$$f_r = \frac{3\times 10^8}{2\sqrt{10}}\sqrt{\frac{1000^2}{14}+\frac{1000^2}{8}+0}$$

$$f_r = 6.82\,\text{GHz}$$

Example 6

$$\epsilon_r = 10$$
$$a = 14\,\text{mm}$$
$$b = 8\,\text{mm}$$
$$d = 16\,\text{mm}$$

Solution

$$f_r = \frac{c}{2\sqrt{\epsilon_r}}\sqrt{\frac{m^2}{a} + \frac{n^2}{b} + \frac{p^2}{d}}$$

$$f_r = \frac{3 \times 10^8}{2\sqrt{10}}\sqrt{\frac{1000^2}{14} + \frac{1000^2}{8} + 0}$$

$$f_r = 6.82\,\text{GHz}$$

4.4 Resonant Frequency of RDRA for Experimentations

The RDRAs can be prototyped with various materials and sizes as per the requirements.

Table 4.1 consists of list of RDRA materials, permittivity, dimensions, and computed resonant frequency.

Example 7 Compute resonant frequency when RDRA dimensions are $10 \times 10 \times 10\ \text{mm}^3$ and dielectric constant of material used is 10.

$$(f_r)m,n,p = \frac{c}{2\pi\sqrt{\epsilon\mu}}\sqrt{\left(\frac{m\pi}{a}\right)^2 + \left(\frac{n\pi}{b}\right)^2 + \left(\frac{p\pi}{d}\right)^2}$$

Resonant frequencies in isolated case are 49.7 and 25.8 GHz with ground plane (Table 4.2).

Table 4.1 RDRA materials, permittivity, dimensions, and computed resonant frequency

S. no.	Material	Permittivity	RDRA dimension ($a \times b \times h$) mm	Resonant frequency simulated by HFSS	Resonant frequency calculated
Countis Laboratories					
1.	$MgO–SiO_2–TiO_2$ (CD-9)	9.8	9 × 6 × 7.6	7.43	7.6757
2.	$MgO–SiO_2–TiO_2$ (CD-9)	9.8	14.3 × 25.4 × 26.1	3.5	3.7430
3.	$MgO–CaO–TiO_2$ (CD-20)	20.0	10.16 × 10.2 × 7.11	4.71	4.6215
4.	$MgO–CaO–TiO_2$ (CD-20)	20.0	10.16 × 7.11 × 10.2	4.55	4.5941
5.	$MgO–CaO–TiO_2$ (CD-20)	20.0	10.2 × 10.2 × 7.89	4.635	4.4833
6.	$MgO–CaO–TiO_2$ (CD-100)	100.0	10 × 10 × 2	4.57	4.2158
7.	$MgO–CaO–TiO_2$ (CD-100)	100.0	10 × 10 × 1	7.97	7.7587
8.	$MgO–CaO–TiO_2$ (CD-100)	100.0	12.7 × 12.7 × 1	7.72	7.6628
9.	$MgO–CaO–TiO_2$ (CD-100)	100.0	5 × 10 × 1	8.85	8.1828
10.	$MgO–CaO–TiO_2$ (CD-100)	100.0	10 × 5 × 1	8.5	8.0147
Emerson & Cuming Microwave Products N.V.					
11.	Magnesium titanate (ECCOSTOCK@)	10.0	14 × 8 × 8	5.5	5.6117
12.	Magnesium titanate (ECCOSTOCK@)	10.0	14.3 × 25.4 × 26.1	3.92	3.7055
13.	Zirconia (ECCOSTOCK@)	20.0	10.16 × 10.2 × 7.11	4.71	4.6215
14.	Zirconia (ECCOSTOCK@)	20.0	10.16 × 7.11 × 10.2	4.55	4.5941
15.	Zirconia (ECCOSTOCK@)	20.0	10.2 × 10.2 × 7.89	4.635	4.4833
16.	Strontium titanate (ECCOSTOCK@)	100.0	10 × 10 × 2	4.57	4.2158
17.	Strontium titanate (ECCOSTOCK@)	100.0	10 × 10 × 1	7.97	7.7587
18.	Strontium titanate (ECCOSTOCK@)	100.0	12.7 × 12.7 × 1	7.72	7.6628
19.	Strontium titanate (ECCOSTOCK@)	100.0	5 × 10 × 1	8.85	8.1828

(continued)

Table 4.1 (continued)

S. no.	Material	Permittivity	RDRA dimension ($a \times b \times h$) mm	Resonant frequency simulated by HFSS	Resonant frequency calculated
20.	Strontium titanate (ECCOSTOCK@)	100.0	10 × 5 × 1	8.5	8.0147
Morgan Advanced Materials					
21.	CaMgTi (Mg, Ca titanate) (D20)	20.0	10.16 × 10.2 × 7.11	4.71	4.6215
22.	CaMgTi (Mg, Ca titanate) (D20)	20.0	10.16 × 7.11 × 10.2	4.55	4.5941
23.	CaMgTi (Mg, Ca titanate) (D20)	20.0	10.2 × 10.2 × 7.89	4.635	4.4833
24.	ZrTiSn (Zr, Sn titanate) (D36)	37.0	18 × 18 × 9	2.45	2.1617
Temex Components & Temex Telecom, USA					
25.	Zr Sn Ti Oxide (E2000)	37.0	18 × 18 × 9	2.45	2.1617
Trans-Tech Skyworks Solutions, Inc.					
26.	BaZnCoNb (D-83)	35.0–36.5	18 × 18 × 6	2.532	2.7081
27.	BaZnCoNb (D-83)	35.0–36.5	18 × 6 × 18	2.835	2.3947
T-CERAM, RF & Microwave					
28.	E-11	10.8	15.2 × 7 × 2.6	11.6	10.379
29.	E-11	10.8	15 × 3 × 7.5	6.88	7.0937
30.	E-11	10.8	15.24 × 3.1 × 7.62	6.21	6.9440
31.	E-20	20.0	10.16 × 10.2 × 7.11	4.71	4.6215
32.	E-20	20.0	10.16 × 7.11 × 10.2	4.55	4.5941
33.	E-20	20.0	10.2 × 10.2 × 7.89	4.635	4.4833
34.	E-37	37.0	18 × 18 × 9	2.45	2.1617
TCI Ceramics, Inc.					
35.	DR-36	36.0	18 × 18 × 6	2.532	2.7081
36.	DR-36	36.0	18 × 6 × 18	2.835	2.3947

Table 4.2 Fringing effect along "*b*" dimensions increased effective along *y*-direction of RDRA

S. no.	Permittivity	Dimension (*a* (length) × *b* (width) × *d* (depth)) mm	Resonant frequency	Effective width (b')	Multiple factor	% change in width
37.	10.0	14.3 × 25.4 × 26.1	3.5	34.22	1.3474	34.7381
38.	10.0	14 × 8 × 8	5.5	14.13	1.7665	76.6535
39.	10.0	15.24 × 3.1 × 7.62	6.21	8.33	2.8872	168.7230
40.	20.0	10.2 × 10.2 × 7.89	4.635	15.31	1.5014	50.1419
41.	20.0	10.16 × 10.2 × 7.11	4.71	15.15	1.4858	48.5797
42.	35.0	18 × 18 × 6	2.532	24.12	1.34	33.9973
43.	35.0	18 × 18 × 9	2.45	25.64	1.4244	42.4423
44.	100.0	10 × 10 × 1	7.97	11.24	1.1242	12.4237

MATLAB program and simulation effective length due to fringing effect:

```
%%Dimensions of RDRA
%%length
d=[14.3,14.0,15.24,10.2,10.16,18,18,10];
%%width
w=[25.4,8,3.1,10.2,10.2,18,18,10];
%%height
h=[26.1,8,7.62,7.89,7.11,6,9,1];
%%Mode
m=1;
n=1;
p=1;
c=3e8;
cons=[10.0,10.0,10,20,20,35,35,100];
syms y real
for i=drange(1:8)
kx(i)=pi/d(i);
kz(i)=pi/2/h(i);
ko=sqrt((kx(i).^2+y.^2+kz(i).^2)/cons(i));
f=real(y.*tan(y*w(i)/2)-sqrt((cons(i)-1)*ko.^2-y.^2));
ky(i)=fzero(inline(f),[0,(pi/w(i))-0.01]);
%%Resonant frequency
fre(i)=c/2/pi*sqrt((kx(i).^2+ky(i).^2+kz(i).^2)/cons(i))*1e3;
Effwidth(i)=pi/ky(i);
factor(i)=Effwidth(i)./w(i);
perchangwidth(i)=((Effwidth(i)-w(i))/w(i))*100;
end
```

Effective increased length computations due to fringing effect:

Program 1

```
%%Dimensions of DRA
%%length
d=[14.3,14.0,15.24,10.2,10.16,18,18,10];
%%width
w=[25.4,8,3.1,10.2,10.2,18,18,10];
%%height
h=[26.1,8,7.62,7.89,7.11,6,9,1];
%%Mode
m=1;
n=1;
```

```
p=1;
c=3e8;
cons=[10.0,10.0,10,20,20,35,35,100];
syms y real
for i=drange(1:8)
kx(i)=pi/d(i);
kz(i)=pi/2/h(i);
ko=sqrt((kx(i).^2+y.^2+kz(i).^2)/cons(i));
f=real(y.*tan(y*w(i)/2)-sqrt((cons(i)-1)*ko.^2-y.^2));
ky(i)=fzero(inline(f),[0,(pi/w(i))-0.01]);
%%Resonant frequency
fre(i)=c/2/pi*sqrt((kx(i).^2+ky(i).^2+kz(i).^2)/cons(i))*1e3;
Effwidth(i)=pi/ky(i);
factor(i)=Effwidth(i)./w(i);
perchangwidth(i)=((Effwidth(i)-w(i))/w(i))*100;
end
```

Results:

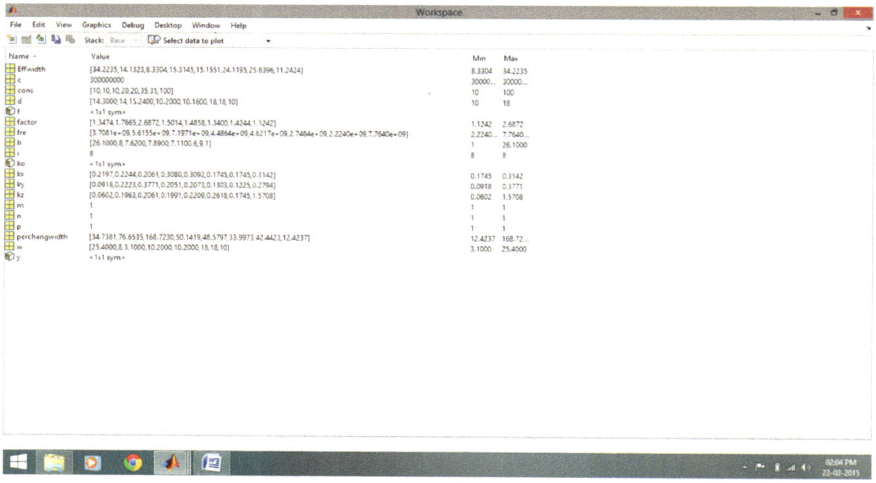

Program 2

```
m=1;
n=1;
p=1;
E_r=10;
a=15.24e-03;
b=3.1e-03;
d=7.62e-03;
c=3e+08;
kx=m*pi/a;
ky=n*pi/b;
kz=p*(pi/d)/2;
ko=sqrt(kx^2+ky^2+kz^2)/sqrt(E_r);
fo=(c*ko/pi)/2;
foghz=fo/(1e+09);
```

Results:

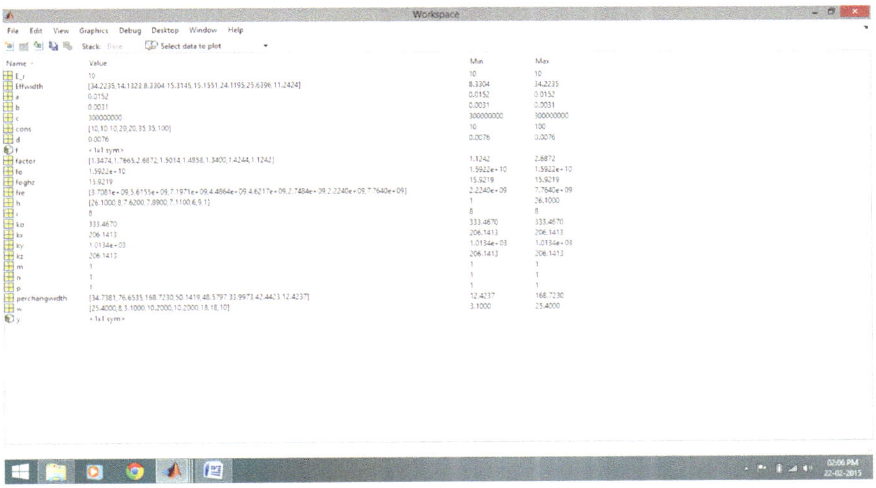

Program 3

MATLAB programs taking parameters a, b, d same and comparing frequency using:

Program 1: Characteristic equation

```
m=1
n=1
p=1
E_r=10
a=14.3e-03
b=25.4e-03
d=26.1e-03
c=3e+08
k_x=m*pi/a
k_y=n*pi/b
k_z=p*(pi/d)/2
k_o=sqrt(k_x^2+k_y^2+k_z^2)/sqrt(E_r)
f_o=(c*k_o/pi)/2
f_oGHz=f_o/1e+09
```

Output:

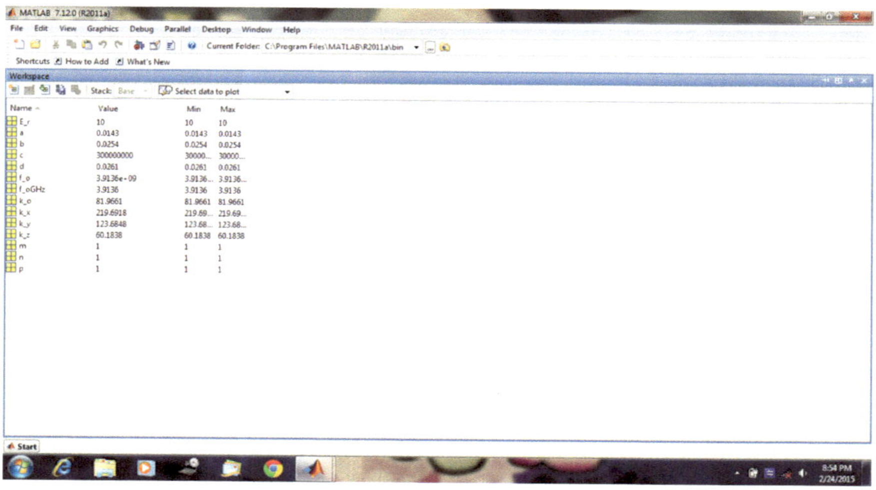

Program 4

Transcendental equation for same dimensions:

```
m=1;
n=1;
p=1;
E_r=10;
a=14.3e-03;
b=25.4e-03;
d=26.1e-03;
c=3e+08;
syms y real
kx=pi/a;
kz=pi/d/2;
ko=sqrt(kx^2+y^2+kz^2)/sqrt(E_r);
f=real(y*tan(y*b/2)-sqrt((E_r-1)*ko^2-y^2));
ky=fzero(inline(f),[0,(pi/b)-0.01]);
fre=c/2/pi*sqrt((kx^2+ky^2+kz^2)/E_r)*1e3;
effwidth=pi/ky;
factor=effwidth/b;
perchangwidth=((effwidth-b)/b)*100;
```

Output:

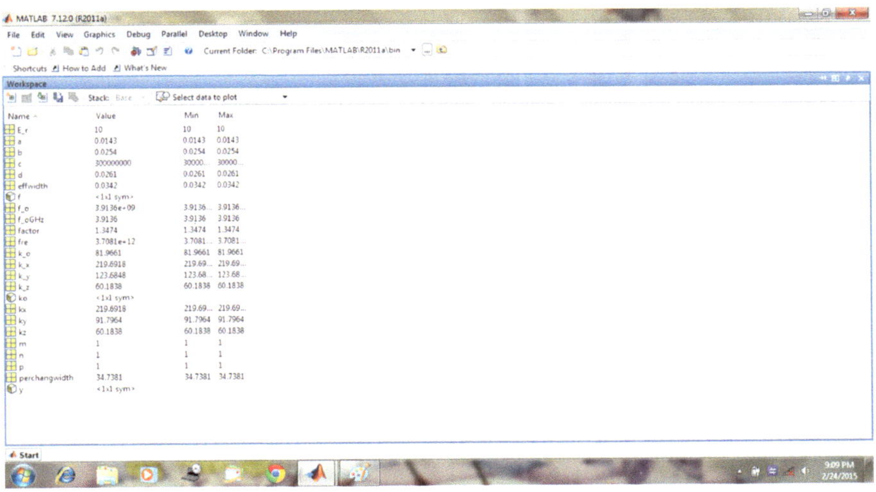

Program 5

```
MATLAB programs taking parameters a, b, d same and comparing
frequency using:
Characteristic equation
      Where a=17mm
            b=25mm
            c=10mm
m=1;
n=1;
p=1;
E_r=10;
a=17e-03;
b=25e-03;
d=10e-03;
c=3e+08;
k_x=m*pi/a;
k_y=n*pi/b;
k_z=p*(pi/d)/2;
k_o=sqrt(k_x^2+k_y^2+k_z^2)/sqrt(E_r);
f_o=(c*k_o/pi)/2;
f_oGHz=f_o/1e+09;
```

Output:

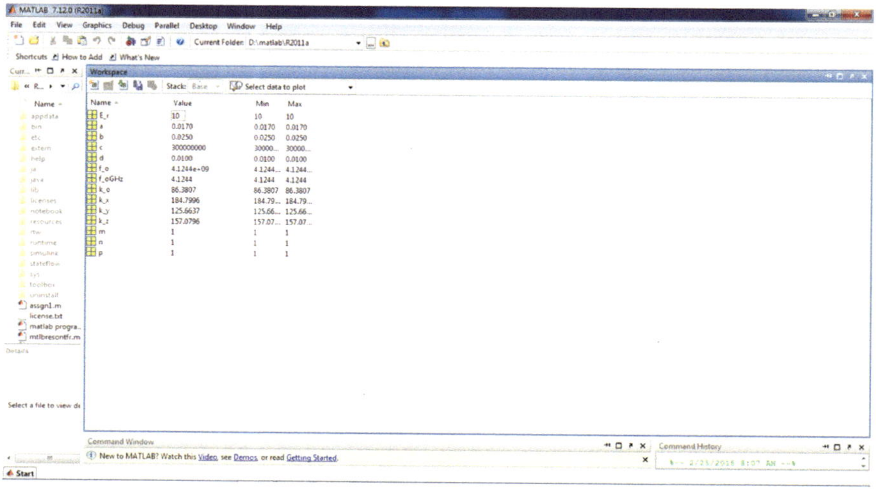

Program 6

Transcendental equation

```
m=1;
n=1;
p=1;
E_r=10;
a=17e-03;
b=25e-03;
d=10e-03;
c=3e+08;
syms y real
kx=pi/a;
kz=pi/d/2;
ko=sqrt(kx^2+y^2+kz^2)/sqrt(E_r);
f=real(y*tan(y*b/2)-sqrt((E_r-1)*ko^2-y^2));
ky=fzero(inline(f),[0,(pi/b)-0.01]);
fre=c/2/pi*sqrt((kx^2+ky^2+kz^2)/E_r)*1e3;
effwidth=pi/ky;
factor=effwidth/b;
perchangwidth=((effwidth-b)/b)*100;
```

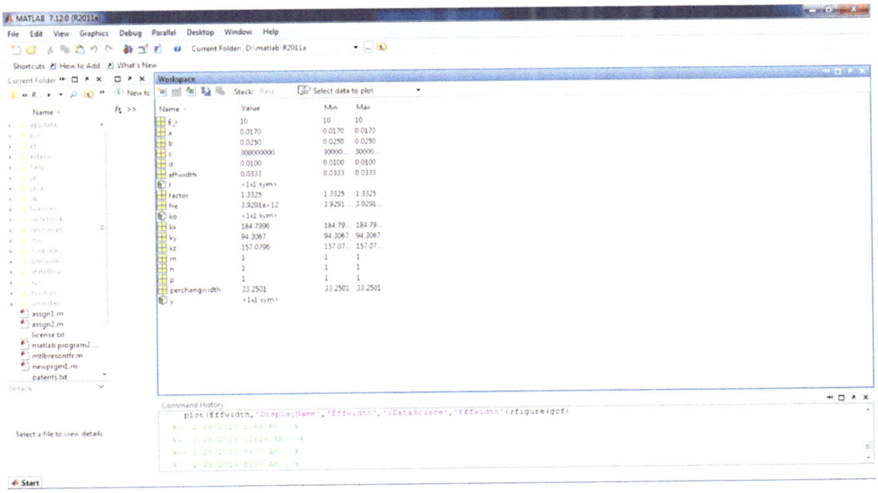

Program 7

```
MATLAB programs taking parameters a,b,d same and comparing
frequency using:
Characteristic equation

m=1
n=1
p=1
E_r=10
a=14.3e-03
b=25.4e-03
d=26.1e-03
c=3e+08
k_x=m*pi/a
k_y=n*pi/b
k_z=p*(pi/d)/2
k_o=sqrt(k_x^2+k_y^2+k_z^2)/sqrt(E_r)
f_o=(c*k_o/pi)/2
f_oGHz=f_o/1e+09
```

Output:

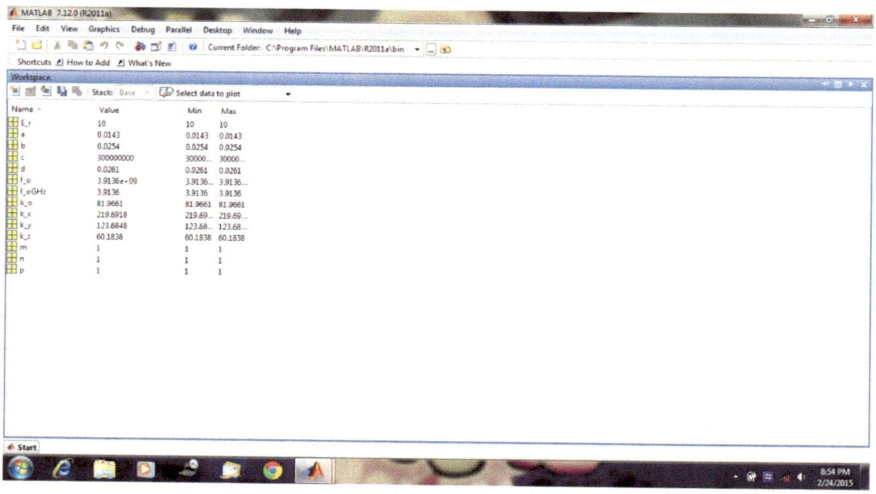

Program 8

Transcendental equation

```
m=1;
n=1;
p=1;
E_r=10;
a=14.3e-03;
b=25.4e-03;
d=26.1e-03;
c=3e+08;
syms y real
kx=pi/a;
kz=pi/d/2;
ko=sqrt(kx^2+y^2+kz^2)/sqrt(E_r);
f=real(y*tan(y*b/2)-sqrt((E_r-1)*ko^2-y^2));
ky=fzero(inline(f),[0,(pi/b)-0.01]);
fre=c/2/pi*sqrt((kx^2+ky^2+kz^2)/E_r)*1e3;
effwidth=pi/ky;
factor=effwidth/b;
perchangwidth=((effwidth-b)/b)*100;
```

Output:

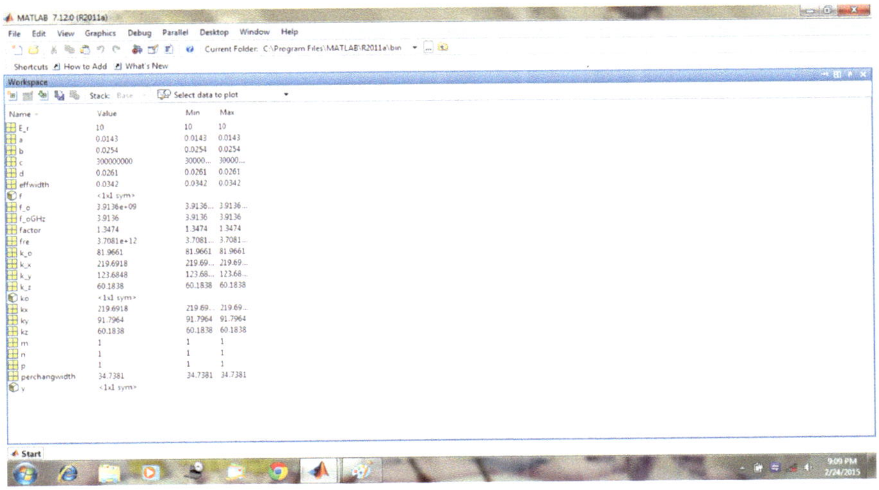

Program 9

```
m=3;
n=3;
a=10;
b=5;
x=linspace(-5,5,51);
y=linspace(-2.5,2.5,51);
[xi,yi]=meshgrid(x,y);
Ez=cos(m*pi*xi/a).*cos(n*pi*yi/b);
Ez=Ez.^2;
Ez=sqrt(Ez);
surf(xi,yi,Ez)
view([-45,60])
%%view([180,0])
drawnow
```

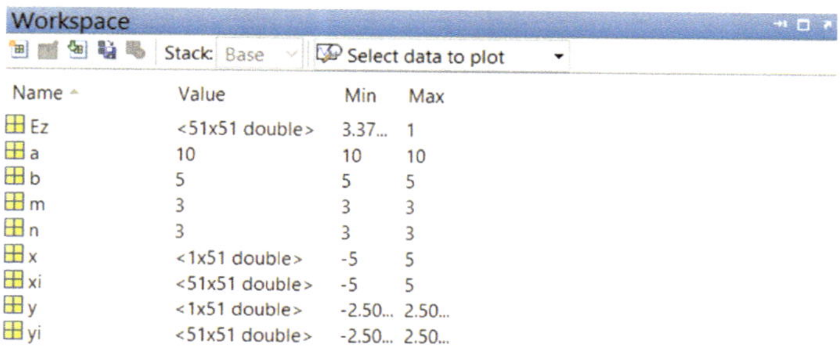

Program 10

```
d=[14.3,14.0,15.24,10.2,10.16,18,18,10];
w=[25.4,8,3.1,10.2,10.2,18,18,10];
h=[26.1,8,7.62,7.89,7.11,6,9,1];
m=1;
n=1;
p=1;
c=3e8;
cons=[10.0,10.0,10,20,20,35,35,100];
syms y real
for i=drange(1:8)
kx(i)=pi/d(i);
kz(i)=pi/2/h(i);
ko=sqrt((kx(i).^2+y.^2+kz(i).^2)/cons(i));
f=real(y.*tan(y*w(i)/2)-sqrt((cons(i)-1)*ko.^2-y.^2));
ky(i)=fzero(inline(f),[0,(pi/w(i))-0.01]);
fre(i)=c/2/pi*sqrt((kx(i).^2+ky(i).^2+kz(i).^2)/cons(i))*1e3;
Effwidth(i)=pi/ky(i);
factor(i)=Effwidth(i)./w(i);
perchangwidth(i)=((Effwidth(i)-w(i))/w(i))*100;
end
```

Workspace

Stack: Base   Select data to plot

Name	Value	Min	Max
Effwidth	[34.2235,14.1323,...	8.33...	34.2...
Ez	<51x51 double>	3.37...	1
a	10	10	10
b	5	5	5
c	300000000	3000...	3000...
cons	[10,10,10,20,20,3...	10	100
d	[14.3000,14,15.24...	10	18
f	<1x1 sym>		
factor	[1.3474,1.7665,2....	1.12...	2.68...
fre	[3.7081e+09,5.61...	2.22...	7.76...
h	[26.1000,8,7.6200...	1	26.1...
i	8	8	8
ko	<1x1 sym>		
kx	[0.2197,0.2244,0....	0.17...	0.31...
ky	[0.0918,0.2223,0....	0.09...	0.37...
kz	[0.0602,0.1963,0....	0.06...	1.57...
m	1	1	1
n	1	1	1
p	1	1	1
perchangwidth	[34.7381,76.6535,...	12.4...	168....
w	[25.4000,8,3.1000...	3.10...	25.4...
x	<1x51 double>	-5	5
xi	<51x51 double>	-5	5
y	<1x1 sym>		
yi	<51x51 double>	-2.50...	2.50...

Q.No. 1 Develop transcendental equation for moat-shaped RDRA.

Q.No. 2 Compute propagation constants in x-, y-, and z-directed propagated RDRAs, when feed probe is given. Compute its resonant frequency when RDRA dimensions are $5 \times 5 \times 3$ mm^3 and dielectric constant used is 20.

Chapter 5
Mathematical Analysis of RDRA Amplitude Coefficients

Abstract Mathematical analysis of amplitude coefficients in rectangular DRA (RDRA) have been evaluated. Rigorous theoretical analysis has been developed for different resonant modes inside RDRA. The resonance phenomenon and its potential use as radiator have been described. The dielectric polarization P is equal to the total dipole moment induced in the volume of the material by the electric field. The discontinuity of the relative permittivity at the resonator surface allows a standing electromagnetic wave to be supported in its interior at a particular resonant frequency, thereby leading to maximum confinement of energy within the resonator. Certain field distributions or modes will satisfy Maxwell's equations and boundary condition. Mathematical solution to get amplitude coefficients C_{mnp} along with its phase coefficients has been obtained. These are also known as eigenvector.

Keywords Amplitude coefficients · Resonant modes · Radiation lobes · Fourier transform · Discrete solution · PMC (perfect magnetic conducting) · PEC (perfect electrical conducting) · Dominant mode · Higher-order modes

5.1 Introduction

Rigorous theoretical analysis has been developed for resonant modes in rectangular DRA (RDRA). RDRA resonance phenomenon and its potential, as a radiator have been long back described. Accordingly, external electric fields bring the charges of the molecules of the dielectric into a certain ordered arrangement in space. The dielectric polarization P is equal to the total dipole moment induced in the volume of the material by the electric field. The discontinuity of the relative permittivity at the resonator surface allows a standing electromagnetic wave to be supported in its interior at a particular resonant frequency, thereby leading to maximum confinement of energy within the resonator. Certain field distributions or modes will satisfy Maxwell's equations and boundary conditions. Resonant modes are field structures that can exist inside the RDRA. The RDRA prototype is shown in Fig. 5.1.

R.S. Yaduvanshi and H. Parthasarathy, *Rectangular Dielectric Resonator Antennas*, DOI 10.1007/978-81-322-2500-3_5

Fig. 5.1 Homogenous dielectric RDRA on ground plane

5.2 Amplitude Coefficients C_{mnp}

Time domain fields can be written as follows: $E_z(x,y,z,t) = \sum_{mnp} \mathrm{Re}(C_{mnp} e^{j\omega(mnp)t} u_{mnp}(x,y,z))$, using orthonormality.

In discrete form,

$$\sum_{m,n,p} |C_{mnp}| u_{mnp}(x,y,z) \cos\left(\omega\left(\omega_{mnp}\right)t + \Psi(mnp)\right)$$

The probe current can be expressed as:

$$E_z\,(x,y,\delta,t) = \int G(x,y) \frac{j\omega\mu I \mathrm{d}l(x^2+y^2)}{4\pi\left(x^2+y^2+\delta^2\right)^{3/2}} e^{\left(j\omega t - \frac{\omega}{c}\sqrt{x^2+y^2+\delta^2}\right)} I(\omega) e^{j\omega t}\,\mathrm{d}\omega$$

where $G(x,y)$ are the constant terms associated with current.

$$\text{Resonator current} = \sum_p |C_{mnp}| \sqrt{\frac{2}{d}} \sin\left(\frac{p\pi\delta}{d}\right) \cos(\omega(mnp)t + \Psi(mnp)) u_{m,n}(x,y);$$

$$\text{Probe current} = \int G(x,y) \frac{j\omega\mu I \mathrm{d}l(x^2+y^2)}{4\pi\left(x^2+y^2+\delta^2\right)^{3/2}} I(\omega) e^{jkt} \mathrm{d}\omega$$
$$\left(e^{\left(j\omega t - \frac{\omega}{c}\sqrt{x^2+y^2+\delta^2} + \psi_{mnp}\right)}\right)(u_{m,n}(x,y)\mathrm{d}x\,\mathrm{d}y\;;$$

The probe current must be equal to the resonator current due to principle of orthonormality.

$$\underline{E}(x,y,z,t) = \sum_{mnp=1}^{\infty} \text{Re}\left\{C(mnp)e^{j\omega(mnp)t}\,\underline{\psi}_{mnp}^{E}(x,y,z)\right\} + \sum_{mnp=1}^{\infty} \text{Re}\left\{D(mnp)e^{j\omega(mnp)t}\,\underline{\bar{\phi}}_{mnp}^{E}(x,y,z)\right\}$$

$$\underline{H}(x,y,z,t) = \sum_{mnp=1}^{\infty} \text{Re}\left\{C(mnp)e^{j\omega(mnp)t}\,\underline{\psi}_{mnp}^{H}(x,y,z)\right\} + \sum_{mnp=1}^{\infty} \text{Re}\left\{D(mnp)e^{j\omega(mnp)t}\,\underline{\bar{\phi}}_{mnp}^{H}(x,y,z)\right\}$$

$$\underline{E}_{\perp} = -\frac{\gamma}{h^2}\underline{\nabla}_{\perp}E_z - \frac{j\omega\mu}{h^2}\underline{\nabla}_{\perp}\,H_z x\,\hat{z}$$

From duality

$$\underline{H}_{\perp} = -\frac{\gamma}{h^2}\underline{\nabla}_{\perp}H_z - \frac{j\omega\epsilon}{h^2}\underline{\nabla}_{\perp}\,E_z x\,\hat{z}$$

From above two equations, we obtain E_x and E_y fields as given below:

$$E_x = \frac{m\pi x}{a}\alpha'(mnp)\cos\left(\frac{m\pi x}{a}\right)\sin\left(\frac{n\pi y}{b}\right) - \frac{n\pi}{b}\beta'(mnp)\cos\left(\frac{m\pi x}{a}\right)\sin\left(\frac{n\pi y}{b}\right);$$
$$E_y = \frac{n\pi y}{b}\alpha'(mnp)\sin\left(\frac{m\pi x}{a}\right)\cos\left(\frac{n\pi y}{b}\right) - \frac{m\pi}{a}\beta'(mnp)\cos\left(\frac{m\pi x}{a}\right)\sin\left(\frac{n\pi y}{b}\right);$$

and

$$E_z = \sum_{m,n,p} \text{Re}[C(mnp)e^{j\omega(mnp)t}\sqrt{\frac{2}{d}}\sin\left(\frac{p\pi\delta}{d}\right)\;u_{m,n}(x,y)$$

$$E_z = \int \frac{j\omega\mu I\text{d}l(x^2+y^2)}{4\pi\left(x^2+y^2+\delta^2\right)^{3/2}}e^{-j\frac{\omega}{c}\sqrt{x^2+y^2+\delta^2}}.\,I(\omega)e^{jkt}\text{d}\omega$$

Here, $I\ (\omega)$ is the Fourier transform of source current, i.e., $I(t)$ is the probe current

$$I(\omega) = \frac{1}{2}\sum_{mnp}|I(mnp)|\left[\delta(\omega-\omega(mnp))e^{j\phi}(mnp) + e^{j\phi}(mnp)\delta(\omega+\omega(mnp))\right]$$

$$E_z(x,y,z,t) = \frac{\mu dl(x^2+y^2)}{4\pi(x^2+y^2+\delta^2)^{3/2}} \omega(mnp)|I(mnp)| \sin\left(\omega(mnp)\left(t - \frac{\sqrt{x^2+y^2+\delta^2}}{c} + \phi(mnp)\right)\right)$$
$$= |C_{mnp}| u_{mn}(x,y) \cos(\omega(mnp)t + \psi(mnp)) \sqrt{\frac{2}{d}} \sin\left(\frac{p\pi\delta}{d}\right)$$

Hence, $C_{mnp} = \sqrt{\frac{\beta(mnp)^2 + \alpha(mnp)^2}{\left[\frac{\sqrt{2}}{\sqrt{d}}\sin\left(\frac{p\pi\delta}{d}\right)\right]^2}}$; amplitude coefficient

$$\psi(mnp) = \tan^{-1}\left[\frac{\alpha_{mnp}\cos(\phi(mnp)) + \beta_{mnp}\sin(\phi(mnp))}{\alpha_{mnp}\sin(\phi(mnp)) - \beta_{mnp}\cos(\phi(mnp))}\right]; \text{Phase}$$

This completely solves the problem of RDRA resonant modes' coefficients in homogeneous medium.

5.3 RDRA Maxwell's Equation-Based Solution

Maxwell's equations with J electric and M magnetic sources:

$$\nabla \times \underline{E} = -j\omega\mu\underline{H} - \underline{M}; \qquad (a)$$

$$\nabla \times \underline{H} = \underline{J} + j\omega\epsilon\underline{E}; \qquad (b)$$

$$\nabla \times E = \frac{\rho_v}{\varepsilon}; \quad \nabla \times H = \frac{\rho_m}{\mu};$$

where ρ_v is the electric charge density, and ρ_m is the magnetic charge density.

$$\text{For consistency, } -j\omega\varepsilon\nabla \times H - \nabla \times M = 0;$$
$$\nabla \times J + j\omega\varepsilon\nabla \times E = 0;$$
$$\text{i.e.} \quad \nabla \times M + j\omega\rho_m = 0, \nabla \times J + j\omega\rho = 0;$$

namely conservation of electric and magnetic charge:

$$\nabla\times(\nabla \times E) = -j\omega\mu\nabla \times H - \nabla \times M;$$

taking curl on both sides

$$\text{or} \quad \nabla(\nabla \times E) - \nabla^2 E = -j\omega\mu\,(J + j\omega\epsilon E) - \nabla \times M;$$

$$\text{or} \quad (\nabla^2 + k^2)\underline{E} = \frac{\nabla\rho}{\varepsilon} + j\omega\mu J + \underline{\nabla} \times \underline{H} = \underline{s}\,(\text{electric source}); \quad (c)$$

i.e., $\underline{E}$ satisfies the Helmholtz equation with source.

$$\text{Likewise,} \quad \nabla \times (\nabla \times \underline{H}) = \nabla \times J + j\omega\varepsilon\nabla \times E$$
$$\text{or} \quad \nabla(\nabla \times H) - \nabla^2 H = \nabla \times J + j\omega\varepsilon(-j\omega\mu H - M)$$
$$\text{or} \quad (\nabla^2 + k^2)\underline{H} = \frac{\underline{\nabla}\rho_m}{\mu} + j\omega\varepsilon M - \underline{\nabla} \times \underline{J} = \underline{f}$$
$$\text{(magnetic source due to probe);} \quad \text{(d)}$$

Hence, $\underline{H}$ also satisfies Helmholtz equation with source. Rectangular cavity resonator sidewalls are the perfect magnetic conductors (PMC) and top and bottom surfaces are the perfect electric conductors (PEC). Applying these boundary conditions, we get the following equation:

$$H_z = 0; \quad \text{where } x = 0, a \text{ or } y = 0, b$$

So,

$$H_z(x, y, z) = \sum_{m,n \geq 1} \phi_{mn}(z) u_{mn}(x, y|a, b) \tag{5.1}$$

where

$$u_{mn}(x, y|a, b) = \frac{2}{\sqrt{ab}} \sin\left(\frac{m\pi x}{a}\right) \sin\left(\frac{n\pi y}{b}\right) \tag{5.2}$$

as we know,

$$H_z^{(0)}(x, y, z) = C_{mn} \frac{2}{\sqrt{ab}} \sin\left(\frac{m\pi x}{a}\right) \sin\left(\frac{n\pi y}{b}\right) \sin\left(\frac{p\pi z}{d}\right)$$

Let

$$f_z(x, y, z) = \sum_{mn \geq 1} f_{zmn}(z) u_{mn}[x, y|a, b] \tag{5.3}$$

$$\begin{aligned}(\nabla^2 + k^2)H_z &= \sum_{mn} \phi''_{mn}(z) + (k^2 - h^2[m, n|a, b]\phi_{mn}(z))u_{mn}[x, y|a, b] \\ &= f_z \Rightarrow \phi''_{mn}(z) + \left(k^2 - h^2[m, n|a, b]\phi_{mn}(z)\right) \\ &= f_{zmn}(z)\end{aligned} \tag{5.4}$$

where $k^2 = \omega^2\mu\epsilon$, and $h^2[x, y|a, b] = \pi^2\left(\left(\frac{m}{a}\right)^2 + \left(\frac{n}{b}\right)^2\right)$;

$$H_z = 0; \quad \text{for} \quad z = 0, d; \quad \text{completely determines}\, \varnothing_{mn}(z)$$

from (1),

Taking Laplace Transform of (5.4);

$$S^2\phi_{mn}(s) - S\phi_{mn}(0) - \phi'_{mn}(0) + \gamma_z^2[m,n]\widehat{\varnothing}_{mn}(s) + \hat{f}_{zmn}(s)$$

So,

$$\widehat{\varnothing}_{mn}(s) = \frac{\hat{f}_{zmn}(s)}{s^2 + \gamma_z^2[m,n]} + \frac{S\varnothing_{mn}(0) - \varnothing'_{mn}}{S^2 + \gamma_z^2[m,n]} \tag{5.5}$$

where $\gamma_z^2[m,n] = k^2 - h^2[m,n|a,b]$.

Thus,

$$\widehat{\varnothing}_{mn}(z) = \frac{1}{\gamma_z[m,n]}\int_0^z \sin(\gamma_z[m,n](z-\xi))\, f_{zmn}(\xi)\mathrm{d}\xi$$
$$+ C_1\sin(\gamma_z[m,n]z) + C_2\cos(\gamma_z[m,n]z)$$

$$\varnothing_{mn}(0) = \varnothing_{mn}(d) = 0 \quad \Rightarrow \quad C_2 = 0,$$

$$C_1 = \frac{-1}{\gamma_z[m,n]\sin(\gamma_z[m,n]d)}\int_0^d \sin(\gamma_z[m,n](d-\xi)f_{zmn}(\xi)\mathrm{d}\xi)$$

So,

$$\varnothing_{mn}(z) = \frac{-1}{\gamma_z[m,n]\times\sin(\gamma_z[m,n]d)}\sin(\gamma_z[m,n]d)\int_0^z \sin(\gamma_z[m,n](z-\xi))f_{zmn}(\xi)\mathrm{d}\xi$$
$$-\sin(\gamma_z[m,n]z)\int_0^d \sin(\gamma_z[m,n](d-\xi))f_{zmn}(\xi)\mathrm{d}\xi$$

In the limit $k^2 \to \pi^2\left(\left(\frac{m}{a}\right)^2 + \left(\frac{n}{b}\right)^2 + \left(\frac{p}{d}\right)^2\right)$, we have, $\gamma_z^2[m,n] \to \left(\frac{\pi p}{d}\right)^2$ and we get,

$$\varnothing_{mn}(z) \to \frac{d}{\pi p}\left\{\int_0^z \sin\left(\frac{\pi p}{d}(z-\xi)\right)f_{zmn}(\xi)\mathrm{d}\xi - \sin\left(\frac{\pi pz}{d}\right)\lim_{\lambda\to\frac{\pi p}{d}}\frac{\int_0^d \sin\left(\frac{\pi p}{d}(d-\xi)\right)f_{zmn}(\xi)\mathrm{d}\xi}{\sin(\lambda d)}\right\} \tag{5.6}$$

The limit in ∞ showing resonance, when

$$k^2 = \pi^2\left(\left(\frac{m}{a}\right)^2 + \left(\frac{n}{b}\right)^2 + \left(\frac{p}{d}\right)^2\right)$$

Let

$$f_{mn}(z) = \sum_{r=1}^{\infty} f_{zmnr}\sqrt{\frac{2}{d}}\sin\left(\sin\frac{r\pi t}{d}\right) \tag{5.7}$$

Then

$$\begin{aligned}&\int_0^d \sin\left(\frac{\pi p}{d}(d-\xi)\right) f_{zmn}(\xi)\mathrm{d}\xi \\ &= \sqrt{\frac{2}{d}}\sum_r f_{zmnr}\int_0^d \sin\left(\frac{\pi p}{d}(d-\xi)\right)\sin\left(\frac{r\pi\xi}{d}\right)\mathrm{d}\xi\end{aligned} \tag{5.8}$$

$$\begin{aligned}&= (-1)^{p+1}\sqrt{\frac{2}{d}}\sum_r f_{zmnr}\int_0^d \sin(\lambda\xi)\sin\left(\frac{r\pi\xi}{d}\right)\mathrm{d}\xi \\ &= (-1)^{p+1}\sqrt{\frac{2}{d}}\sum_r f_{zmnr}\times\frac{1}{2}\int_0^d \cos\left(\left(\lambda-\frac{r\pi}{d}\right)\xi\right) - \cos\left(\left(\lambda+\frac{r\pi}{d}\right)\xi\right)\mathrm{d}\xi \\ &= \frac{1}{2}(-1)^{p+1}\sqrt{\frac{2}{d}}\sum_r f_{zmnr}\left[\frac{\sin\left(\left(\lambda-\frac{r\pi}{d}\right)d\right)}{\left(\lambda-\frac{r\pi}{d}\right)} - \frac{\sin\left(\left(\lambda+\frac{r\pi}{d}\right)d\right)}{\left(\lambda+\frac{r\pi}{d}\right)}\right]\end{aligned} \tag{5.9}$$

Here λ propagation parameter $= kz \approx \frac{\pi p}{d}$

Thus,

$$\begin{aligned}&\frac{1}{\sin(\lambda d)}\int_0^d \sin(\lambda(d-\xi))\; f_{zmn}(\xi)\mathrm{d}\xi \\ &= \frac{1}{2}(-1)^{p+1}\sqrt{\frac{2}{d}}\sum_r f_{zmnr}\;\frac{1}{\sin(\lambda d)}\left[\frac{\sin(\lambda d)(-1)^r}{\left(\lambda-\frac{r\pi}{d}\right)} - \frac{\sin(\lambda d)(-1)^r}{\left(\lambda+\frac{r\pi}{d}\right)}\right] \\ &= \frac{1}{\sqrt{2d}}\;(-1)^{p+1}\sum_r (-1)^r f_{zmnr}\left[\frac{1}{\left(\lambda-\frac{r\pi}{d}\right)} - \frac{1}{\left(\lambda+\frac{r\pi}{d}\right)}\right] \\ &= \frac{(-1)^{p+1}}{\sqrt{2d}}\;\sum_r (-1)^r f_{zmnr}\left[\frac{\frac{2r\pi}{d}}{\left(\lambda^2-\left(\frac{r\pi}{d}\right)^2\right)}\right]\end{aligned} \tag{5.10}$$

Writing $\lambda = \frac{p\pi}{d} + \delta\ (\delta \to 0),$ we get,

$$\begin{aligned}
&\frac{1}{\sin(\lambda d)} \int_0^d \sin(\lambda(d-\xi)) \quad f_{zmn}(\xi)\mathrm{d}\xi \\
&\approx \frac{(-1)^{p+1}}{\sqrt{2\mathrm{d}}}(-1)^p \frac{f_{zmnp}}{\delta} \\
&= -\frac{1}{\delta\sqrt{2\mathrm{d}}} f_{zmnp} \qquad \text{(Dominant term)}
\end{aligned} \tag{5.11}$$

Hence,

$$\varnothing_{mn}(z) \approx \frac{d}{\pi p}\left\{\int_0^z \sin(\lambda(z-\xi)) \sum_{r\geq 1} f_{zmnr}\sqrt{\frac{2}{d}} \sin\left(\frac{r\pi\xi}{d}\right) \mathrm{d}\xi \ + \sin\left(\frac{p\pi z}{d}\right)\frac{1}{\delta\sqrt{2d}} f_{zmnp}\right\} \tag{5.12}$$

Now

$$\begin{aligned}
\int_0^z \sin(\lambda(z\ \xi)) \sin\left(\frac{r\pi\xi}{d}\right) \mathrm{d}\xi &= \frac{1}{2}\int_0^z \left[\cos\left(\lambda z - \left(\lambda + \frac{r\pi}{d}\right)\xi\right) - \cos\left(\lambda z - \left(\lambda - \frac{r\pi}{d}\right)\xi\right)\right]\mathrm{d}\xi \\
&= \frac{1}{2}\left[\frac{\left(\sin\left(\frac{r\pi z}{d}\right) + \sin(\lambda z)\right)}{\left(\lambda + \frac{r\pi}{d}\right)} - \frac{\left(\sin\left(\frac{r\pi z}{d}\right) - \sin(\lambda z)\right)}{\left(\lambda - \frac{r\pi}{d}\right)}\right] \\
&= \frac{1}{2}\left[\frac{\left(\sin\left(\frac{r\pi z}{d}\right) + \sin\left(\frac{p\pi z}{d} + \delta z\right)\right)}{\frac{\pi(p+r)}{d} + \delta} - \frac{\left(\sin\left(\frac{r\pi z}{d}\right) - \sin\left(\frac{p\pi z}{d} + \delta z\right)\right)}{\left(\frac{\pi(p-r)}{d} + \delta\right)}\right]
\end{aligned} \tag{5.13}$$

There is no dominant term here, i.e., if $\geq \mathrm{O}(\frac{1}{\delta})$, where O-order.

Hence, for $k^2 = \pi^2\left(\left(\frac{m}{a}\right)^2 + \left(\frac{n}{b}\right)^2\right) + \left(\frac{\pi p}{d} + \delta\right)^2$

$$\begin{aligned}
\varnothing_{mn}(z) &\approx \frac{d}{\pi p}\left(\frac{1}{\delta\sqrt{2d}} f_{zmnp} \sin\left(\frac{\pi p z}{d}\right)\right) \\
&= \frac{1}{\pi p\delta}\sqrt{\frac{d}{2}} f_{zmnp} \sin\left(\frac{p\pi z}{d}\right)
\end{aligned} \tag{5.14}$$

Likewise, propagation in x direction can be taken as:

$$\left(\nabla^2 + (k)^2\right)H_x = f_x$$
$$H_x(x, y, z) = \sum_{m,n} \text{ø}_{xmn}\tilde{u}_{mn}(y, z|b, d)$$
$$\text{Let, } f_x(\vec{x}, y, z) = \sum_{m,n} f_{xmn}(\mathrm{x})\tilde{u}_{mn}(y, z|b, d)$$

where $\tilde{u}_{mn}(y, z|b, d) = \frac{2}{\sqrt{bd}}\sin\left(\frac{m\pi y}{b}\right)\cos\left(\frac{n\pi y}{b}\right)$; orthogonal 2D half wave Fourier basis function.

Then,

$$\text{ø}''_{xmn}(x) + \left(k^2 - h^2[m, n|b, d]\right)\text{ø}_{xmn}(x) = f_{xmn}(x)$$

Hence, general solution can be given as follows:

$$\text{ø}_{xmn}(x) = \frac{1}{\gamma_x[m, n]}\int_0^x \sin(\gamma_x[m, n](x - \xi))f_{xmn}(\xi) + C_1\cos(\gamma_x[m, n]x)C_2\sin(\gamma_x[m, n]x) \tag{5.15}$$

Likewise,

$$H_y(x, y, z) = \sum_{m,n} \text{ø}_{ymn}(y)u_{mn}(x, z|a, d)$$
$$f_y(x, y, z) = \sum_{m,n} f_{ymn}(y)u_{mn}(x, z|a, d)$$

with

$$\text{ø}''_{ymn}(y) + \left(k^2 - h^2[m, n|a, d]\right)\varphi_{ymn}(y) = f_{ymn}(y)$$

and with the boundary conditions:

$$E_x = 0 \text{ where } x = 0, a \text{ or } z = 0, d;$$
$$E_y = 0 \text{ where } y = 0, b \text{ or } z = 0, d;$$

The general solution for $\text{ø}_{ymn}(y)$ is given as follows:

$$\text{ø}_{ymn}(y) = \frac{1}{\gamma_y[m, n]}\int_0^y \sin\left(\gamma_y[m, n](z - \xi)\right)f_{ymn}(\xi)\mathrm{d}\xi + D_1\cos(\gamma_y[m, n]y) + D_2\sin(\gamma_y[m, n]y) \tag{5.16}$$

Here, $\gamma_x[m,n] = (k^2 - h^2[m,n|b,d])^{1/2}$

$$\gamma_y[m,n] = (k^2 - h^2[m,n|a,d])^{1/2}$$

The equation

$$\nabla \times H = J + j\omega\epsilon E$$

gives

$$j\omega\epsilon E_x = H_{z,y} - H_{y,z} - J$$
$$j\omega\epsilon E_y = H_{x,z} - H_{z,x} - J$$

We assume that J on the walls is zero. Then, the boundary conditions yields

$$H_{z,y} - H_{y,z} = 0; \quad \text{where} \quad x = 0, a;$$
$$H_{x,z} - H_{z,x} = 0; \quad \text{where} \quad y = 0, b$$

Recall that H_z has been completely determined.

5.4 RDRA Inhomogeneous Permittivity and Permeability

$$\varepsilon = \varepsilon_0\left(1 + \delta_p X_e(x,y)\right) \tag{5.17}$$

$$\mu = \mu_0\left(1 + \delta_p X_m(x,y)\right) \tag{5.18}$$

At some known frequency ω and δ_{p} as perturbation parameter, the solution has been worked out using perturbation techniques to determine shift in the frequency.

As per Maxwell's equation,

$$\nabla \times \underline{E} = -j\omega\mu\underline{H}$$
$$\nabla \times \underline{H} = j\omega\varepsilon\underline{E}$$

where boundary conditions are given as follows:

$$0 \le x \le a = W$$
$$0 \le y \le b = L$$
$$0 \le z \le d = h$$

Due to duality $E \to H, H \to -E$, and $\mu \leftrightarrow \varepsilon$.

Sidewalls have been taken as PMC (magnetic conductor walls) and top and bottom as PEC (perfect electrical conductor).

$$
\begin{aligned}
&H_{\tan} = 0; \quad \text{on side walls} \\
&H_x, H_y = 0; \quad \text{when } \; y = 0, b \\
&H_y, H_x = 0; \quad \text{when } \; x = 0, d \\
&E_x, E_y = 0; \quad \text{when } \; z = 0, d
\end{aligned}
$$

$$\underline{E} = \underline{E}(x, y)e^{-\gamma z} \tag{5.19}$$

$$\underline{H} = \underline{H}(x, y)e^{-\gamma z} \tag{5.20}$$

Propagation constant is given as:
$h_0^2 = \gamma^2 + \omega^2 \mu_0 \epsilon_0, = \gamma^2 + k^2$; when $k^2 = \omega^2 \mu_0 \epsilon_0$

$$E_X = \frac{\gamma}{-h^2} E_{z'x} - \frac{j\omega\mu}{h^2} H_{z'y} \tag{5.21}$$

$$E_Y = \frac{\gamma}{-h^2} E_{z'y} - \frac{j\omega\mu}{h^2} H_{z'x} \tag{5.22}$$

$$H_X = \frac{\gamma}{-h^2} H_{z'x} - \frac{j\omega\epsilon}{h^2} E_{z'y} \tag{5.23}$$

$$H_Y = \frac{\gamma}{-h^2} H_{z'y} - \frac{j\omega\epsilon}{h^2} E_{z'x} \tag{5.24}$$

Top and bottom walls are perfect electric conditions so that

$$E_x, E_y = 0; \quad \text{when} \quad z = 0, d$$

$$\underline{E} \sim \underline{E}(x, y)\exp(-\gamma z), \qquad \sim \underline{H} - \underline{H}(x, y)\exp(-\gamma z)$$

$$\underline{E} = \underline{E}_\perp + E_z \hat{z}, \quad \underline{H} = \underline{H}_\perp + H_z \hat{z}.$$

$$\underline{\nabla} = \underline{\nabla}_\perp + \hat{z}\frac{\partial}{\partial z} = \underline{\nabla}_\perp - \gamma \hat{z}$$

$$\underline{\nabla}_\perp E_z \times \hat{z} - \gamma \hat{z} \times E_\perp = -j\omega\mu(\underline{H}_\perp) \tag{5.25}$$

$$\underline{E} = \underline{E}_\perp + E_z \hat{z}, \quad \underline{H} = \underline{H}_\perp + H_z \hat{z}. \tag{5.26}$$

$$\underline{\nabla}_\perp H_z \times \hat{z} - \gamma \hat{z} \times \underline{H}_\perp = j\omega\epsilon \underline{E}_\perp, \tag{5.27}$$

$$\underline{\nabla}_{\perp} \times \underline{H}_{\perp} = j w \epsilon \, E_z \hat{z} \tag{5.28}$$

Taking $\hat{z} \times$ of (5.25) gives

$$\underline{\nabla}_{\perp} E_z + \gamma \underline{\mathrm{E}}_{\perp} = -j\omega\mu\hat{z} \times \underline{\mathrm{H}}_{\perp} \tag{5.29}$$

Equations (5.11) and (5.13) can be changed as:

$$\begin{bmatrix} j\omega\epsilon & \gamma \\ -\gamma & j\omega\mu \end{bmatrix} \begin{bmatrix} E_{\perp} \\ \hat{Z} \times \underline{H}_{\perp} \end{bmatrix} = \begin{bmatrix} \underline{\nabla}_{\perp} H_z \times \hat{z} \\ \underline{\nabla}_{\perp} \quad E_z \end{bmatrix} \tag{5.30}$$

$$\begin{bmatrix} E_{\perp} \\ \hat{z} \times \underline{H}_{\perp} \end{bmatrix} = \frac{\begin{bmatrix} -j\omega\mu & -\gamma \\ \gamma & j\omega\epsilon \end{bmatrix} \begin{bmatrix} \underline{\nabla}_{\perp} H_z \times \hat{Z} \\ \underline{\nabla}_{\perp} E_z \end{bmatrix}}{\omega^2 \mu\epsilon + \gamma^2}$$

$$\underline{E}_{\perp} = \frac{-\gamma}{h^2} \underline{\nabla}_{\perp} E_z - \frac{j\omega\mu}{h^2} \underline{\nabla}_{\perp} H_z \times \hat{z} \tag{5.31}$$

$$\hat{z} \times \underline{H}_{\perp} = \frac{\gamma}{h^2} \underline{\nabla}_{\perp} H_z \times \hat{z} + \frac{j\omega\mu}{h^2} \underline{\nabla}_{\perp} E_z \tag{5.32}$$

$$h^2 = h^2(x, y) = \gamma^2 + \omega^2 \mu(x, y)\varepsilon(x, y) = h_0^2 + k_0^2 \delta\chi(x, y),$$

where $h_0^2 = \gamma^2 + \omega^2 \mu_0 \varepsilon_0, k^2 = \omega^2 \mu_0 \varepsilon_0 = \gamma^2 + k^2$

$$\chi(x, y) = \chi_e(x, y) + \chi_m(x, y) + \delta \times \chi_e(x, y)\chi_m(x, y)$$

Taking $\hat{z}$ of (5.32) gives

$$-\left(\frac{\gamma}{h^2}\right) \nabla_{\perp} H_z + \frac{j\omega\varepsilon}{h^2} \hat{z} \times \nabla_{\perp} E_z = \underline{H}_{\perp} \tag{5.33}$$

from Eqs. (5.31) and (5.33),

$$\begin{aligned} E_X &= \frac{\gamma}{-h^2} E_{z,x} - \frac{j\omega\mu}{h^2} H_{z,y} \\ E_Y &= \frac{\gamma}{-h^2} E_{z,y} + \frac{j\omega\mu}{h^2} H_{z,x} \\ H_X &= \frac{\gamma}{-h^2} H_{z,x} + \frac{j\omega\epsilon}{h^2} E_{z,y} \\ H_y &= \frac{\gamma}{-h^2} H_{z,y} - \frac{j\omega\epsilon}{h^2} E_{z,x} \end{aligned}$$

From Eqs. (5.25) and (5.26),

$$\nabla_\perp \times \left(\frac{\gamma}{h^2}\nabla_\perp E_z + \frac{j\omega\mu}{h^2}\nabla_\perp H_z \times \hat{Z}\right) - j\omega\mu H_z\hat{z} = 0 \quad (5.34)$$

$$\text{or} \quad \left(\hat{z}\times\underline{\nabla}_\perp\left(\frac{\gamma}{h^2}\right), \underline{\nabla}_\perp E_z\right) - \left(\underline{\nabla}_\perp, \frac{j\omega\mu}{h^2}\underline{\nabla}_\perp H_z\right) - j\omega\mu H_z = 0$$

or

$$\underline{\nabla}_\perp^2 H_z + h^2 H_z + \left(\underline{\nabla}_\perp\left(\frac{j\omega\mu}{h^2}\right), \underline{\nabla}_\perp H_z\right) \times \frac{h^2}{j\omega\mu} - \frac{h^2}{j\omega\mu}\left(\hat{Z}\times\underline{\nabla}_\perp\left(\frac{\gamma}{h^2}\right), \underline{\nabla}_\perp E_z\right) = 0$$

or

$$(\underline{\nabla}_\perp^2 + h_0^2)H_z + \delta\left\{k^2 X H_z \delta^{-1}\log\left(\frac{\mu}{h^2}\right) + (\underline{\nabla}_\perp, \underline{\nabla}_\perp H_z) + \frac{\gamma k^2}{j\omega\mu h^2}(\underline{\nabla}_\perp X_1 \underline{\nabla}_\perp E_z)\right\} = 0 \quad (5.35)$$

Now, we retain only $0\,(\delta)$ terms.

$$\chi \sim \chi_e + \chi_m$$
$$\frac{\chi_m}{h^2} \sim \frac{\chi_m}{h_0^2}$$
$$\frac{\gamma k^2}{j\omega\mu h^2} \sim \frac{\gamma k^2}{j\omega\mu h_0^2}$$

and (5.35) becomes

$$\begin{aligned}(\underline{\nabla}_\perp^2 + h_0^2)H_z + \delta\{k^2\,\chi\,H_z + (\underline{\nabla}_\perp\chi_m - \frac{\epsilon^2\underline{\nabla}_\perp\chi}{h_0^2}, \underline{\nabla}_\perp H_z)\\ + \frac{\gamma k^2}{j\omega\mu_0 h_0^2}(\underline{\nabla}_\perp, \chi\underline{\nabla}_\perp E_z)\} = 0\end{aligned} \quad (5.36)$$

By duality

$$E \to H, \quad H \to -E, \quad \chi_e \leftrightarrow \chi_m$$

$$\epsilon_0 \leftrightarrow \mu_0, \quad \chi \leftrightarrow \chi$$

we get from (5.36)

$$
\begin{aligned}
&(\underline{\nabla}_{\perp}^2 + h_0^2)E_z + \delta\Bigg\{k^2\chi E_z + (\underline{\nabla}_{\perp}\chi_e - \frac{k^2}{h_0^2}\underline{\nabla}_{\perp}\chi, E_z) + (\underline{\nabla}_{\perp}\chi_e - \frac{k^2}{h_0^2}\underline{\nabla}_{\perp}\chi, E_z) \\
&\qquad - \frac{\gamma k^2}{j\omega\epsilon_0 h_0^2}(\underline{\nabla}_{\perp}\chi, \underline{\nabla}_{\perp}H_z)\Bigg\} = 0
\end{aligned}
\tag{5.37}
$$

Boundary conditions are given as:

$$H_z = 0,\ \ x = 0, a \ \text{ and } \ Y = 0, b \quad H_z = 0,\ \ Z = 0, d$$

$$H_X = 0,\ \ Y = 0, b\ \ H_Y = 0,\ \ x = 0, a\ \ E_X = 0,\ \ x = 0, a$$

$$E_X = E_Y = 0,\ \ Z = 0, d\ \ E_Y = 0\ \ y = 0, b$$

Equations (5.28) and (5.29) are the own fundamental equations, let $h_0^2 = \lambda$.

Let

$$\lambda = \lambda_{m,n}^{(0)} + \delta \times \lambda^{(1)} + 0(\delta^2)$$

$$E_z = E_z^{(0)} + \delta E_z^{(1)} + 0(\delta^2) \tag{5.38}$$

$$H_z = H_z^{(0)} + \delta H_z^{(1)} + 0(\delta^2) \tag{5.39}$$

if there is non-homogeneity $\rightarrow$

$$\lambda_{n,m}^{(0)} = \pi^2\left(\frac{m^2}{a^2} + \frac{n^2}{b^2}\right)$$

$$\left(\underline{\nabla}_{\perp}^2 + \lambda_{n,m}^{(0)}\right)E_z^{(0)} = 0$$

$$\left(\underline{\nabla}_{\perp}^2 + \lambda_{n,m}^{(0)}\right)H_z^{(0)} = 0 \tag{5.40}$$

By Eqs. (5.36) and (5.37) $\left(0\left(\delta^0\right)\right)$

$$H_x^{(0)} = \frac{-\gamma}{\lambda_{m,n}^0} H_{z,x}^{(0)} + \frac{j\omega\epsilon 0}{\lambda_{m,n}^0} E_{z,y}^{(0)}$$

Since

$$H_Z^{(0)} = 0,\ \ \text{when}\ \ Y = 0, b \quad H_{z,x}^{(0)} = 0,\ \ \text{when}\ \ y = 0, b$$

Then

$$H_x^{(0)} = 0, \quad \text{when } y = 0, b \quad \rightarrow \quad E_{z,Y}^{(0)} = 0, \quad \text{when } Y = 0, b$$

Likewise

$$E_{z,X}^{(0)} = 0, \quad \text{when } X = 0, a$$

Thus,

$$H_z^{(0)} = C_{mn} \sin\left(\frac{m\pi x}{a}\right) \sin\left(\frac{n\pi y}{b}\right) \times \frac{2}{\sqrt{ab}} \tag{5.41}$$

$$E_z^{(0)} = D_{mn} \cos\left(\frac{m\pi x}{a}\right) \cos\left(\frac{n\pi y}{b}\right) \times \frac{2}{\sqrt{ab}} \tag{5.42}$$

If z-dependent is taken into account, then $H_z^{(0)}, E_z^{(0)}$ must be multiplied by $\exp(\pm\gamma z)$ according to Eq. (5.34),

$$E_x^{(0)} = \mp \frac{-\gamma}{\lambda_{m,n}^{0}} - \frac{j\omega\mu_0}{\lambda_{m,n}^{0}} \quad H_{z,Y}^{(0)}$$

and $E_x^{(0)} = 0$, when $z = 0, d$, and $H_z^{(0)} = 0$, when $z = 0, d$.

We get $E_{z,X}^{(0)} = 0$, when $z = 0, d$ then,

$$E_z^{(0)}(x, y, z) = D_{mn} \frac{2}{\sqrt{ab}} \cos\left(\frac{m\pi x}{a}\right) \cos\left(\frac{n\pi y}{b}\right) \sin\left(\frac{p\pi\delta}{d}\right) \tag{5.43}$$

$$\gamma = \frac{jp\pi}{d}, \quad p = 1, 2, 3.$$

Since $H_z^{(0)} = 0$, when $z = 0, d$,

$$H_z^{(0)}(x, y, z) = C_{mn} \frac{2}{\sqrt{ab}} \sin\left(\frac{m\pi x}{a}\right) \sin\left(\frac{n\pi y}{b}\right) \sin\left(\frac{p\pi z}{d}\right) \tag{5.44}$$

Frequency of oscillations:

$$\omega = \omega_{mnp}$$

$$\gamma^2 + \omega^2 \mu_0 \epsilon_0 = \lambda_{mn}^{(0)}$$

$$\text{or } \omega = \frac{\pi}{\sqrt{\mu_0 \epsilon_0}} \left(\left(\frac{m}{a}\right)^2 + \left(\frac{n}{a}\right)^2 \left(\frac{p}{d}\right)^2 \right)^{1/2}.$$

5.5 RDRA with Probe Current Excitation

The rectangular cavity has dimensions a, b, and d as shown in Fig. 5.2. Sidewalls are taken as magnetic conductors (PMC), and top and bottom surfaces are as PEC; theoretical fields (modes) solution has been worked under boundary conditions with a square-type feed probe for excitation.

$E_X, E_y = 0$, top and bottom plane being electric walls.

$E_X, E_y = 0$, sidewalls being magnetic walls.

$$H_z(x,y,z,t) = \sum_{mnp} \mathrm{C}(m,n,p)\frac{2}{\sqrt{ab}}\sin\left(\frac{m\pi x}{a}\right)\sin\left(\frac{n\pi y}{b}\right)\sin\left(\frac{p\pi z}{d}\right)\{\cos(\omega(m,n,p)t + \phi(m,n,p))$$

where m, n, and p are the integers (half wave variations in particular direction, i.e., x, y, z directions, respectively); a, b, and d are the dimensions (width, length, and height) of the RDRA, $C(m,n,p)$ and $\phi(m,n,p)$ are the magnitude and phase coefficients of H_z and $D(m,n,p)$ and $\psi(m,n,p)$ for E_z.

Let, orthogonal 2D half wave Fourier basis function $= \frac{2}{\sqrt{ab}}\sin\left(\frac{m\pi x}{a}\right)\sin\left(\frac{n\pi y}{b}\right) = u_{mn}(x,y)$ for convenience.

$$E_z(x,y,z,t) = \sum_{mnp} d(m,n,p)\frac{2}{\sqrt{ab}}\cos\left(\frac{m\pi x}{a}\right)\cos\left(\frac{n\pi y}{b}\right)\cos\left(\frac{p\pi z}{d}\right)\{\cos\omega(m,n,p)t + \psi(m,n,p)\}$$

Let, $\frac{2}{\sqrt{ab}}\cos\left(\frac{m\pi x}{a}\right)\cos\left(\frac{n\pi y}{b}\right) = v_{mn}(x,y)$ for convenience = orthogonal 2D half wave Fourier basis function.

From Lorentz Gauge conditions, $E_z = -j\omega A_z - \frac{\partial\bar{\phi}}{\partial z}$

Therefore, the magnetic vector potential can be given as below in discrete form after taking Fourier transform of A_z.

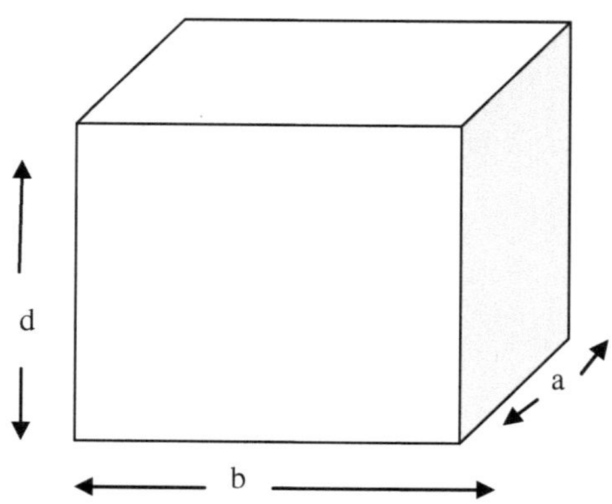

Fig. 5.2 RDRA with square feed probe inserted in $a \times b$ plane

$\hat{A}_z(x, y, z, \omega) = \frac{\mu}{4\pi}\hat{I}(\omega)\frac{\delta l e^{-j\frac{\omega r}{c}}}{r}$; where $\delta l =$ probe length

Div $\underline{\hat{A}} = \frac{\partial \hat{A}_Z}{\partial z}$; need to be computed.

Now, if we insert this probe at the location defined below into the cavity to find the fields pattern, we get:

$$\frac{l_2}{2} < |x - \frac{a}{2}| < \frac{a}{2}, \quad \frac{l_2}{2} < |y - \frac{b}{2}| < \frac{b}{2}$$

Then, the magnetic vector potential will be

$$\frac{\partial \hat{A}_z}{\partial z} = \frac{\mu \hat{I} \delta l}{4 \times \pi}\left[\frac{\partial}{\partial z}\frac{e^{-jkr}}{r}\right) = \frac{\mu \hat{i} \delta l}{4 \times \pi}\left(\frac{\cos\theta}{r^2} - \frac{jk\cos\theta}{r}\right)e^{-jkr} = -\frac{j\omega}{c^2}\underline{\hat{\phi}}$$

and scalar potential will be

$$\underline{\hat{\phi}} = \frac{\mu \hat{I} \delta l j c^2}{4 \times \pi\omega}\cos\theta\left(\frac{1}{r^2} - \frac{jk}{r}\right)e^{-jkr} = \frac{\mu \hat{i} \delta l c^2}{4 \times \pi\omega}\left(\frac{z}{r^3} - \frac{jkz}{r^2}\right)e^{-jkr}$$

Differentiating $\underline{\hat{\phi}}$ w.r.t. z

$$\frac{\partial \underline{\hat{\phi}}}{\partial z} = \frac{\mu \hat{i} \delta l c^2}{4\pi\omega}\left(\frac{1}{r^3} + \frac{3z^2}{r^5} - \frac{jk}{r^2} - \frac{2jkz^2}{r^4} + \left(\frac{z}{r^3} - \frac{jkz}{r^2}\right)\left(\frac{-jkz}{r}\right)\right)e^{-jkr} \tag{5.45}$$

when, $\hat{E}_z = -j\omega\hat{A}_z - \frac{\partial\hat{\phi}}{\partial z}$, substituting $\frac{\partial\hat{\phi}}{\partial z}$ in $\hat{E}_z$,

$$\hat{E}_z = \left[\frac{-j\omega\mu\hat{I}\delta l}{4\pi r} - \frac{\mu\hat{I}\delta l c^2}{4\pi\omega}\left(\frac{1}{r^3} + \frac{3z^2}{r^5} - \frac{jk}{r^2} - \frac{2jkz^2}{r^4} - \frac{jkz^2}{r^4} - \frac{k^2z^2}{r^3}\right)\right]e^{-jkr} \tag{5.46}$$

If we take $\theta = \frac{\pi}{2}, \quad z = 0$

$$\text{If we take} \quad \theta = \frac{\pi}{2}, \quad z = 0$$

$$\hat{E}_z = \frac{\omega c \mu \hat{I} \delta l}{4\pi k}\left[\frac{-jk^2}{r} - \left(\frac{1}{r^3} + \frac{3\,\delta^2}{r^5} - \frac{jk}{r^2} - \frac{2jk\delta^2}{r^4} - \frac{jk\delta^2}{r^4} - \frac{k^2\delta^2}{r^3}\right)\right]$$

Also,

$$r = \sqrt{x^2 + y^2 + \delta^2}$$

$\frac{k^2 r^2}{r} = \frac{1}{r}$, for $r = \lambda = 2\pi/k$.

Coulomb component of electric field is dominant in this inductive zone $r^2 \approx \delta^2$ given that $\delta \ll r$.

Minimum of r $\approx l_2$ and Maximum of $r = (a, b)$;

$$kr \ll 1$$

Hence,

$$\hat{E}_z \approx \frac{\mu c \hat{I} \delta l}{4\pi k r^3} \quad \text{and} \quad \approx \frac{\mu c^2 \delta l}{4\pi r^3 j\omega} \hat{I}(\omega);$$

$$E_z(t, x, y\delta) \approx \frac{\delta l}{4\pi\epsilon\left(\sqrt{x^2 + y^2}\right)^3} \int_0^t I(\tau)\mathrm{d}\tau \approx \frac{Q(t)\delta l}{4\pi\epsilon r^3} \tag{5.47}$$

Charge flowing through the resonator is $Q(t) = \int_0^t I(\tau)\mathrm{d}\tau$ or equivalently $= \frac{\hat{I}(\omega)}{j\omega} = \hat{Q}(\omega)$,

Here,

$$\frac{Q(t)\delta l}{4\pi\epsilon(x^2 + y^2)^{\frac{3}{2}}} \approx \sum_{mnp} D(m, n, p) v_{mn}(x, y) \sin\left(\frac{\pi p \delta}{d}\right) \cos(\omega(m, n, p)t + \psi(m, n, p))$$

and

$$\frac{Q(t)}{4\pi\epsilon(x^2 + y^2)^{\frac{3}{2}}} = \sum_{mnp} C(m, n, p) v_{mn}(x, y) \frac{\pi p}{d} \cos(\omega(m, n, p)t + \gamma(m, n, p))$$

For complete solution, we need to compute $D(m, n, p)$ and $\psi(m, n, p)$ coefficients for H_z fields and $C(m, n, p)$ and $\gamma(m, n, p)$ for E_z fields. The $D(m, n, p)$ and $C(m, n, p)$ are the desired resonant modes. For region,

$$\frac{l_2}{2} < |x - \frac{a}{2}| < \frac{a}{2}$$

$$\frac{l_2}{2} < |y - \frac{b}{2}| < \frac{b}{2}$$

$$\frac{Q(t)}{4\pi\epsilon} \int\int \frac{v_{mn}(x, y)}{(x^2 + y^2)^{\frac{3}{2}}} \mathrm{d}x\mathrm{d}y = \sum_p \frac{D(m, n, p)\pi p}{d} \cos(\omega(m, n, p)t + \psi(m, n, p)) \tag{5.48}$$

$$\left\{\frac{a + l_2}{2} < x < a;\ 0 < x < \frac{a - l_2}{2}\right\} \cap \left\{\frac{b + l_2}{2} < y < b \cup 0 < y \frac{b - l_2}{2}\right\}$$

$$\langle \cos(\omega t)\cos(\omega t)\rangle = \frac{1}{2}$$
$$\langle \sin(\omega t)\cos(\omega t)\rangle = 0$$

$$\frac{D(m,n,p)\pi p}{2d}\cos(\psi(m,n,p)) = \frac{1}{4\pi\epsilon}\langle Q(t)\cos(\omega(m,n,p)t)\rangle \int \frac{v_{mn}(x,y)}{(x^2+y^2)^{\frac{3}{2}}}\mathrm{d}x\mathrm{d}y$$
$$-\frac{D(m,n,p)\pi p}{2d}\sin(\psi(m,n,p)) = \frac{1}{4\pi\epsilon}\langle Q(t)\sin(\omega(m,n,p)t)\rangle \int \frac{v_{mn}(x,y)\mathrm{d}x\mathrm{d}y}{(x^2+y^2)^{\frac{3}{2}}}$$

Hence,

$$D(m,n,p) = 2\mathrm{d}/\pi p\sin(\psi(m,n,p))\left(\frac{1}{4\pi\epsilon}\langle Q(t)\sin(\omega(m,n,p)t)\rangle \int \frac{v_{mn}(x,y)\mathrm{d}x\mathrm{d}y}{(x^2+y^2)^{\frac{3}{2}}}\right). \tag{5.49}$$

5.6 RDRA Resonant Modes Coefficients in Homogeneous Medium

The basic Maxwell's theory can be applied with boundary conditions to express RDRA resonant fields as superposition of these characteristics frequencies. RDRA is shown in Fig. 5.3. u_{mn} depends on input excitation = orthogonal Fourier basis function, h_{mn} resonant mode (cut off frequency), k propagation constant. The generation of modes or characteristics frequencies $\omega(mnp)$ due to electromagnetic fields oscillations inside the cavity resonator has been described. Orthogonal Fourier basis function $u_{m,n}(x,y) = \frac{2}{\sqrt{ab}}\sin\left(\frac{m\pi x}{a}\right)\sin\left(\frac{n\pi y}{b}\right)$; $\omega(mnp)$ is the characteristic frequency and $\psi(mnp)$ is the phase of current applied. The rectangular cavity resonator is excited at the center with an antenna probe carrying current $i(t)$ of some known

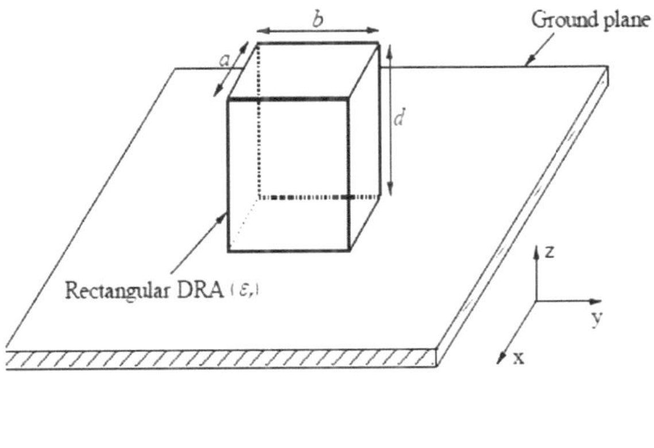

Fig. 5.3 RDRA with ground plane

frequency $\omega(mnp)$. This generates the field E_z inside the cavity of the form given below:

$$k^2 + \gamma_{mn}^2 = h_{mn}^2$$

hence,

$$k^2 = h_{mn}^2 + \frac{\pi^2 p^2}{d^2}$$

$$E_z(x, y, z, t) = \sum_{m,n,p} \mathrm{Re} \int C_{mnp}\, e^{j\omega(mnp)t}\, u_{mnp}(x, y, z);$$

or $\sum\limits_{m,n,p} |C_{mnp}| u_{mnp}(x, y, z) \cos(\omega(mnp)t + \psi(mnp));$

$$E_z\ (x, y, \delta, t) = \int G(x, y) \frac{j\omega\mu I \mathrm{d}l(x^2 + y^2)}{4\pi (x^2 + y^2 + \delta^2)^{3/2}} e^{\left(j\omega t - \frac{\omega}{c}\sqrt{x^2+y^2+\delta^2}\right)} \times I(\omega) e^{j\omega t} \mathrm{d}\omega$$

where $G(x,y)$ are the constant terms associated with the current.

Equating RDRA probe current fields with the antenna-radiated current fields at $z = \delta$;

Radiated currents:

$$= \sum_{p} |C_{mnp}| \sqrt{\frac{2}{d}} \sin\left(\frac{p\pi\delta}{d}\right) \cos(\omega(mnp)t + \phi(mnp)\ u_{m,n}(x, y));$$

Due to orthonormality, probe currents will be equal to radiated fields.

Probe currents:

$$= \int G(x, y) \frac{j\omega\mu I \mathrm{d}l(x^2 + y^2)}{4\pi (x^2 + y^2 + \delta^2)^{3/2}} I(\omega) e^{jkt} \mathrm{d}\omega \left(e^{\left(j\omega t - \frac{\omega}{c}\sqrt{x^2+y^2+\delta^2} + \psi_{mnp}\right)} u_{m,n}(x, y) \right) \mathrm{d}x\mathrm{d}y$$

It is clear that these two expressions have to be equal due to energy conservation. The probe current can be defined as:

$$I(\omega) = \frac{1}{2} \sum_{mnp} |I(mnp)| \left[\delta(\omega - \omega(mnp) e^{j\varnothing(mnp)}) + e^{j\varnothing(mnp)} \delta(\omega - \omega(mnp)) \right]$$

The antenna probe current must contain only the resonator characteristics frequencies $\omega(mnp)$. The radiated and input currents are equated as:

$$\sum_p = |C_{mnp}|\sqrt{\frac{2}{d}}\sin\left(\frac{p\pi\delta}{d}\right)\cos((\omega(mnp)t + \phi(mnp))u_{m,n}(x,y))$$

$$= \int G(x,y)\frac{j\omega\mu I \mathrm{d}l(x^2+y^2)}{4\pi(x^2+y^2+\delta^2)^{3/2}}.I(\omega)e^{jkt}\mathrm{d}\omega(e^{\left(j\omega t-\frac{\omega}{c}\sqrt{x^2+y^2+\delta^2}+\psi_{mnp}\right)}u_{m,n}(x,y)\mathrm{d}x\mathrm{d}y;$$

probe current = radiated current; thus C_{mnp} can be completely determined.

Hence, we can conclude that modes generation is due to the dipole moment in cavity resonator, mostly depend on size, dimensions of device, excitation type, coupling, and point of excitation.

5.7 RDRA Modes with Different Feed Position

Let us take $z = \delta$, i.e., very small probe length inserted into RDRA resonator at point of insertion $\left({}^{a}\!/_{2}, {}^{b}\!/_{2}, \delta\right)$ or $\left(x - {}^{a}\!/_{2}, y - {}^{b}\!/_{2}, \delta\right)$; where δ—length of insertion.

(H_x, H_y, E_x, E_y), transverse fields; (E_z and H_z) longitudinal fields

$$E_z = \sum_{mnp} u_{mnp}(x,y,z)R_e\left(C_{mnp}e^{j\omega_{mnp}t}\right)$$

where $u_{mnp}(x,y,z) = \frac{2^{3/2}}{\sqrt{abd}}\sin\left(\frac{m\pi x}{a}\right)\sin\left(\frac{n\pi y}{b}\right)\sin\left(\frac{p\pi z}{d}\right) = E_z$, (when top and bottom walls are PMC, rest all four walls are PEC).

Appling boundary conditions on transparent sidewalls (or all four sides of RDRA or resonator) and top and bottom planes as electrical walls, we get $H_z = 0$, for magnetic walls; and $E_z = 0$, for electrical walls; fields to be computed are

(E_z, H_z)—longitudinal fields;

$$E_z(x,y,z,t) = \sum_{mnp}\mathrm{Re}\int C_{mnp}e^{j\omega(mnp)t}u_{mnp}(x,y,z)$$

At $z = 0$; E_z, E_x, E_y all will be zero

$E_z = \sum_{mnp}\mathrm{Re}\left[C_{mnp}e^{j\omega(mnp)t}\right]\sqrt{\frac{2}{d}}\sin\left(\frac{p\pi\delta}{d}\right)u_{mn}(x,y)$; this is the E_z field in the resonator at $z = \delta$. It must be equated to the corresponding field generated by the antenna probe, i.e., for the above two expressions to be equal, the antenna probe currents must contain frequencies only from the set $\{\omega(mnp)\}$.

Where $u_{mn}(x,y) = \frac{2}{\sqrt{ab}}\sin\left(\frac{m\pi}{a}\right)\sin\left(\frac{n\pi}{b}\right)$

E_z will exist little above from $z = 0$ plane; $E_z = \int \frac{j\omega\mu I \mathrm{d}l e^{-jk\sqrt{x^2+y^2+\delta^2}}}{4\pi\sqrt{x^2+y^2+\delta^2}} I(\omega)\mathrm{d}\omega$; where $I(\omega)$ is the Fourier transform of $i(t)$

$$E_z = -j\omega A_z - \frac{\partial \phi}{\partial z}$$

$$\mathrm{div}{\cdot}A = \frac{\mu I \mathrm{d}l}{4\pi}(-jk\cos\theta)e^{-jkr} \tag{5.50}$$

$$\frac{kc^2}{\omega} = -\frac{j\omega}{c^2}\phi$$

Hence, scalar potential $\phi = \frac{\mu I \cos}{4\pi r} e^{jkr}$

$$\begin{aligned} \frac{\partial\phi}{\partial z} &= \frac{\mu I\cos\theta}{4\pi r}(-jk\cos\theta)e^{-jkr} \\ &= \frac{jk\mu I\cos 2\theta}{4\pi r}e^{-jkr} \end{aligned} \tag{5.51}$$

$$E_z = -\frac{j\omega\mu I \mathrm{d}l}{4\pi r}e^{-jkr} + \frac{j\omega c\mu I dl\cos 2\theta}{4\pi r}e^{-jkr}$$

Hence,

$$E_z \text{ will be } \frac{j\omega\mu I \mathrm{d}l\sin^2\theta}{4\pi r}\ e^{-jkr}$$

where

$$\cos\theta = \frac{\delta}{\sqrt{x^2+y^2+\delta^2}}$$

$$\sin^2\theta = \frac{x^2+y^2}{x^2+y^2+\delta^2}$$

$$E_z|_{z=} = \delta\sum_{mnp}\mathrm{Re}[C(mnp)e^{j\omega(mnp)t} \times \sqrt{\frac{2}{d}}\sin\left(\frac{p\pi\delta}{d}\right)u_{m,n}(x,y) \tag{5.52}$$

$$E_z = \int \frac{j\omega\mu I \mathrm{d}l(x^2+y^2)}{4\pi\left(x^2+y^2+\delta^2\right)^{3/2}}e^{-j\frac{j\omega}{c}\sqrt{x^2+y^2+\delta^2}} \times I(\omega)e^{jkt}\mathrm{d}\omega \tag{5.53}$$

Here, $I(\omega)$ is the Fourier transform of source current, i.e., $I(t)$ probe current

$$I(\omega) = \frac{1}{2}\sum_{mnp} |I(mnp)|[\delta(\omega - \omega(mnp))e^{j\o}(mnp) + e^{j\o}(mnp)\delta(\omega - \omega(mnp))]$$

$$I(\omega) = \frac{1}{2}\sum_{mnp} |I(mnp)|[\delta(\omega - \omega(mnp))e^{j\o}(mnp) + e^{j\o}(mnp)\delta(\omega - \omega(mnp))]$$

$$I(\omega) = \int \cos(\omega(mnp)t)e^{-j\omega t}\mathrm{d}t$$

When $\omega(mnp) = \pi^2\sqrt{\frac{m^2}{a^2} + \frac{n^2}{b^2} + \frac{p^2}{d^2}}$, probe current magnitude and phase $I(\omega) = \sum_{m,np}|I(mnp)|\cos(\omega(mnp)t + \phi(mnp))$ $\o(mnp)$ is the phase of current at frequency $\omega(mnp)$.

$$\begin{aligned} &E_z(x, y, z, t) \\ &= \frac{\mu \mathrm{d}l(x^2 + y^2)}{4\pi(x^2 + y^2 + \delta^2)^{3/2}}\omega(mnp)|I(mnp)|\sin\left(\omega(mnp)\left(t - \frac{\sqrt{x^2 + y^2 + \delta^2}}{c} + \phi(mnp)\right)\right) \\ &= |C_{mn}|u_{mn}(x, y)\cos\omega((mnp)t + \psi(mnp))\sqrt{\frac{2}{d}}\sin\left(\frac{p\pi\delta}{d}\right). \end{aligned} \tag{5.54}$$

5.8 *R*, *L*, *C* Circuits and Resonant Modes

The information contained in eigenvalue or eigenvector of modes can impart the knowledge of antenna radiation behavior, surface current distribution, input impedance, and its feeding point location. Combinations of feeding configuration and dimensions can generate or excite various modes. Thus, modes can be effectively used in design control of an antenna. Surface current and geometry of an antenna give eigenfunctions or eigenvectors. Closed-loop currents of eigenvectors that present inductive nature are the magnetic fields. Horizontal and vertical eigenvectors are noninductive are electric fields. These electric fields are produced by supplied probe currents. Number of lobes in radiation pattern gets increased if mode number or order of mode is increased and vice versa. The modal excitation coefficients shall depend on position, magnitude, and phase of the applied probe current. The effective current is superposition of all modes excited. The eigenvalue is most important because its magnitude tells effectiveness of radiation or reactive power and modes are the solution of characteristics equation. Smaller magnitude of

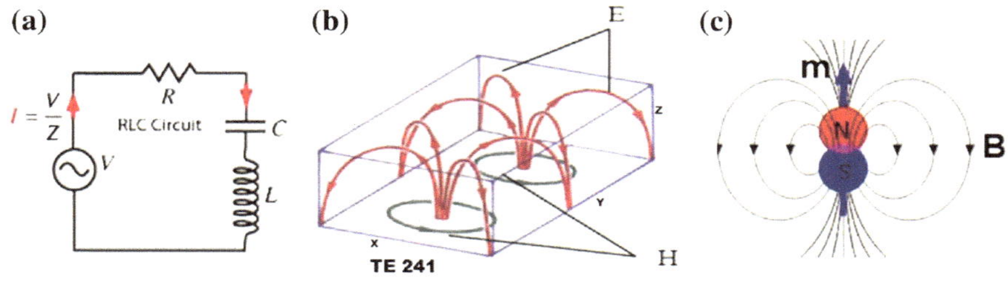

Fig. 5.4 a *RLC* circuit, b resonance higher modes, c magnetic dipoles

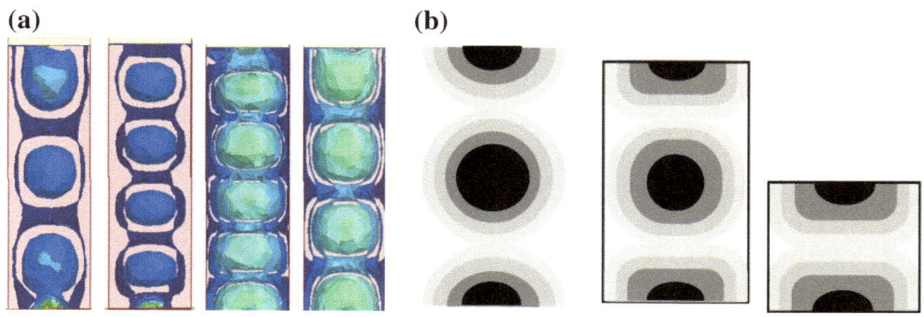

Fig. 5.5 Higher-order even and odd modes

eigenvalue is more efficient. Positive eigenvalue is the magnetic energy storing mode, and if modes are negative, it stores electric energy. The eigenvalue variation versus frequency gives information about resonance and radiation nature. Excitation angle can have impact on antenna quality factor. The excited mode will adjust the phase of the reflected currents. Orthogonality of modes can be used to produce circular polarization in the RDRA. Figure 5.4 represents the equivalent *RLC* circuit of RDRA, resonant modes excited, and corresponding magnetic dipoles. Figure 5.5 depicts the even and odd modes generation. Figure 5.6 presents RDRA HFF model along with its equivalent *RLC* circuit. Figures 5.7 and 5.8 are *RLC* circuits which are used for derivation of resonant frequency and impedance.

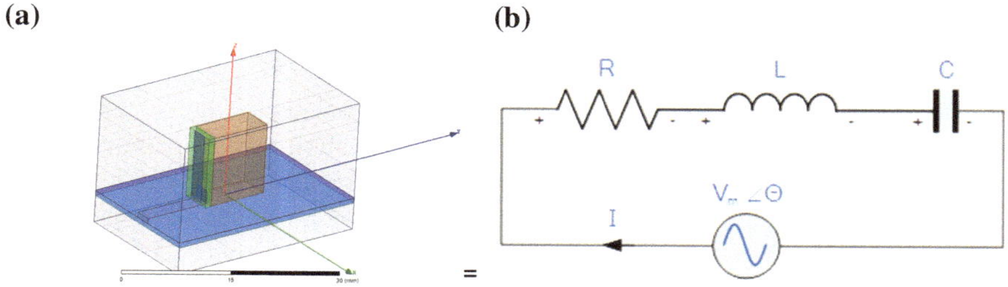

Fig. 5.6 a RDRA model and b equivalent *RLC* circuit

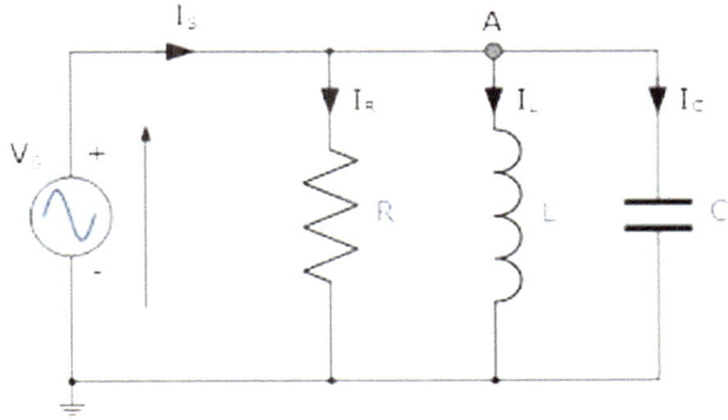

Fig. 5.7 *RLC* circuit

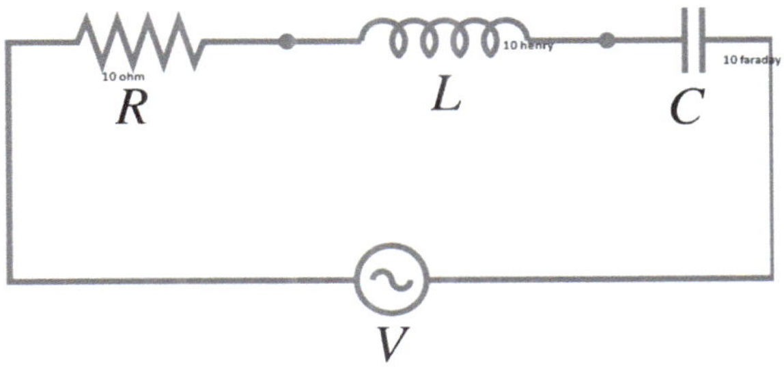

Fig. 5.8 Series *RLC* circuit

R, *L*, *C* equivalent circuit: An antenna can be represented as *R*, *L*, *C* circuitry with natural frequencies ω_c and forced resonance due to excitation eigen-valued (ω_{mnp}) has been determined along with eigenvector J_{mnp}. Separation of all frequencies will be the out come of modes. The second-order differential equation is the general solution of equivalent antenna (*R*, *L*, *C*) circuit. Fourier solution will provide a discrete solution of resonance. J_e is excitation current or probe current and γ is an propagation constant $(\gamma = \alpha + j\beta)$. *L*, *C* circuit will introduce non-homogeneous or inhomogeneous matter, ω^2 will be replaced in this case by $\omega^2\mu\epsilon \times -\gamma$ is replaced by $\tilde{\gamma}$ introducing decay. $H_z^{(f)}$ represents forced resonance mode.

$$L\ddot{q} + \frac{q}{C} + R\dot{q} = v_s(t)$$

where $\ddot{q} = \frac{d^2q}{\mathrm{d}t^2}$

$$X_L = j\omega L, \quad X_C = \frac{1}{j\omega C}$$

Taking Fourier Transform

$$\left((j\omega L)^2 + \frac{1}{C} + j\omega R\right)Q(\omega) = V_s(\omega)$$

$$Q\,(\omega) = \frac{V_s(\omega)}{(j\omega L)^2 + \frac{1}{c} + j\omega R}$$

$$q(t) = \int_{-\infty}^{+\infty} \frac{Q(\omega)e^{j\omega t}\mathrm{d}\omega}{2\pi}$$

$$R = 0$$

$$(\omega L)^2 = \frac{1}{\omega c}$$

$$\omega = \frac{1}{\sqrt{LC}}$$

$$\underline{J}_{se}(x, y, z) = J_{sx}(x, y)\delta(z - d_0)\hat{x} + J_{sy}(x, y)\delta(z - d_0)\hat{y}$$

where J_s is the current surface density, and J_e is the electron current

$$\int J_s \mathrm{d}z = J_{sx}(x, y, \omega)\hat{x} + J_{sy}(x, y, \omega)$$

From Maxwell's equation,

$$\underline{\nabla} \times \underline{H} = \underline{J}_e + (\sigma + j\omega\epsilon)E$$

$$\underline{\nabla} \times \underline{E} = -j\omega\mu\underline{H}$$

$$-\nabla^2 E = -j\omega\mu(\sigma + j\omega\epsilon)E + \underline{J}_e$$

$$\nabla^2 E_z = \gamma^2(\omega)E_z$$

When

$$\gamma(\omega) = \sqrt{j\omega\mu(\sigma + j\omega\epsilon)} = \alpha(\omega) + j\beta(\omega)$$

Similarly, we can compute

$$\nabla^2 H_z = \gamma^2(\omega)H_z$$

Boundary conditions are applied

$$H_z = 0, \;\; x = 0, a, \;\; \text{or} \;\; y = 0, b, \;\; z = 0, d$$

$$E_x = 0, \;\; x = 0, a, \;\; \text{when} \;\; z = 0, d, E_y = 0, y = 0, \;\; b, z = 0, d;$$

Fields propagating is H_z for TE mode

$$H_z(x,y,z,\omega) = \sum \frac{2\sqrt{2}}{\sqrt{abd}} \sin\left(\frac{n\pi x}{a}\right) \sin\left(\frac{m\pi y}{b}\right) \sin\left(\frac{p\pi d}{d}\right) \mathrm{Re}\left(C(mnp)\exp^{(j\omega(nmp)t)}\right)$$

$$(\nabla^2 - \gamma^2(\omega))E_z = 0$$

γ and $\tilde{\gamma}$ are two propagation constants

$$\begin{aligned}
&(\nabla^2 - \gamma^2(\omega))E_z = 0 \\
&(\nabla^2 - \gamma^2(\omega))\underline{E}_\perp = \underline{J}_e \\
&-\nabla^2 \underline{H} = \underline{\nabla} \times \underline{J}_e + (\sigma + j\omega\epsilon)(-j\omega\mu j\underline{H}) \\
&(\nabla^2 - \gamma^2(\omega))\underline{H} = -\underline{\nabla} \times \underline{J}_e \\
&(\nabla^2 - \gamma^2(\omega))\underline{E}_\perp = \underline{J}_e \\
&(\nabla^2 - \tilde{\gamma}^2)H_x = J_{sy}\delta'(z - d_0) \\
&(\nabla^2 - \tilde{\gamma}^2)H_y = -J_{sx}\delta'(z - d_0) \\
&\tilde{\gamma}^2(\omega) + \pi^2\left(\frac{n^2}{a^2} + \frac{m^2}{b^2} + \frac{p^2}{d^2}\right) = 0 \\
&j\omega\mu(\sigma + j\omega\epsilon) + \pi^2\left(\frac{n^2}{a^2} + \frac{m^2}{b^2} + \frac{p^2}{d^2}\right) = 0 \\
&-\tilde{\gamma}^2\left(\omega_{(mnp)}\right) = \omega^2\mu\epsilon - j\omega\mu\sigma = \pi^2\left(\frac{n^2}{a^2} + \frac{m^2}{b^2} + \frac{p^2}{d^2}\right) \\
&\omega(mnp) = \omega_{\mathrm{Real}}(mnp) + j\omega_{\mathrm{Img}}(mnp) \\
&e^{j\omega(mnp)t} = e^{j\omega_{\mathrm{Real}}(mnp)t} e^{-\omega_{\mathrm{Img}}(mnp)t} \\
&\omega^2\mu\epsilon \rightarrow \omega^2\mu\epsilon - j\omega\mu\sigma = -\tilde{\gamma}^2(\omega) \\
&(J_{s_{xy}}(x, y\ \omega) - J_{s_{yx}}(x, y\ \omega)\delta(z - d_0) = \sum J_z(n,m,p,\omega)u_{nmp}(x,y,z)
\end{aligned}$$

$$J[n,m,p,\omega] = \int_0^a \int_0^b (J_{s_{xy}}(x,y,\omega) - J_{s_{yx}}(x,y,\omega) \times \frac{2\sqrt{2}}{\sqrt{abd}} \sin\left(\frac{n\pi x}{a}\right) \sin\left(\frac{m\pi y}{b}\right) \sin\left(\frac{d_0\pi p}{d}\right) \mathrm{d}x\mathrm{d}y \tag{5.55}$$

$$(\nabla^2 - \tilde{\gamma}^2(\omega))H_z = \sum_{mnp} J_z[mnp,\omega]u_{mnp}(x,y,z)$$

$$H_z^{(f)} = \sum_{mnp} H_z[mnp,\omega]u_{mnp}(x,y,z);$$

When $H_z^{(f)}$ is the forced resonant mode, then

$$J_z[nmp,\omega] = [\pi^2\left(\frac{n^2}{a^2}+\frac{m^2}{b^2}+\frac{p^2}{d^2}\right)+\tilde{\gamma}^2(\omega)]H_z[nmp,\omega]$$

$$H_z^{(f)}(x,y,z\ \omega) = \sum_{mnp}\frac{J_z[nmp,\omega]u_{mnp}(x,y)}{\tilde{\gamma}^2(\omega)-\tilde{\gamma}^2(\omega(nmp))}$$

where

$$v_{mnp} = \frac{2\sqrt{2}}{\sqrt{abd}}\cos\left(\frac{n\pi x}{a}\right)\cos\left(\frac{m\pi y}{b}\right)\cos\left(\frac{p\pi z}{d}\right) = H_z$$

$$J_{sy}(x,y,\omega)\delta'(z-d_0) = \sum J_y[mnp,\omega]v_{mnp}(x,y,z)$$

Hence, current density

$$J_y[n,m,p,\omega] = \int_0^a\int_0^b\int_0^d J_{Sy}(x,y,\omega)\delta'(z-d_0)v_{mnp}(x,y,z)\mathrm{d}x\mathrm{d}y\mathrm{d}z$$

This completes the general solution of *R*, *L*, *C* circuit.

5.9 Resonant Modes Based on *R*, *L*, *C* Circuits

$$\underline{\nabla}_\perp H_z\times\hat{z}-\gamma\ \hat{z}\times\underline{H}_\perp = J_e+(\sigma+j\omega\epsilon)E_\perp \tag{5.56}$$

$$\underline{\nabla}_\perp E_z\times\hat{z}-\gamma\ \hat{z}\times\underline{E}_\perp = -j\omega\mu\underline{H}_\perp \tag{5.57}$$

$$\hat{z}\times(\underline{\nabla}_\perp E_z\times\hat{z}-\gamma\ \hat{z}\times\underline{E}_\perp) = \hat{z}\times(-j\omega\mu\underline{H}_\perp)$$

$$\underline{\nabla}_\perp E_z+\gamma\underline{E}_\perp = -j\omega\mu\ \hat{z}\times\underline{H}_\perp \tag{5.58}$$

Eliminate $\hat{z}\times\underline{H}_\perp$ from Eqs. (5.56) and (5.58)

$$\underline{\nabla}_\perp E_z+\gamma\underline{E}_\perp = \frac{-j\omega\mu}{\gamma}(\underline{\nabla}_\perp H_z\times\hat{z}-J_e-(\sigma+j\omega\varepsilon)E_\perp).$$

$$(\gamma^2-\tilde{\gamma}^2(\omega))E_\perp = -j\omega\mu\ \underline{\nabla}_\perp H_z\times\hat{z}+j\omega\mu\ J_e-\gamma\ \underline{\nabla}_\perp E_z$$

Hence,

$$E_{\perp} = \frac{-j\omega\mu}{\gamma^2 - \tilde{\gamma}^2(\omega)} \nabla_{\perp} H_z \times \hat{z} + \frac{j\omega\mu}{\gamma^2 - \tilde{\gamma}^2(\omega)} J_e - \frac{\gamma \nabla_{\perp} E_z}{\gamma^2 - \tilde{\gamma}^2(\omega)}$$

Parallel *RLC* Circuits solution:

$$Ii + L\frac{\mathrm{d}i}{\mathrm{d}t} + \frac{1}{c}\int i \,\mathrm{d}t = V$$

On differentiating

$$L\frac{\mathrm{d}i^2}{\mathrm{d}^2 t} + R\frac{\mathrm{d}i}{\mathrm{d}t} + \frac{1}{c} = 0$$

Second-order linear, homogeneous differential equation dividing by *L* both sides gives the following:

$$\frac{\mathrm{d}i^2}{\mathrm{d}^2 t} + \frac{R}{L}\frac{\mathrm{d}i}{\mathrm{d}t} + \frac{1}{LC} = 0$$

Taking Laplace transform

$$S^2 + \frac{R}{L}S + \frac{1}{LC} = 0$$

$$S = -\frac{R}{L} \pm \frac{\sqrt{\left(\frac{R}{L}\right)^2 - 4 \times \frac{1}{LC}}}{2}$$

$$S_1 = -\frac{R}{L} + \frac{\sqrt{\left(\frac{R}{L}\right)^2 - \left(\frac{2}{\sqrt{LC}}\right)^2}}{2}$$

$$S_2 = -\frac{R}{L} - \frac{\sqrt{\left(\frac{R}{L}\right)^2 - \left(\frac{2}{\sqrt{LC}}\right)^2}}{2}$$

Series *RLC* circuit

Let

$$\lambda = -\frac{R}{2L}$$
$$\omega_2 = \sqrt{\left(\frac{R}{2L}\right)^2 - \left(\frac{1}{Lc}\right)^2}$$
$$S_1 = \lambda + \omega_2$$
$$S_2 = \lambda - \omega_1$$

Hence, solution of differential equation can be written as:

$$I = A_1 e^{s_1 t} + A_z e^{s_2 t}$$

Here, A_1 and A_2 are the magnitude of currents

Now

$$\text{Care 1} \quad \left(\frac{R}{2L}\right)^2 > \frac{1}{Lc}$$

$$\text{Care 2} \quad \left(\frac{R}{2L}\right)^2 < \frac{1}{Lc}$$

$$\text{Care 3} \quad \left(\frac{R}{2L}\right)^2 = \frac{1}{Lc}$$

$$V = Ri + \frac{L\mathrm{d}i}{\mathrm{d}t} + \frac{1}{C}\int i\,\mathrm{d}t$$

Taking Laplace transformation

$$\frac{V}{s} = RI(s) + LsI(s) + \frac{I(s)}{sC}$$
$$I(s) = \frac{V}{s\left[R + Ls + \frac{1}{sC}\right]}$$
$$I(s) = \frac{V}{Rs + Ls^2 + \frac{1}{C}}$$

Since,

$$s_{1,2} = \frac{-R/L \pm \sqrt{\left(\frac{R}{L}\right)^2 - 4/LC}}{2}$$

$$I(s) = \frac{V}{L\left[s^2 + \frac{Rs}{L} + \frac{1}{LC}\right]} = \frac{V}{L(s_1 - s_2)}\left[\frac{1}{s - s_1} - \frac{1}{s - s_2}\right]$$

Taking Laplace inverse of equation

$$I(t) = \frac{1}{L\sqrt{\frac{R}{L^2} - \frac{4}{LC}}} e^{s_1 t} - \frac{1}{L\sqrt{\frac{R}{L^2} - \frac{4}{LC}}} e^{s_2 t}$$

Example 5.1 Series *RLC* circuit solution

$$v = R \times i(t) + L\frac{\mathrm{d}i(t)}{\mathrm{d}t}\frac{1}{C}\int_0^t i(t) \times \mathrm{d}t$$

Taking Laplace transform on both the sides gives

$$\frac{v}{s} = I(s) \times R + L[s \times I(s) - i(0)] + \frac{1}{C \times s} I(s) \quad [i(0) = 0]$$

$$\frac{v}{s} = I(s)\left[R + L \times s + \frac{1}{C \times s}\right]$$

$$v = I(s)\left[R \times s + L \times s^2 + \frac{1}{C}\right]$$

Roots of the equation are as follows:

$$L \times s^2 + R \times s + C^{-1} = 0$$

$$S = \frac{-R}{2L} \pm \sqrt{\left(\frac{R}{2L}\right)^2 - \frac{1}{LC}}$$

$$\text{Let} \quad s_1 = s = \frac{-R}{2L} + \sqrt{\left(\frac{R}{2L}\right)^2 - \frac{1}{LC}}$$

$$\text{and} \quad s_2 = \frac{-R}{2L} - \sqrt{\left(\frac{R}{2L}\right)^2 - \frac{1}{LC}}$$

$$\text{Now,} \quad s_1 + s_2 = \frac{-R}{L} \text{ and } s_1 s_2 = \frac{1}{LC}$$

$$v = I(s) \times L[s^2 - s \times (s_1 + s_2) + s_1 s_2]$$

$$v = I(s) \times L[s(s - s_1) - s_2(s - s_1)]$$

$$v = I(s) \times L[(s - s_1)(s - s_2)]$$

$$I(s) = \frac{v}{L} \times \frac{1}{(s - s_1)(s - s_2)}$$

Using partial fraction solution, we get

$$I(s) = \frac{v}{L(s_1 - s_2)} \times \left[\frac{1}{(s - s_1)} - \frac{1}{(s - s_2)}\right]$$

Taking inverse Laplace transform on both the sides

$$i(t) = \frac{v}{L(s_1 - s_2)} \times [e^{s_1 t} - e^{s_2 t}]$$

$$s_1 - s_2 = \frac{R}{L}\sqrt{1 - \frac{4L}{C}}$$

$$i(t) = \frac{v}{R\sqrt{1 - \frac{4L}{C}}} \times [e^{s_1 t} - e^{s_2 t}]$$

$$\text{Let,} \quad A_1 = -A_2 = \frac{v}{R\sqrt{1 - \frac{4L}{C}}}$$

$$i(t) = A_1 e^{s_1 t} + A_2 e^{s_2 t}$$

Chapter 6
Mathematical Analysis of Radiation Pattern of RDRA

Abstract In this chapter, detailed study using mathematical analysis for radiation pattern of RDRA has been described. RF excitation with proper impedance match can generate *J*-current density into surfaces of RDRA, which leads to produce *A*-magnetic vector potential and finally *E*-electric intensity or *H*-magnetic field intensity. Acceleration or deceleration of charge carriers causing current is mandatory phenomenon for radiations. Wave can only propagate if wave vector $k > k_c$, where k_c is cutoff frequency. The lowest resonance can be termed as dominant mode and second and third resonances are higher-order modes. Propagation constant $k_x = n\pi/a$, and propagation takes place if $k_x > n\pi/a$, while no propagation takes place if $k_x < n\pi/a$. Thus, standing waves inside the resonator are formed and energy storing will take place. Hence, mode spectrum will result into corresponding resonant frequency generation. Wave propagation can be well defined by Helmholtz equation. The Maxwell's equations describe the behavior of electromagnetic fields and form the basis of all EM classical phenomena. P_{rad} (power radiated) can be evaluated using Parseval's power theorem. The radiated power is produced by oscillating dipole moments. The current varying in time can be analyzed by Fourier analysis. If medium is inhomogeneous, wave possesses exponential growth or decay in some direction. Thus, Poynting vector "*S*" shall give the magnitude and phase of the radiated fields in particular direction.

Keywords Impedance match · Current density · Magnetic vector potential · Power radiated · Poynting vector · Persvals power theorem · Moat-shaped DRA

6.1 Introduction

RF excitation with proper impedance match can generate *J*-current density into surfaces of RDRA, which leads to produce *A*-magnetic vector potential and finally *E*-electric intensity. Acceleration or deceleration of charge carriers causing current is mandatory phenomenon for radiations. Wave can only propagate if wave vector

R.S. Yaduvanshi and H. Parthasarathy, *Rectangular Dielectric Resonator Antennas*, DOI 10.1007/978-81-322-2500-3_6

$k > k_c$, where k_c is cutoff frequency and the lowest resonance can be termed as dominant mode and second and third resonances are higher-order modes. Propagation constant $k_x = n\pi/a$. Propagation takes place if $k_x > n\pi/a$, while no propagation takes place $k_x < n\pi/a$. Standing waves inside the resonator are formed and energy storing will take place. Hence, mode spectrum will result into corresponding resonant frequency generation due to equivalent RLC circuit formation. Wave propagation can be well defined by Helmholtz equation. The Maxwell's equations describe the behavior of electromagnetic fields and form the basis of all EM classical phenomenon. P_{rad} (power radiated) can be evaluated using Parseval's power theorem. The radiated power is produced by oscillating dipole moments. The current varying in time can be analyzed by Fourier analysis. If medium is inhomogeneous, wave possesses exponential growth or decay in some direction. Thus, Poynting vector "*S*" shall give the magnitude and phase of the radiated fields in particular direction.

Finally, the radiation pattern produced by the surface electric and magnetic current densities on the RDRA surfaces is computed. PEC walls, the surface electric current density is $J_s = \hat{n} \times E$.

Then, the far-field magnetic vector and electric vector potentials are determined by the usual reactance potential formulae as follows:

$$\underline{A}(\omega, \underline{r}) = \frac{\mu}{4\pi} \frac{e^{-jkr}}{\sigma} \int_S \underline{J}_s(\omega, \underline{r}') \exp(jk\hat{r} \cdot \underline{r}') \, \mathrm{d}s(\underline{r}'); \tag{6.1a}$$

and

$$\underline{F}(\omega, \underline{r}) = \frac{\epsilon}{4\pi} \frac{e^{-jkr}}{\sigma} \int_S \underline{M}_s(\omega, \underline{r}') \exp(jk\hat{r} \cdot \underline{r}') \, \mathrm{d}s(\underline{r}'). \tag{6.1b}$$

Lorentz force conditions are applied to determine the far-field electric scalar and magnetic scalar potentials as follows:

$$\begin{aligned} \underline{\overline{\phi}}_e(\omega, \underline{r}) &= \frac{j}{\omega\epsilon\mu} \operatorname{div} \underline{A}(\omega, \underline{r}) \\ &= \frac{k}{\omega\epsilon\mu} (\hat{r}, \underline{A}(\omega, \underline{r})) \end{aligned} \tag{6.2a}$$

$$\begin{aligned} \underline{\overline{\phi}}_m(\omega, \underline{r}) &= \frac{j}{\omega\epsilon\mu} \operatorname{div} \underline{F}(\omega, \underline{r}) \\ &= \frac{k}{\omega\epsilon\mu} (\hat{r}, \underline{F}(\omega, \underline{r})) \end{aligned} \tag{6.2b}$$

The far-field electric and magnetic fields (i.e., up to Order (r^{-1})) are then determined as follows:

$$\underline{E} = -\underline{\nabla}\underline{\bar{\phi}}_e - j\omega\underline{A} + \frac{1}{\epsilon}\underline{\nabla} \times \underline{F}; \tag{6.3}$$

$$\begin{aligned} \underline{H} &= -\frac{1}{\mu}\underline{\nabla} \times \underline{A} - \underline{\nabla}\underline{\bar{\phi}}_m - j\omega\underline{F}; \\ &= \frac{jk^2}{\omega\epsilon\mu}\hat{r}(\hat{r}, \underline{A}) - j\omega\underline{A}\underline{\phi}\frac{jk}{\epsilon}\hat{r} \times \underline{F}; \\ &= -j\omega\underline{A}_\perp\underline{\phi_m}\frac{jk}{\epsilon}\hat{r} \times \underline{F}; \end{aligned} \tag{6.4}$$

where

$$\begin{aligned} \underline{A}_\perp &= A_\theta\hat{\theta} + A_\phi\hat{\phi}; \\ \underline{H}_\perp &= \frac{jk}{\mu}\hat{r} \times \underline{A} - j\omega\underline{F}_1; \end{aligned}$$

Finally, we derive expression for the Poynting vector as follows:

$$\underline{S} = \frac{1}{2}\mathrm{Re}\{E \times H^*\}.$$

Up to order $\left(\frac{1}{r^2}\right)$ i.e., value $1/r^2$ is taken into account from where, the RDRA radiation resistance is evaluated:

$$\frac{1}{2}I^2R_r = \lim_{r\to\infty}\int S \cdot \hat{r} \cdot r^2 \cdot \mathrm{d}\Omega;$$

when I is the input current to the RDRA, R_r or $R_r(\omega)$ is radiation resistance and depends on the frequency.

6.2 Radiation Pattern of RDRA Due to Probe Current *i*(*t*) and Probe Length d*l*

$$\frac{\mu I\overrightarrow{dl}e^{-jkr}}{4\pi r} = \underline{\overrightarrow{A}}; \quad \text{where } A \text{ is magnetic vector potential} \tag{6.5}$$

From Helmholtz equation $|\underline{\vec{A}}|$

$$\underline{E} = -j\omega\underline{\vec{A}}$$

Radiated power can be given as follows:

$$\frac{|\underline{E}|^2}{2\eta} = \frac{\omega^2 |\vec{\underline{A}}|^2}{2\eta}, \quad \sqrt{\frac{\mu}{\epsilon}} = \eta = \text{characteristic impedance.}$$

$$\vec{\underline{A}} = \frac{\mu}{4\pi} \int\limits_{\text{Volume}} \frac{J(\underline{r}', \omega) e^{-jk|\underline{r}-\underline{r}'|}}{|\underline{r} - \underline{r}'|} \mathrm{d}^3 r'; \text{ at source.} \tag{6.6}$$

We know that radiation pattern can be defined by the electrical field intensity E_θ, E_ϕ:

$$E_\theta = -j\omega A_\theta \quad \text{and} \quad \underline{A}_\theta = \hat{\theta} \cdot \underline{A}$$

Antenna surface current density can be expressed as follows:

$$J(\underline{r}', \omega) = \sum_{mnp} \underline{J}_s[mnp, \underline{r}'] e^{j\omega(mnp)t}; \quad \text{where, } r = (x, y, z) \tag{6.7}$$

The magnetic vector potential in terms of J can be written as follows:

$$\begin{aligned} \underline{A} &= \frac{\mu}{4\pi} \sum_{mnp} \int \frac{\underline{J}_s[mnp, \underline{r}'] e^{j\omega(mnp)\left(t - \frac{|\underline{r}-\underline{r}'|}{c}\right)}}{|\underline{r} - \underline{r}'|} \mathrm{d}s(\underline{r}'); \quad \text{where, d}s \text{ is surface of RDRA} \\ &= \frac{\mu}{4\pi} \frac{e^{jkn}}{|\underline{r} - \underline{r}'|} \sum_{mnp} \int\limits_s \underline{J}_s[mnp, \underline{r}'] e^{j\omega(mnp)\hat{\underline{r}}\cdot\underline{r}'} \mathrm{d}s(\underline{r}') \end{aligned} \tag{6.8}$$

$$H_\phi = E_\theta / \eta, \quad H_\theta = -{}^{E_\phi}\!/_{\eta}.$$

Hence radiated power can be given as:

$$P_{\text{rad}} = \frac{1}{2\eta} \left(|E_\theta|^2 + |E_\phi|^2 \right)$$

$$\begin{cases} \hat{\theta} = \hat{x} \cos\varphi \cos\theta + \hat{y} \sin\varphi \cos\theta - \hat{z} \sin\theta \\ \hat{\phi} = -\hat{x} \sin\varphi + \hat{y} \cos\varphi \end{cases}$$

$$\begin{aligned} E_\theta = &\frac{\mu}{4\pi r^2} \mathrm{Re} \sum_{mnp} \int\limits_s \left\{ J_{sx}[mnp, \underline{r}'] \cos\varphi \cos\theta + J_{sy[mnp,\underline{r}']} \sin\phi \cos\theta - J_{sz[mnp,\underline{r}']} \sin\theta \right\} \\ &\exp(j\omega \frac{(mnp)}{c} \}(x' \cos\phi \sin\theta + y' \sin\phi \sin\theta + z' \cos\theta) \, \mathrm{d}s(\underline{r}')] \, e^{j\omega(mnp)t}; \end{aligned} \tag{6.9a}$$

$$E_{\varphi} = \mathrm{Re}\sum_{mnp}\int_{s}\left\{-J_{sx}(mnp,\underline{r}')\sin\phi + J_{sy[mnp,\underline{r}']}\cos\phi\right\}e^{j\frac{\omega(mnp)}{c}}$$
$$(x'\cos\varphi\sin\theta + y'\sin\phi\sin\theta + z'\cos\theta)\,\mathrm{d}s(\underline{r}')\,e^{j\omega(mnp)t}\mathrm{d}s(\underline{r}'). \tag{6.9b}$$

Radiated power P_{rad}, x, y, z component wise, can thus be defined as follows:

$$P_x[\hat{r}|mnp] = \int_s J_{sx}(mnp,\underline{r}']e^{j\frac{\omega(mnp)\hat{r}\cdot\underline{r}'}{c}}\mathrm{d}s(\underline{r}') \tag{6.10a}$$

$$P_y[\hat{r}|mnp] = \int_s J_{sy}(mnp,\underline{r}']e^{j\frac{\omega(mnp)\hat{r}\cdot\underline{r}'}{c}}\mathrm{d}s(\underline{r}') \tag{6.10b}$$

$$P_z[\hat{r}|mnp] = \int_s J_{sz}(mnp,\underline{r}']e^{j\frac{\omega(mnp)\underline{r}\cdot\underline{r}'}{c}}\mathrm{d}s(\underline{r}') \tag{6.10c}$$

$$\hat{r}(\theta,\phi) = \hat{x}\cos\phi\sin\theta + \hat{y}\sin\phi\sin\theta + \hat{z}\cos\theta.$$

Let $s = mnp$ for convenience then

$$\begin{aligned} E_\theta &= \mathrm{Re}\sum_s\left\{P_x[\hat{r}|s]\cdot\cos\phi\cos\theta + P_y[\hat{r}|s]\sin\phi\cos\theta - P_z[\hat{r}|s]\sin\theta\right\}e^{j\omega(s)t} \\ &= \mathrm{Re}\sum_s E_{s\theta}e^{j\omega(s)t} \end{aligned} \tag{6.11}$$

where $s = (mnp) = \begin{bmatrix}000\\001\\010\end{bmatrix}$ and so on till $s = [111]$, similarly

$$E_\phi = \mathrm{Re}\sum_s(-P_x[\hat{r}|s]\sin\phi + P_y[\hat{r}|s]\cos\phi)e^{j\omega(s)t} \tag{6.12}$$

6.2.1 Radiation Pattern

Now, power radiation pattern can be defined as follows:

$$\begin{aligned} \frac{|E_\theta|^2+|E_\phi|^2}{2\eta} &= \frac{1}{2}\left\{\sum_s\left(E_{s\theta}e^{j\omega(s)t} + E_{s\phi}e^{j\omega(s)t}\right)\right\} \\ &\quad\times\frac{1}{2}\left\{\sum_s H_{s\theta}e^{j\omega(s)t} + \sum_s H_{s\phi}e^{j\omega(s)t}\right\} \end{aligned} \tag{6.13}$$

$$
\begin{aligned}
&= \frac{1}{4}\left(\sum_{s} E_s \times H_m^* \, e^{j(\omega_s-\omega_m)t} + \sum_{s} E_s^* \times H_m \, e^{j(\omega_m-\omega_s)t}\right) \\
&= \frac{1}{4}\sum_{s}\left[E_{s\theta} \times H_{s\o}^* + E_{s\o}^* \times H_{s\theta}\right] \\
&= \frac{1}{2}\mathrm{Re}\sum_{s}\left(E_{s\theta} \times H_{s\o}^*\right) \\
&= \frac{1}{2}\left(E_{s\theta}\hat{\theta} + E_{\o}\hat{\o}\right) \times \left(\frac{E_{s\theta}^*}{\eta}\hat{\phi} - \frac{E_{s\o}}{\eta}\hat{\theta}\right) \\
&= \sum_{s}\frac{|E_{s\theta}|2}{2\eta}\hat{r} + \frac{|E_{s\o}|2}{2\eta}\hat{r};
\end{aligned}
\tag{6.14}
$$

6.3 Poynting Vector

Poynting vector is defined as radiated power flux per unit solid angle or power radiated in particular direction in specified angular zone.

$H = \nabla \times A$

$E = -\nabla \o - \frac{\mathrm{d}_A}{\mathrm{d}_t}$; scalar and magnetic vector potential from Lorentz gauge conditions.

$S = (E \times H^*)$; S is Poynting vector (energy flow or flux).

$$
Z = \frac{P_{\mathrm{rad}}}{|I|^2} = \text{Input impedance}
$$

$$
\begin{aligned}
\underline{S}\cdot\hat{r} = \frac{1}{2\eta}\sum_{mnp}&\left[\omega(s)^2|P_x(\hat{r}|s)\cos\phi\cos\theta + P_y(\hat{r}|s)\sin\phi\cos\theta - P_z(\hat{r}|s)\sin\theta|^2\right. \\
&\left. + \,\omega(s)^2|P_x(\hat{r}|s)\sin\phi - P_y(\hat{r}|s)\cos\phi|^2\right]
\end{aligned}
\tag{6.15}
$$

$$
\begin{aligned}
\underline{S}\cdot\hat{r}(r,\theta,\phi) = \frac{1}{2\eta}\sum_{mnp}&\omega(mnp)^2\{|P_x(\theta,\phi|mnp)\cos\phi\cos\theta \\
&+ P_y(\theta,\phi|mnp)\sin\phi\cos\theta - P_z(\theta,\phi|mnp)\sin\theta|^2 \\
&+ |P_x(\theta,\phi|mnp)\sin\phi \pm P_y(\theta,\phi|mnp)\cos\phi|^2\}
\end{aligned}
\tag{6.16}
$$

6.4 Moat-Shaped RDRA Radiation Pattern

Moat-shaped RDRA is shown in Fig. 6.1a with *x*, *y*, and *z* coordinates, and feed is given at *a*/2 position.

In Fig. 6.1b, rectangular moat-shaped RDRA is covered with *r* copper plate to reduce resonant frequency.

$E(t, x, y, z)$ is electric field intensity of RDRA to be computed in time domain and $E(\omega, x, y, z)$ in frequency domain having *a*, *b*, and *d* dimensions, excited with feed probe at $\frac{a}{2}, \frac{a}{2}, 0$ point by $I_0 \cos \omega t$ RF current.

$A = A_z \hat{z}$ (due to RF excitation current $I_0 \cos \omega t$ along length *d* inserted into the RDRA).

Hence, magnetic vector potential can be written as follows:

$$A_z(\omega, x, y, z) = \frac{\mu I_0}{4\pi} \int_0^d \frac{e^{-jk|r - a/2\hat{x} - b/2\hat{y} - \xi\hat{z}|}}{|r - a/2\hat{x} - b/2\hat{y} - \xi\hat{z}|} \mathrm{d}\xi, \tag{6.17}$$

(a)

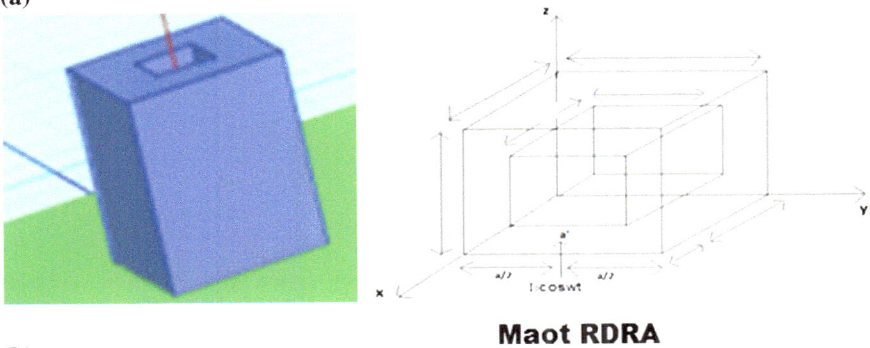

(b)

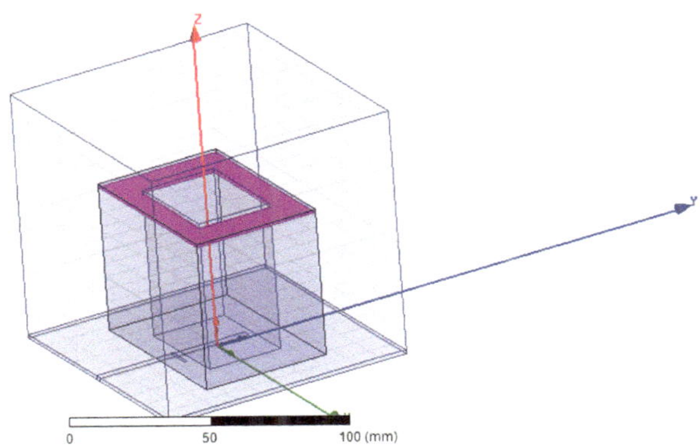

Fig. 6.1 **a** Moat-shaped RDRA. **b** RDRA moat cover with rectangular copper plate to reduce resonant frequency

Let $C = \frac{\mu}{4\pi}$, $k = \omega/c$ and ξ = variable probe length.

$$A_z = CI_0 \int_0^d \left\{ \frac{\exp\{-jk\left((x - a/2)^2 + (y - a/2)^2 + (z - \xi)^2\right)}{\left((x - a/2)^2 + (y - a/2)^2 + (z - \xi)^2\right)^{1/2}} \right\} d\xi \quad (6.18)$$

Far-field approximation can be determined as follows:

$$A_z = \frac{CI_0 e^{-jkr}}{r} P(\theta_0, \varnothing_0), \quad (6.19)$$

where $P(\theta_0, \varnothing_0)$ is radiation pattern.

Here, it is assumed that probe is very small as compared to RDRA.

$$\begin{aligned}
&(x - a/2)^2 + (y - a/2)^2 + (z - \xi)^2 \\
&\quad = (x - a/2)^2 + (y - a/2)^2 + z^2 - 2z\xi \\
&(x - a/2)^2 + (y - a/2)^2 + z^2 \gg d^2 \\
&\quad r = \left((x - a/2)^2 + (y - a/2)^2 + z^2\right)^{1/2}
\end{aligned}$$

where r = distance from the points (x, y, z) in the center of the feed probe $(a/2, a/2, 0)$

$$\begin{aligned}
&\left((x - a/2)^2 + (y - a/2)^2 + (z)^2\right)^{1/2} \\
&\quad = (r^2 - 2z\xi)^{1/2} = r(1 - z\xi/r_0{}^2) = r - z\xi/r_0.
\end{aligned}$$

Hence, magnetic vector potential due to source inside RDRA can be computed as follows:

$$\begin{aligned}
A_z &= \frac{CI_0 e^{-jkr_0}}{r_0} \int_0^d \exp\left\{\frac{jkz\xi}{r_0}\right\} d\xi, \quad \text{where } I_0 \text{ probe RF current.} \\
&= \frac{CI_0 e^{-jkr_0}}{r_0} \frac{\exp\left(jkz\xi/r_0\right)}{\left(jkz/r_0\right)} \left| \begin{matrix} \xi = d \\ \xi = 0 \end{matrix} \right| \quad \text{i.e., variable probe length.} \\
&= \frac{CI_0}{r_0} e^{-jkr_0} \frac{\exp\left(jkzd/r_0\right) - 1}{\left(jkz/r_0\right)} \\
&= \frac{CI_0}{r_0} e^{-jkr_0} \frac{\exp\left(jkzd/2r_0\right) 2j \sin\left(kzd/2r_0\right)}{\left(jkz/r_0\right)}
\end{aligned} \quad (6.20)$$

$$A_z = 2CI_0 \exp\left\{-jk\left(r_0 - {zd}/{2r_0}\right)\right\}\frac{\sin\left({kzd}/{2r_0}\right)}{kz}$$

where, $z = r\cos\theta$.

$$A_z(\omega, x, y, z) = CI_0 \exp(-jkr_0)\exp\frac{\left(\frac{jkd}{2}\cos\theta_0\right)\sin\left(\frac{kd\cos\theta_0}{2}\right)}{kr_0\cos\theta_0} \tag{6.21}$$

here, (r, θ, Φ) are spherical polar coordinates of (x, y, z) so as to relate $\left({a}/{2}, {a}/{2}, 0\right)$, the probe insertion point. Hence, magnetic vector potential can be expressed as follows:

$$\begin{aligned} E(t, x, y, z) = &\frac{\omega}{r_0^3}|P(\theta_0)|\sin(\omega t - kr_0 + \Psi(\theta_0))\left(\left(x - {a}/{2}\right)^2 + \left(y - {a}/{2}\right)^2\right) + \left(x - \frac{a}{2}\right)z\hat{x} \\ &+ \left\{-\frac{\omega\left(\left(y - {a}/{2}\right)z\right)}{r_0^3}|P(\theta_0)|(\sin(\omega t - kr_0 + \Psi(\theta_0)))\right\}\hat{y} \end{aligned} \tag{6.22}$$

$$\begin{aligned} B(t, x, y, z) &= -\frac{k\left(x - \frac{a}{2}\right)}{r_0^2}|P(\theta_0)|\sin(\omega t - kr_0 + \Psi(\theta_0))\hat{y} \\ &+ \frac{k\left(y - \frac{a}{2}\right)}{r_0^2}|P(\theta_0)|\sin(\omega t - kr_0 + \Psi(\theta_0))\hat{x} \\ &= k\frac{P(\theta_0)}{r_0^2}\sin(\omega t - kr_0 + \Psi(\theta_0))\left(\left(y - \frac{a}{2}\right)\hat{x} + \left(x - \frac{a}{2}\right)\hat{y}\right) \end{aligned} \tag{6.23}$$

Finally, we derive expression for the Poynting vector as follows:

$$\underline{S} = \frac{1}{2}\mathrm{Re}\{E \times H^*\}$$

Up to $\mathrm{O}\left(\frac{1}{r^2}\right)$ from where the radiator resistance is evaluated as

$$\frac{1}{2}I^2R_r = \lim_{r\to\infty}\int S\cdot\hat{r}\cdot r^2\cdot \mathrm{d}\Omega$$

where I is the input current to the RDRA. R_r or $R_r(\omega)$ depends on the frequency. Hence, this completes the solution for radiation pattern of RDRA.

6.5 Quality Factor of RDRA

The quality factor Q of the RDRA can be evaluated by comparing the power radiated $P_{\text{rad}} = \frac{1}{2} I^2 R_r$ with the average electromagnetic energy (W) stored with the RDRA as follows:

$$W(\omega) = \frac{1}{4} \int_{[0,a]\times[0,b]\times[0,c]} (\epsilon(\underline{E}, E^*)) + \mu(\underline{H}, H^*)\, dx\, dy\, dz \tag{6.24}$$

The average energy stored per unit cycle with the RDRA is

$$P(\omega) = \frac{W(\omega)}{2\pi/\omega} = \frac{\omega}{2\pi} W(\omega) \tag{6.25}$$

The quality field factor of the RDRA is thus

$$Q(\omega) = \frac{2\omega W(\omega)}{|I(\omega)|^2 \cdot R_r(\omega)};$$

where ω corresponds to resonant frequency.

The quality factor of a resonant mode measures how sharp its resonance is. As per conservation of energy,

$$\int |E|^2 \mathrm{d}v = \int |H|^2 \mathrm{d}v$$

(time) average magnetic energy will be equal to electric energy inside the resonator.

The time-averaged energy dissipated in the walls of RDRA in unit time can be calculated as of energy into walls from the electromagnetic fields in the cavity normal component of energy based on the boundary conditions as energy flux density as follows:

$$\underline{S} = \left(\frac{C}{8\pi}\right) \mathrm{Re}(E \times H^*) \tag{6.27}$$

Hence, total energy dissipated is given by

$$\frac{c}{8\pi} \oint \mathrm{Re}|H|^2 \mathrm{d}f$$

Change in resonant frequency due to dielectric material used in RDRA:
The resonant frequency is reduced by $\sqrt{\mu\epsilon}$
If $\omega \to \omega\sqrt{\mu\epsilon}$

$\omega a, \omega b$ are orthogonal frequencies, and Ea and Eb are orthogonal fields.
$\frac{\omega'}{2|\omega''|}$ = quality factor (Q), ω' is real frequency, and ω'' is imaginary frequency.
Complex freq $\omega = \omega' + j\omega''$

$$\int Ea \cdot Eb^* \mathrm{d}v = \int Ha \cdot Hb^* \mathrm{d}v = 0$$

Resonator filled with non-absorbing dielectric, for which ϵ and μ differ from unity by replacing ω by $\omega\sqrt{\mu\epsilon}$ and E by ϵE, and H by μH.

The (time) average energy flux through surface is

$$\underline{S} = \frac{c}{8\pi}\mathrm{Re}\left(E_t \times H_t^*\right) \tag{6.28}$$

where $S = \frac{c}{4\pi}(E \times H)$.

If Q of heat evolved per unit time and volumes

$$Q = \frac{\omega}{4\pi}\left(\epsilon''\overline{E^2} = +\mu''\overline{H^2}\right) \tag{6.29}$$

Bar denotes time-average exciting frequency, must be exactly equal to the chosen resonance frequency, and is required to establish field configuration inside resonator. This results in dissipation of energy in the cavity walls and dielectric filling of the cavity resonator. A measure of the sharpen of response of the cavity to external excitation is quality of the cavity. This is defined as 2π times the ratio of the time-averaged energy stored in the cavity to the energy dissipated.

$$Q = \omega_0 \frac{\text{stored energy}\{W(\omega)\}}{\text{power loss}(I \cdot I \cdot R_r)} \tag{6.30}$$

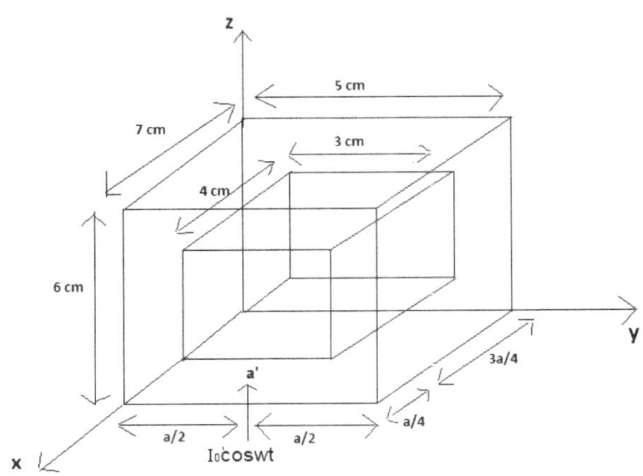

Fig. 6.2 Rectangular RDRA moat

where ω_0—Resonant frequency oscillations of fields are damped and time dependent. Change in frequency $\Delta\omega$ to occur based on superposition of frequencies:

$$\omega = \omega_0 + \Delta\omega$$

$$E(t) = \frac{1}{\sqrt{2\pi}} \int_{-\infty}^{\infty} E(\omega)e^{-j\omega t} \mathrm{d}\omega$$

Q. No. 1 Compute resonant frequency and propagation constant of given RDRA shown in Fig. 6.2 and also compute quality factor of a RDRA having dimensions $10 \times 10 \times 10\ \text{mm}^3$ with dielectric constant 10 and probe current 10 mA.

Chapter 7
Rectangular DRA Higher-Order Modes and Experimentations

Abstract In this chapter, rectangular DRA higher-order modes have been realized by mathematical modeling. Resonant modes are seen with experimentations in anechoic chamber. These resonant modes impart physical insight into the radiating phenomenon of the antenna. Knowledge of modes can be boon to the antenna designer. If antenna resonant modes are known, radiation parameters can be steered. There are two types of modes and they are dominant and higher-order modes. The dominant mode corresponds to the lowest resonant frequency. These higher-order modes can be generated either by increasing electrical length of RDRA or by applying higher excitation frequency. The resonant frequencies of the modes are represented by eigenvalues and currents by eigenvectors. Radiating behavior of the antenna can be predicted by modes. They can also help to determine input excitation point. Moreover, having in mind the current distribution of the modes, the geometry of the antenna can be modified. The aspect ratio is the important parameter in RDRA. Devising control on aspect ratio can alter resonant frequency, gain, and bandwidth. RDRA has two fold design flexibility because of two aspect ratios. The resonant modes of RDRA can be described with an equivalent sequential RLC circuit having different sequential LC values. Thus, they form many series-tuned resonant circuits. The superposition of these modes generally give rise to resulting or weighted resonant frequency. The top-loading RDRA has been completely modeled. Antenna gain and bandwidth enhancement techniques have been worked out with examples.

Keywords Aspect ratio · RLC circuit · Tuned cavity · Weighted sum · Resonant frequency · Eigen frequency · Design flexibility · Top-loading RDRA · Gain and bandwidth enhancement

R.S. Yaduvanshi and H. Parthasarathy, *Rectangular Dielectric Resonator Antennas*, DOI 10.1007/978-81-322-2500-3_7

7.1 Introduction to Higher Modes

Resonant modes impart physical insight into the radiating phenomenon of the antenna. Knowledge of modes can be boon to the antenna designer. If antenna resonant modes are known, radiation parameters can be steered. Any of the antennas have two types of modes. They are dominant and higher-order modes. The dominant mode corresponds to the lowest resonant frequency. Other than dominant frequency, all higher resonant frequencies are higher-order modes. These higher-order modes can be generated either by increasing electrical length of RDRA or by applying higher excitation frequency.

The resonant frequencies of the modes are represented by eigenvalues and currents by eigenvectors. Radiating behavior of the antenna can be predicted by modes. They can also help to determine input excitation point. Moreover, having in mind the current distribution of the modes, the geometry of the antenna can be modified. The aspect ratio is the important parameter in RDRA. Devising control on aspect ratio can alter resonant frequency, gain, and bandwidth. RDRA has two fold design flexibility because of two aspect ratios.

The resonant modes of RDRA can be described with an equivalent sequential RLC circuit having different sequential LC values. Thus, they form many series-tuned resonant circuits. The superposition of these modes generally gives rise to resulting or weighted resonant frequency. The modes are defined as E and H fields pattern inside a device, whose EM wave propagation is governed by Maxwell's equations under certain boundary conditions. RF input excitation currents get distributed on RDRA surfaces. Thus, weighted sum of eigen currents or superposition of all these currents inside device is the resultant mode at any instant of time.

The resonant modes are state of excited fields at any instant inside the device, generally classified as transverse electric (TE), transverse magnetic (TM) and hybrid electromagnetic (HEM), dominant modes and higher modes. TE modes will have only H_z component as propagating fields. TM modes will have only E_z component as propagating fields. These propagating fields are longitudinal fields. HEM has hybrid mode and will have both E_z and H_z components simultaneously as propagating fields at any instant of time. These field perturbations form a particular excited resonant mode in the device.

In the literature, stacking of the RDRA has been used for enhancement in the directivity of the antenna by Petosa [1]. This can be achieved by devising proper control on higher-order modes. Higher modes correspond to higher resonant frequency and higher antenna gain. RDRA higher-order modes and hybrid modes are useful and provide design space to antenna designers, but configure complex fields structure. The generation of higher modes mainly depends on RF excitation, device dimensions, dielectric material, perturbation, and coupling used in RDRA.

An aperture-coupled microstrip slot feed RDRA is discussed in this chapter. This has the advantage of isolating the feeding network from the radiating element. Aspect ratio can be changed by changing the RDRA dimensions a, b, and d. This will have impact on resonant modes, and thus, change in resonant frequency will take place.

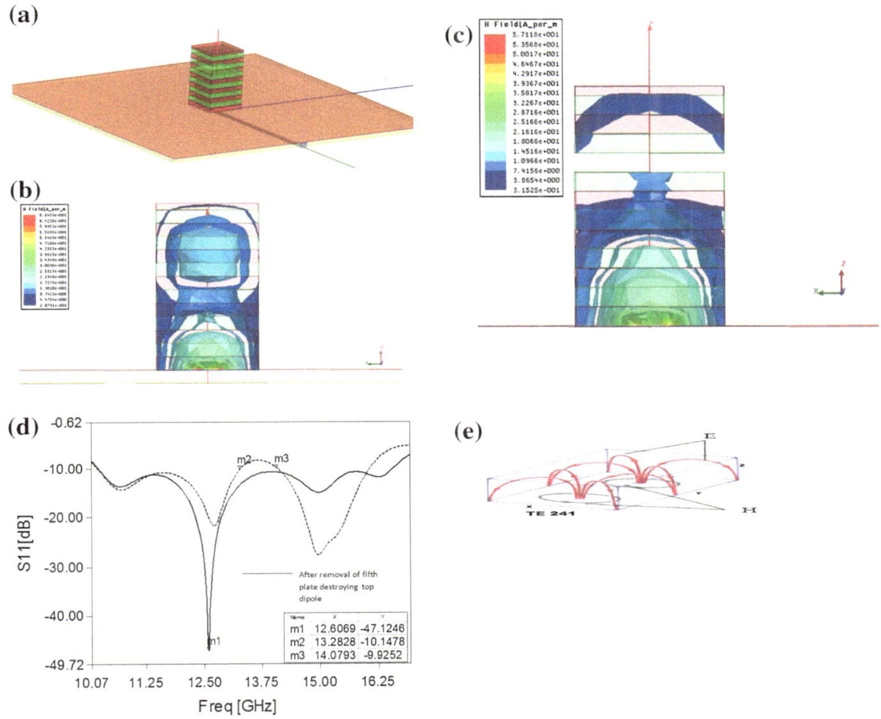

Fig. 7.1 **a** RDRA higher modes. **b** RDRA mode generated. **c** RDRA mode control due to dipole moment. **d** S11 of RDRA with mode merging. **e** Higher-order modes field configuration

The number of half-wave variations corresponding to *x*, *y*, and *z* directions can alter the operating mode. Once the dimensions of RDRA are fixed, modes can also be altered by excitation frequency. Operating frequency of RDRA has inverse relationship with permittivity of the material. The device size can be minimized by using higher permittivity material. Figure 7.1 shows the resonant mode structure.

RDRA shown in Fig. 7.1a has been made with alternate layers of RT-Duroid and FR-4 dielectric materials having permittivity 10.2 and 4.4. These dielectric materials are easily available. The fabrication is also simple. The dimensions of these sheets are $6 \times 6 \times 10\ \text{mm}^3$ and $6 \times 6 \times 0.8\ \text{mm}^3$, respectively. Figure 7.1b shows resonant modes. Figure 7.1c shows how these dipoles are broken by introducing air between these stacking layers. Figure 7.1d presents mode merging due to proper dipole control, i.e., merging these modes by removing one stacking layer and creating airspace between these dipoles. This has enhanced the antenna gain. Figure 7.1e shows higher-order modes in RDRA. Figure 7.2 presents prototype RDRA with VNA feed probe. Figure 7.3 shows modes or field pattern. Figures 7.4 and 7.5 show odd and even modes. Figure 7.6 represents structure, and Fig. 7.7 represents gain plot. Figure 7.8 shows microstripline used in RDRA feed. Figure 7.9 shows RDRA with top loading. Figure 7.10 shows *H* and *E* fields

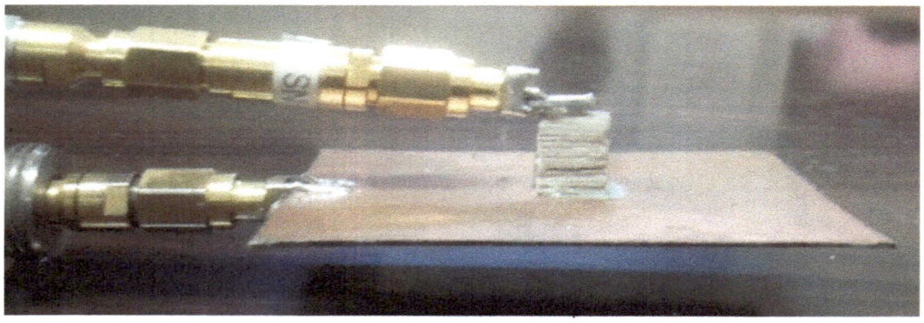

Fig. 7.2 RDRA excitation by aperture-coupled slot

Fig. 7.3 *EH* fields pattern or resonant modes

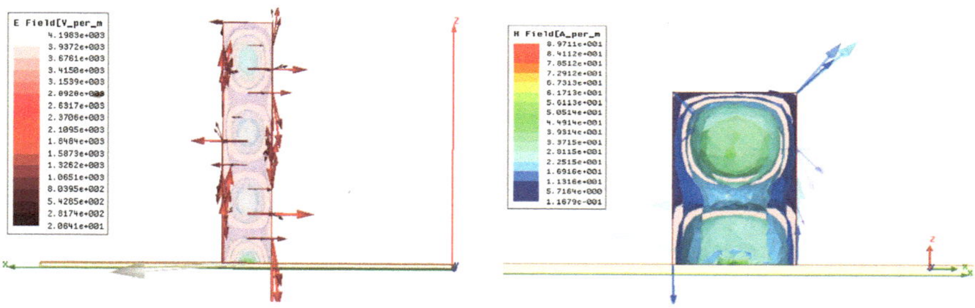

Fig. 7.4 Odd number of modes

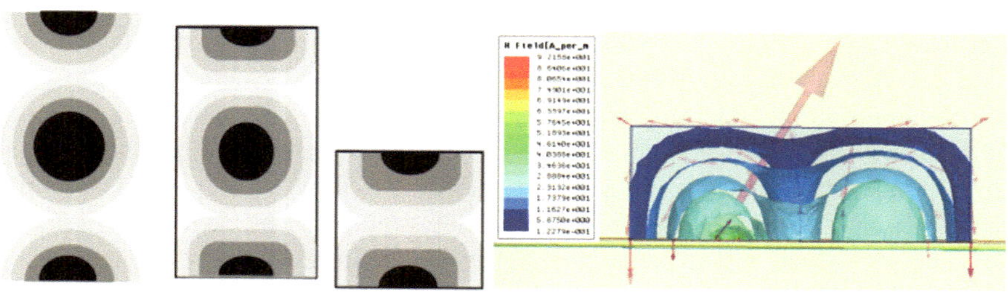

Fig. 7.5 Even number of modes

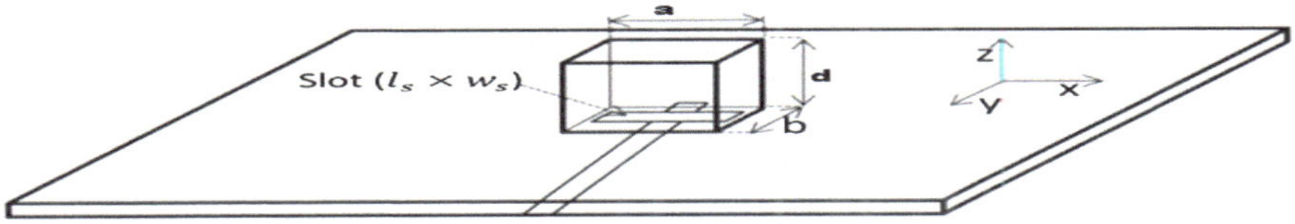

Fig. 7.6 Structure of RDRA

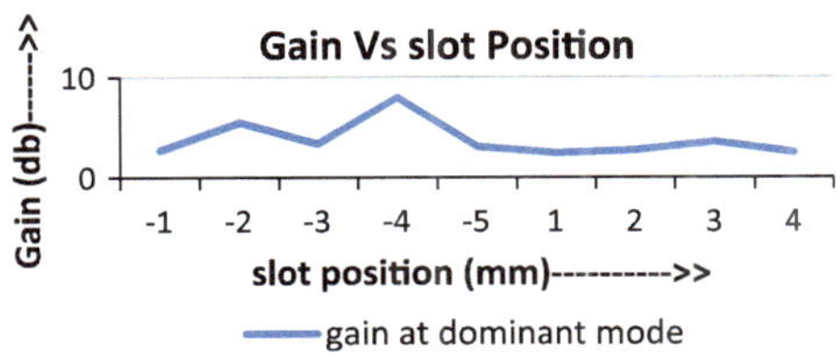

Fig. 7.7 Slot position versus gain

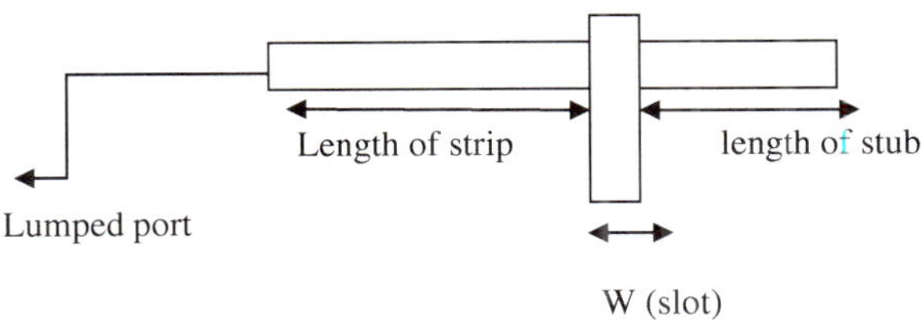

Fig. 7.8 Slot and microstrip feed

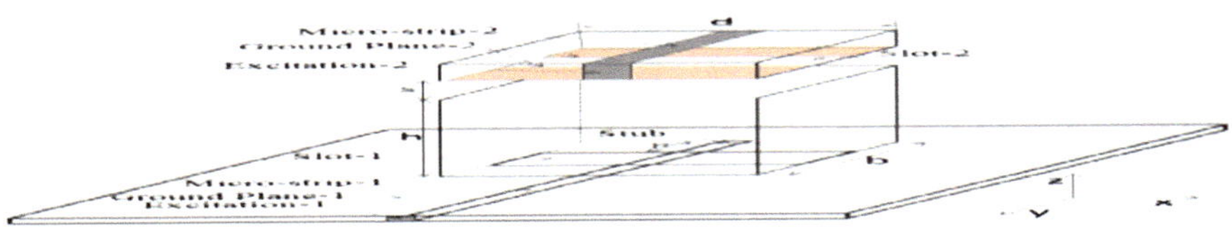

Fig. 7.9 Top-loaded RDRA with *a*, *b*, and *d* = *h* dimensions

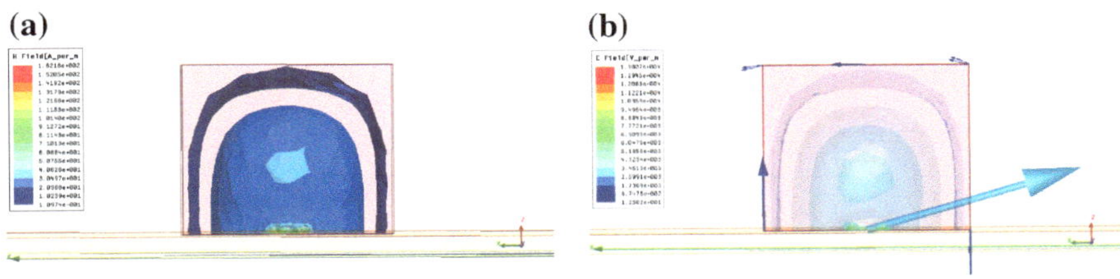

Fig. 7.10 **a** *H* field distribution inside RDRA and **b** *E* field distribution inside RDRA

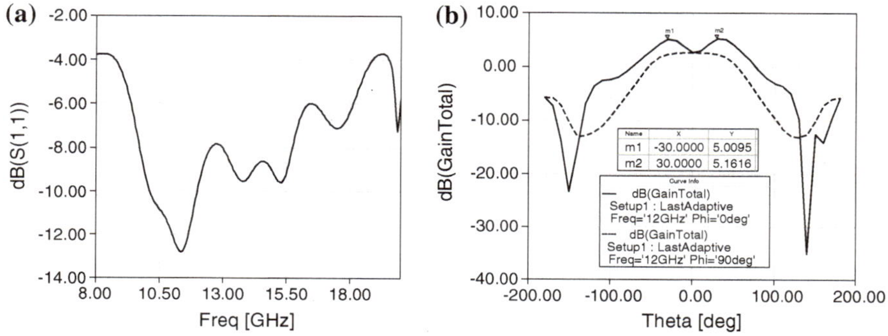

Fig. 7.11 **a** Return loss at $f = 11$ GHz and **b** radiation pattern showing 5.16 dB gain of the antenna at $f = 12$ GHz

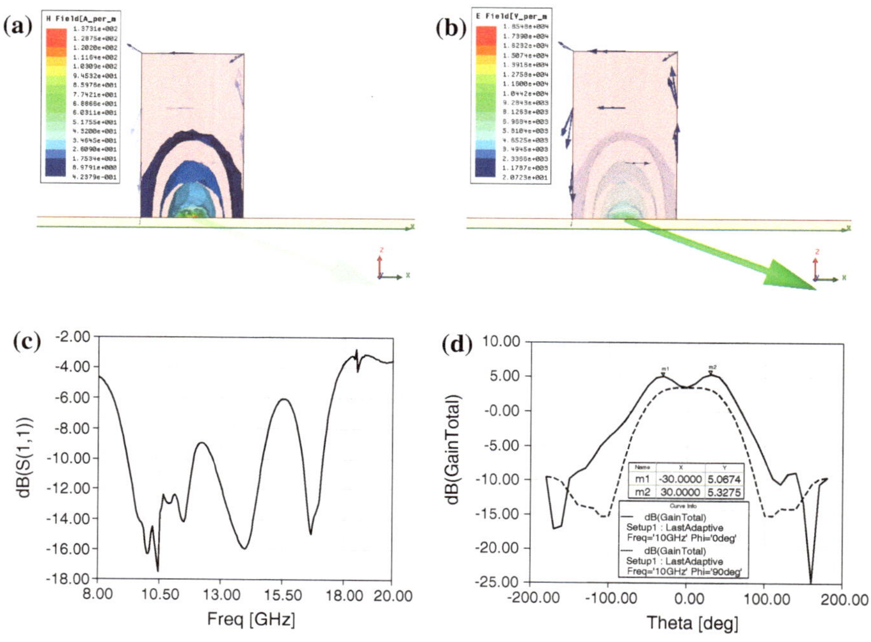

Fig. 7.12 TE_{111} mode at frequency 10 GHz. **a** *H* field distribution, **b** *E* field distribution, **c** return loss, **d** gain

distributions inside RDRA. Figure 7.11 shows the plot of return loss at f = 11 GHz and radiation pattern. Figure 7.12 shows TE_{111} mode at a frequency of 10 GHz. Figures 7.13, 7.14, 7.15 and 7.16 show resonant modes and Fig. 7.17 TE_{113} mode at a frequency of 10 GHz. Figure 7.18 shows TE_{115} mode at a frequency of 12 GHz. Figure 7.19 depicted TE_{117} resonant mode at frequency 15 GHz. Figures 7.20, 7.21,

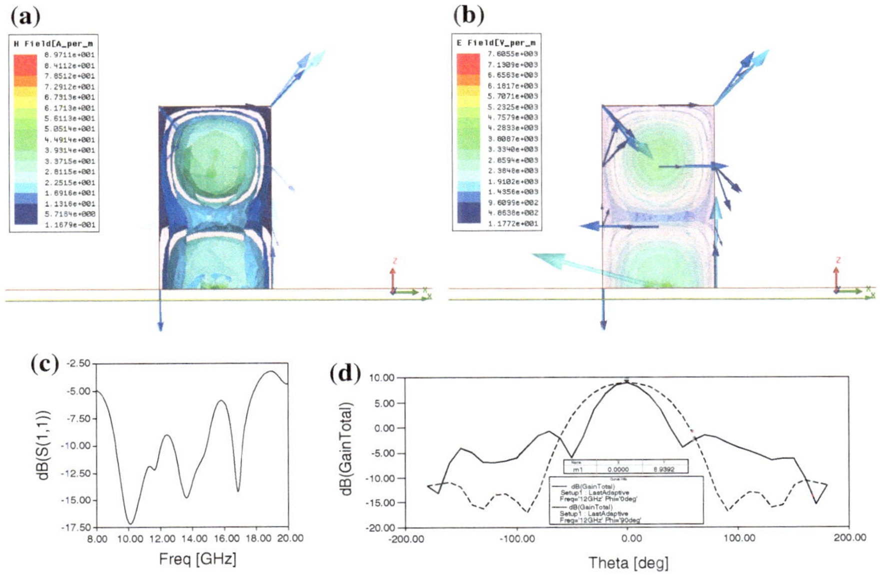

Fig. 7.13 TE_{113} mode at frequency 12 GHz. **a** *H* field distribution, **b** *E* field distribution, **c** return loss, **d** gain

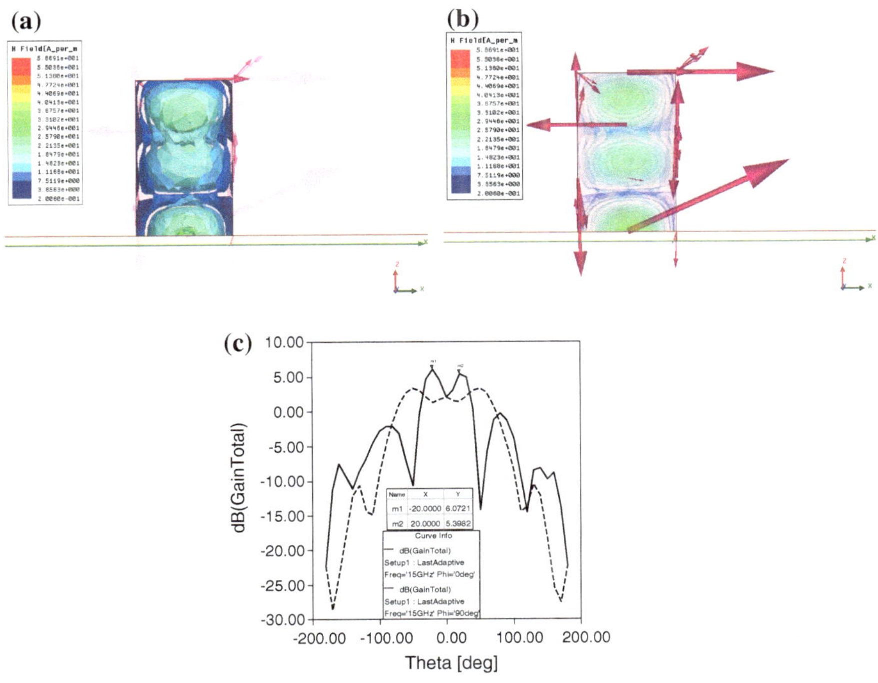

Fig. 7.14 TE_{115} mode at frequency 15 GHz. **a** *H* field distribution, **b** *E* field distribution, **c** gain

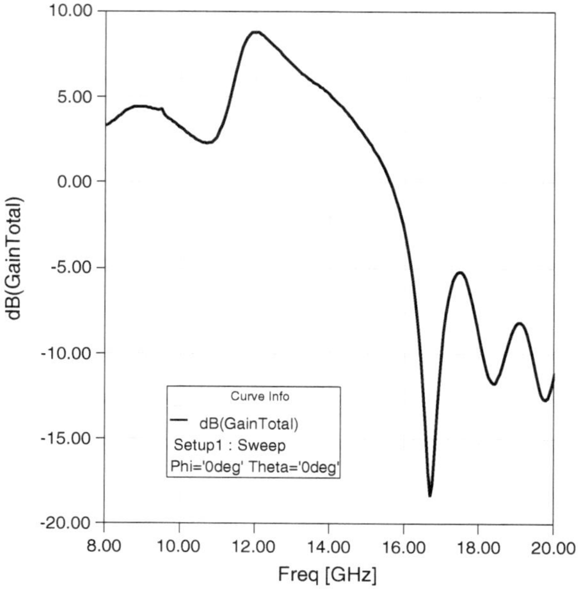

Fig. 7.15 Gain versus frequency plot

7.22, 7.23, 7.24, 7.25, 7.26, 7.27, 7.28, 7.29, 7.30, 7.31 and 7.32 show parameter measured in anechoic chamber and HFSS simulated results. Working is mentioned below each figures.

Resonant modes take the real orthogonal basis for currents on the antenna surfaces.

In this chapter, mechanism for mode generation and their possible control in RDRA are discussed. These are validated with simulated and experimental results using prototype models. Figure 7.2 shows the prototype of RDRA. VNA probes are connected in order to take measurements. Top-loading RDRA is used for generation of higher-order modes. These higher-order modes are of even and odd types. If RDRA design is isolated even as well as odd modes will be available, i.e., both even as well as odd modes will be present in isolated DRA case. RDRA once

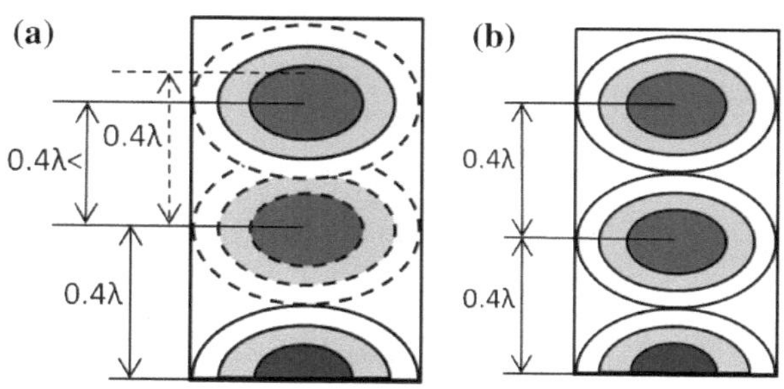

Fig. 7.16 Spacing adjustment between short magnetic dipoles placed at the center of each mode

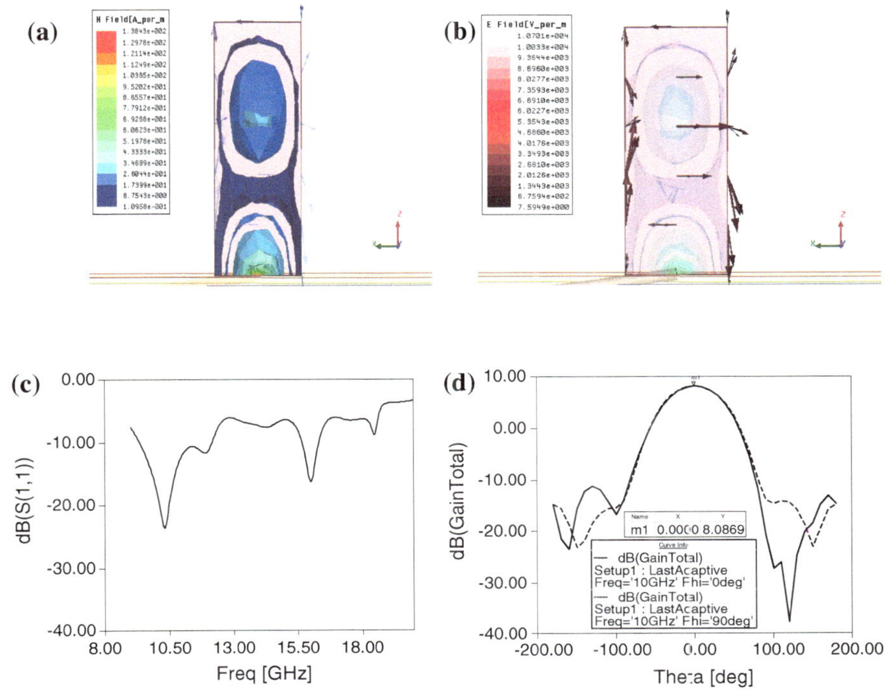

Fig. 7.17 TE_{113} mode at frequency 10 GHz. **a** *H* field distribution, **b** *E* field distribution, **c** return loss, **d** gain

extended with ground plane, only odd modes will be generated, because even modes get grounded. Thus, ground plane canceled out even modes. The *E* and *H* fields patterns are shown in Fig. 7.3.

Higher-order even and odd modes are shown in Figs. 7.4 and 7.5. These modes can further be identified as TE/TM/HEM.

If *H* field is propagating, then it is TE mode. By contrary, if *E* field is propagating, then it is TM mode. Also, when both types of fields, *E* and *H*, are excited simultaneously, then it is HEM mode. HEM modes are most advantageous but have complex structure. The detailed analysis of hybrid modes is described later in Chap. 10. Resonant modes can be shifted, merged, and independently controlled by different techniques. Increasing RDRA electrical length and input excitation frequency can generate higher-order modes into RDRA.

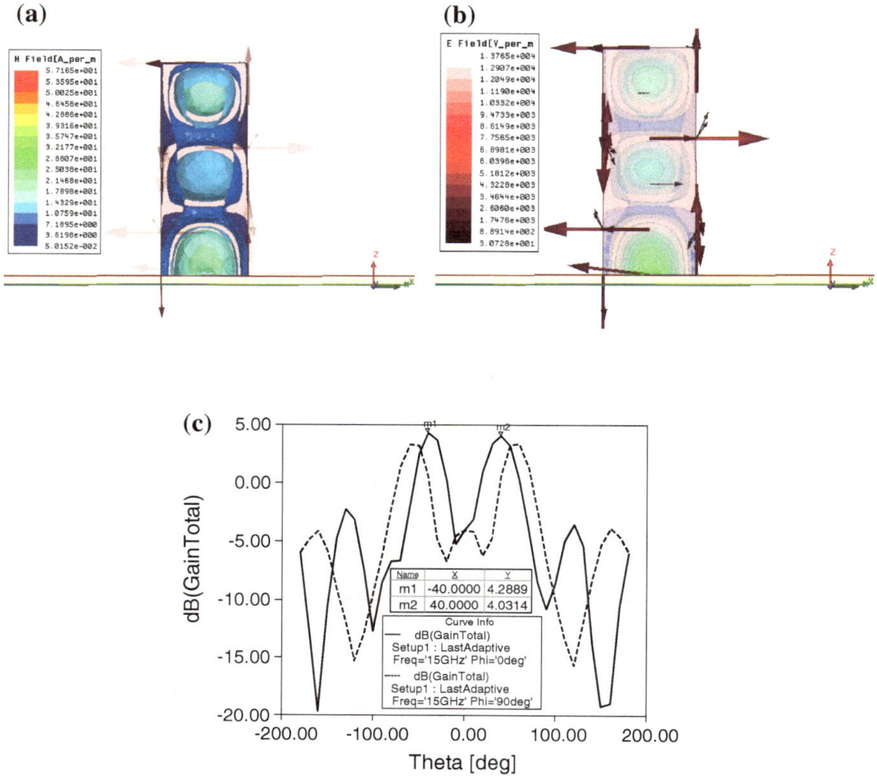

Fig. 7.18 TE_{115} mode at frequency 12 GHz. **a** *H* field distribution, **b** *E* field distribution, **c** gain

7.2 Resonant Frequency and RDRA Structure

Table 7.1 consists of RDRA specifications for prototype.

The structure of the antenna is shown in Fig. 7.6. Slot and microstrip are shown in Figs. 7.7 and 7.8. The feed is aperture-coupled. The substrate rectangular plane of 50 × 50 mm with a thickness of 0.6 mm was used. FR4 was used as RDRA substrate, and RDRA with a dielectric constant (permittivity) of 10.2 was placed on top of the substrate. The width of the microstrip used was 1.15 mm. Slot dimensions were 6 mm in length and 1 mm in width. RDRA dimensions were $6 \times 6 \times 5$ mm^3.

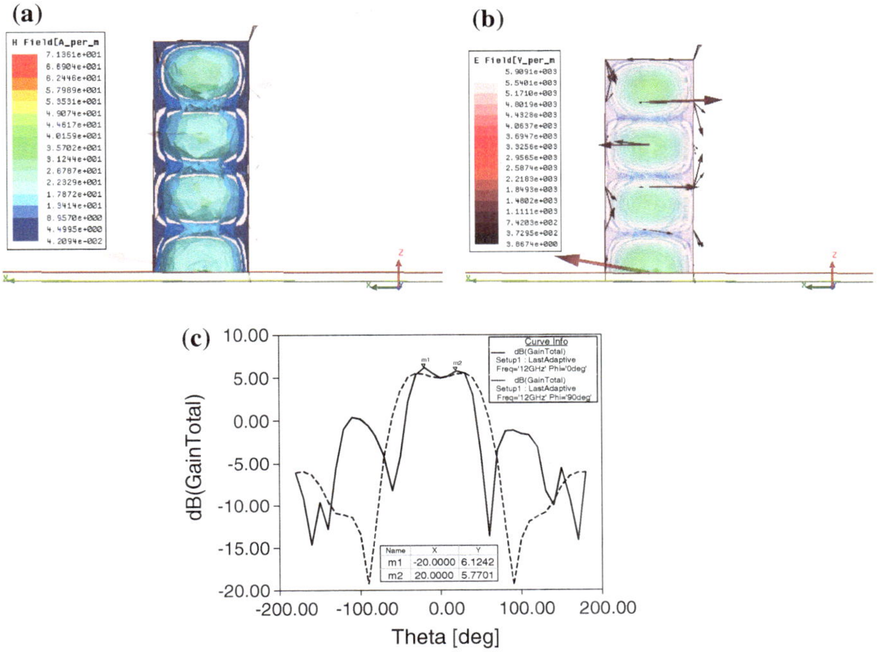

Fig. 7.19 TE_{117} mode at frequency 15 GHz. **a** *H* field distribution, **b** *E* field distribution, **c** gain

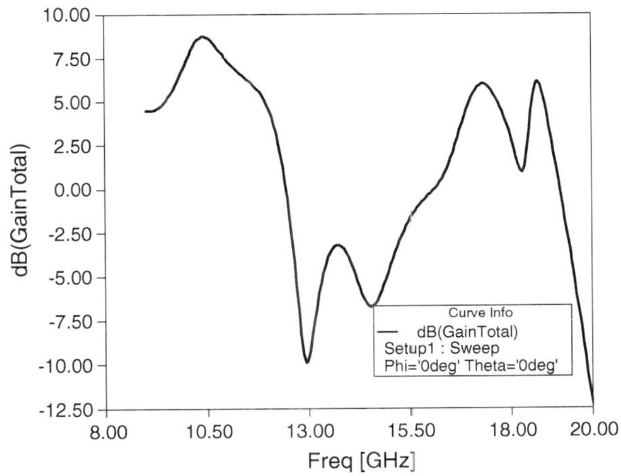

Fig. 7.20 Gain versus frequency plot at h = 15 mm

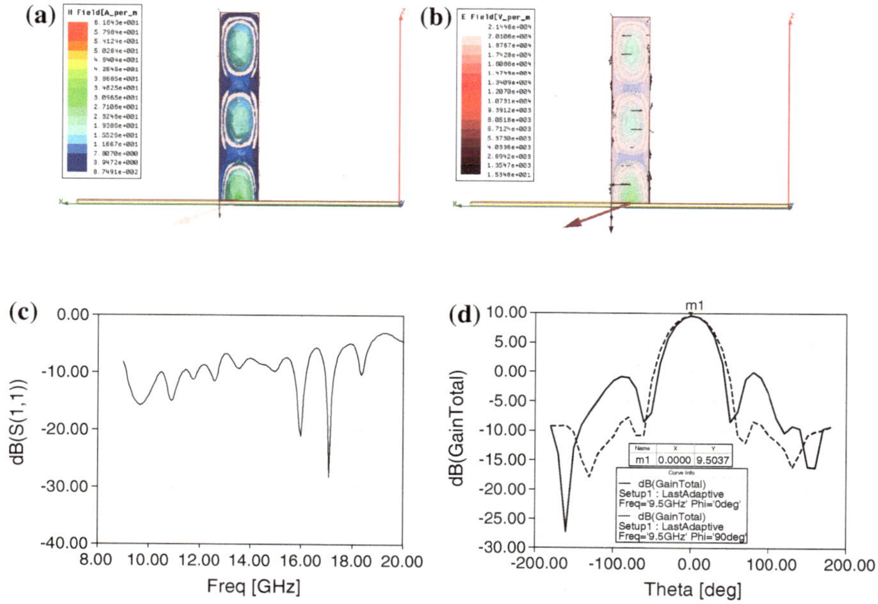

Fig. 7.21 TE_{115} mode at frequency 9.5 GHz. **a** *H* field distribution, **b** *E* field distribution, **c** return loss, **d** gain

7.2.1 Fields in Rectangular DRA

$TE_{\delta 11}$ resonant mode in rectangular DRA the fields can be defined using dielectric waveguide model depending upon given boundry conditions

$$H_x = \frac{(k_x^2 + k_y^2)}{j\omega\mu_0} \cos(k_x x)\ \cos(k_y y)\ \cos(k_z z) \tag{7.1}$$

$$H_y = \frac{(k_y k_x)}{j\omega\mu_0} \sin(k_x x)\ \sin(k_y y)\ \cos(k_z z) \tag{7.2}$$

$$H_z = \frac{(k_z k_x)}{j\omega\mu_0} \sin(k_x x)\ \cos(k_y y)\ \sin(k_z z) \tag{7.3}$$

$$\begin{aligned} E_x &= 0 \\ E_y &= k_z\ \cos(k_x x)\ \cos(k_y y)\ \sin(k_z z) \end{aligned} \tag{7.4}$$

$$E_z = -k_y\ \cos(k_x x)\ \sin(k_y y)\ \cos(k_z z) \tag{7.5}$$

So by solving these equations, we get transcendental equation as follows:

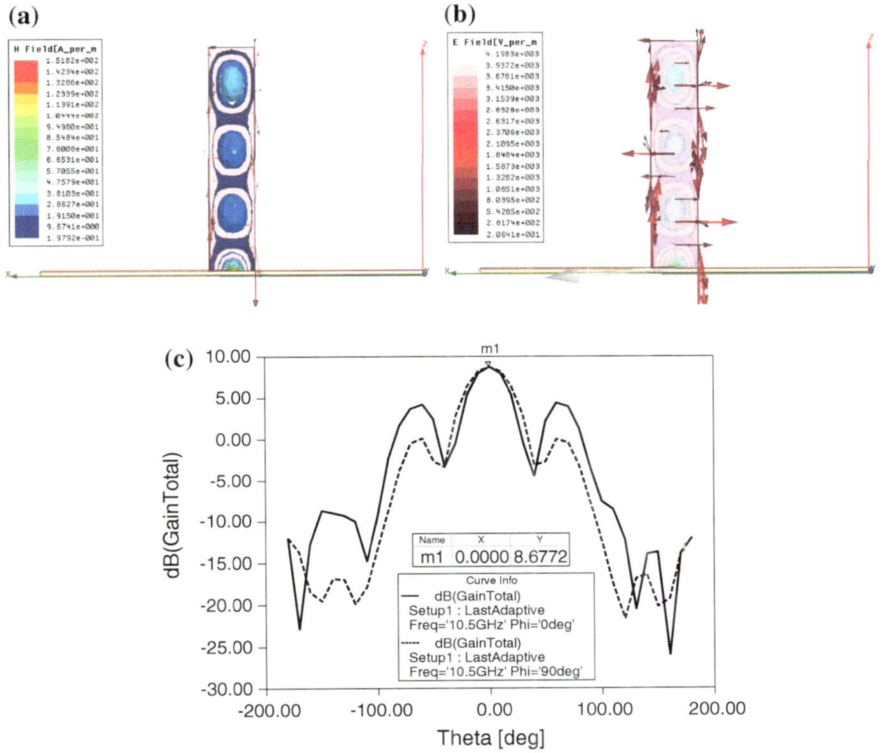

Fig. 7.22 TE_{117} mode at frequency 10.5 GHz. **a** *H* field distribution, **b** *E* field distribution, **c** gain

$$k_x \tan\left(\frac{k_x d}{2}\right) = \sqrt{(\varepsilon_r - 1)k_0^2 - k_x^2} \tag{7.6}$$

The resonant frequency and propagation constant can be determined from the transcendental equation.

The characteristic equation is as follows:

$$k_x^2 + k_y^2 + k_z^2 = \varepsilon_r k_0^2 \tag{7.7}$$

So, the resonant frequency can be obtained for grounded RDRA as follows:

$$f_{mnp} = \frac{c}{2\sqrt{\varepsilon_r}} \sqrt{\left(\frac{m}{a}\right)^2 + \left(\frac{n}{b}\right)^2 + \left(\frac{p}{2d}\right)^2}. \tag{7.8}$$

Frequency for isolated RDRA is given as follows:

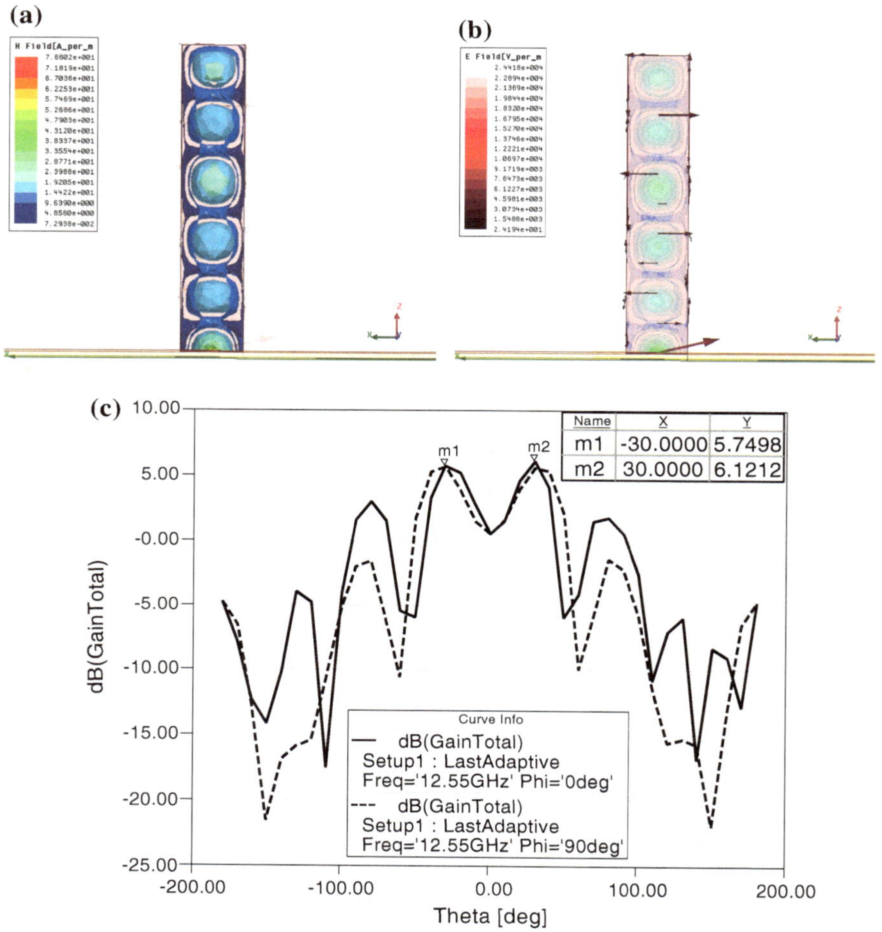

Fig. 7.23 $TE_{1,1,11}$ mode at frequency 15 GHz. **a** *H* field distribution, **b** *E* field distribution, **c** gain

Fig. 7.24 Gain versus RDRA height

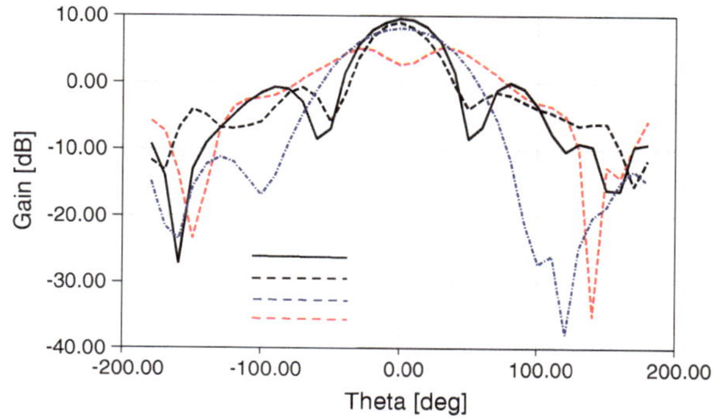

Fig. 7.25 Prototype development of RDRA

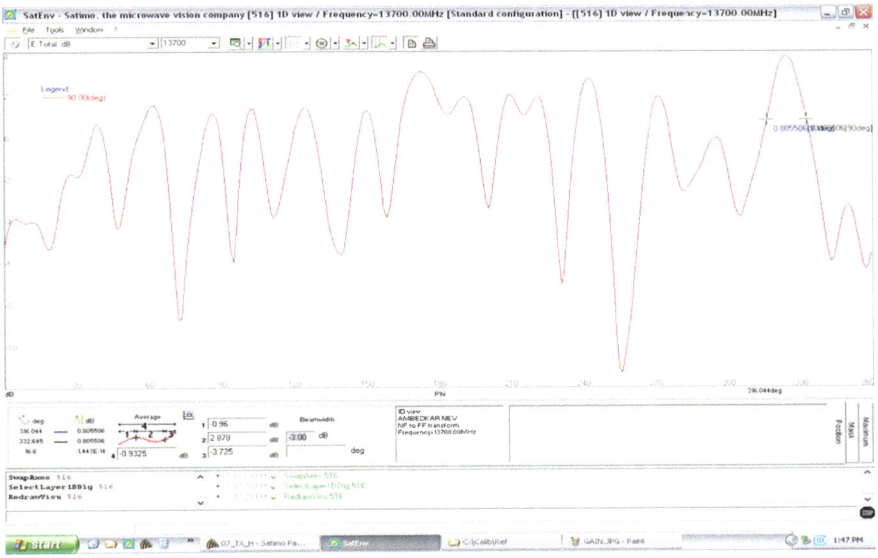

Fig. 7.26 Radiation pattern *H* field measurements under anechoic chamber

$$f_{mnp} = \frac{c}{2\sqrt{\varepsilon_r}}\sqrt{\left(\frac{m}{a}\right)^2 + \left(\frac{n}{b}\right)^2 + \left(\frac{p}{d}\right)^2}; \tag{7.9}$$

k_x, k_y, and k_z can be determined by using characteristic equation. Propagation constant k_0 can be determined in terms of k_x. Transcendental equation can be solved for k_z' using k_x. This solution can be obtained using MATLAB for fixed value of n, p, and d. k_x will now contain a' in place of a. a' is the extended electrical length due to fringing effects. Hence, k_x is the complete solution of transcendental equation.

Example 1 Let us determine the resonant frequency for dominant and higher-order modes of RDRA with given dimensions and dielectric constant:

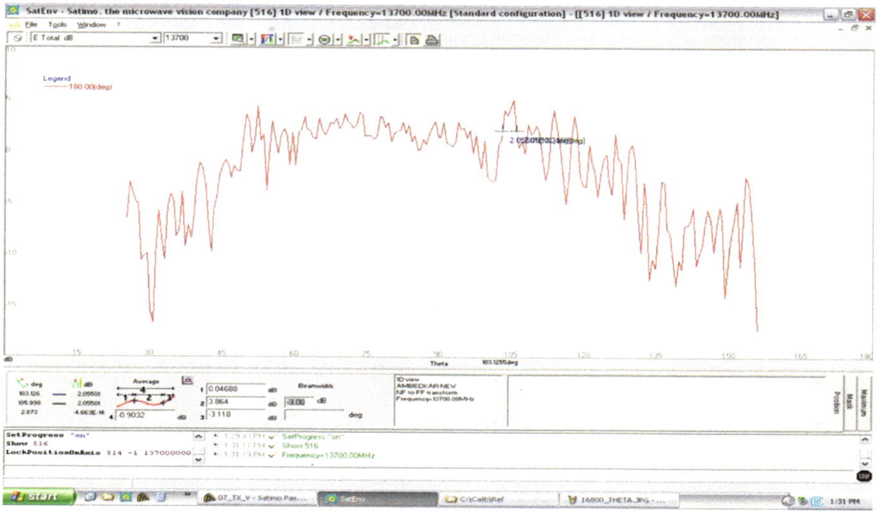

Fig. 7.27 Radiation pattern *E* field measurements under anechoic chamber

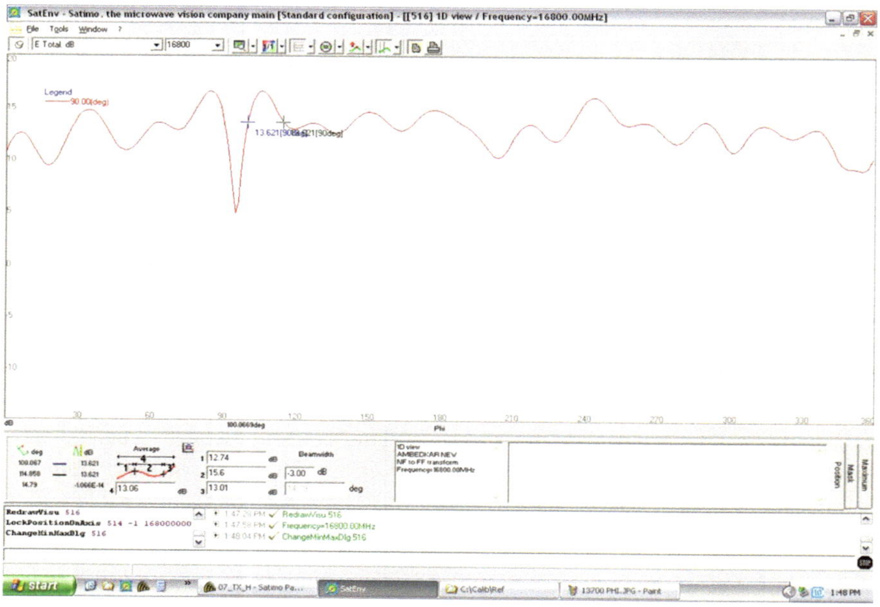

Fig. 7.28 Azimuthal radiation pattern measurements inside anechoic chamber

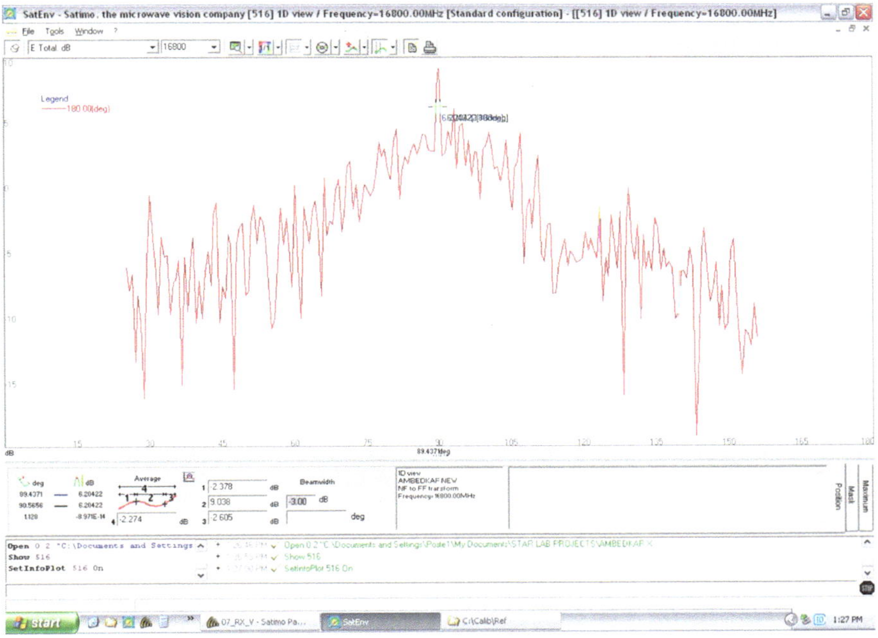

Fig. 7.29 Elevation radiation pattern

Fig. 7.30 Ready for measurement

Fig. 7.31 RDRA mounted with wooden block for measurement

Fig. 7.32 RDRA inside anechoic chamber for measurement

Table 7.1 Specifications of RDRA and design dimensions

S. No.	Element	Dimension (mm)
1	Ground plane	20 × 30
2	Substrate	20 × 30 × 0.8
3	RDRA	4.6 × 9 × 10.8
4	Width of microstrip	2.4
5	Length of stub and microstrip	18.693
6	Ground slot (l × *w*)	3.743 × 0.404

$$\begin{aligned} f_r &= \frac{c/2}{\sqrt{\varepsilon_r}}\sqrt{\left(\frac{m}{a}\right)^2+\left(\frac{n}{b}\right)^2+\left(\frac{p}{d}\right)^2} \\ &= \frac{3\times 10^8}{2\sqrt{10}}\sqrt{\left\{\left(\frac{1}{10}\right)+\left(\frac{1}{10}\right)+\left(\frac{1}{10}\right)\right\}10^6} \\ &= \frac{3\times 10^8}{2\sqrt{10}}\times 10^3\sqrt{\frac{3}{10}} = \frac{3\sqrt{3}\times 10^{11}}{2\times 10} = \frac{3\sqrt{3}}{2}\times 10^{10} \\ f_r &= 2.598\times 10^{10}\,\text{Hz} = 25.98\,\text{GHz} \end{aligned}$$

$$\begin{aligned} \text{Higher-order mode:}\quad f_{r_{113}} &= \frac{3\times 10^8}{2\sqrt{10}}\sqrt{\left(\frac{1}{10}+\frac{1}{10}+\frac{9}{10}\right)10^6} \\ &= \frac{3\times 10^8\times 10^3}{2\sqrt{10}}\times\sqrt{11} = \frac{3\times 10^{11}}{2\times 10}\sqrt{11} - \frac{3\sqrt{11}}{2}\times 10^{10} \\ f_r &= 4.9749\times 10^{10}\ \text{Hz} = 49.749\ \text{GHz} \end{aligned}$$

7.3 Modes (Resonant) Mathematical Solution

Rectangular dielectric resonator antennas (RDRAs) have received lots of attention in the last two decades due to several attractive characteristics, such as design flexibility, high gain, and wide bandwidth. RDRA has two different aspect ratios (*b*/*a*, *d*/*a*), high radiation efficiency, light weight, and low profile. In contrast, patch antenna has low gain, less bandwidth, and design flexibility.

The resonant modes are *E* and *H* fields pattern. They can be expressed as follows:

$$E_z = \sum_{mnp} u_{mnp}(x,y,z)\ \mathrm{Re}\left(C_{mnp}\ \mathrm{e}^{j\omega_{mnp}t}\right);\ \text{where,}\ C_{mnp}\ \text{are amplitude coefficients.} \tag{7.10}$$

where

$$u_{mnp}(x,y,z) = \frac{2^{3/2}}{\sqrt{abd}}\cos\left(\frac{m\pi x}{a}\right)\cos\left(\frac{n\pi y}{b}\right)\sin\left(\frac{p\pi z}{d}\right);\ \text{orthogonal Fourier basis function.} \tag{7.11}$$

$$C_{mnp} = \sqrt{\frac{\beta(mnp)^2 + \alpha(mnp)^2}{\left[\frac{\sqrt{2}}{\sqrt{d}}\sin\left(\frac{p\pi\delta}{d}\right)\right]^2}};\ \text{amplitude coefficient.} \tag{7.12}$$

$$\psi(mnp) = \tan^{-1}\left[\frac{\alpha_{mnp}\cos(\phi(mnp)) + \beta_{mnp}\sin(\phi(mnp))}{\alpha_{mnp}\sin(\phi(mnp)) - \beta_{mnp}\cos(\phi(mnp))}\right],\ \text{Phase.} \tag{7.13}$$

7.4 Top-Loading RDRA

This chapter is developed based on new approach using a top-loading excited RDRA as shown in Fig. 7.9, for generating higher-order modes. The even as well as odd modes can be generated even with ground plane. It has also been studied that even modes (in z-direction) were short-circuited, when RDRA was placed on a ground plane. Short magnetic dipoles are basis for generation of these resonant modes. Nearly, $\lambda/3$, $\lambda/2$, and λ heights of the dielectric resonator generated $TE_{\delta 11}$, $TE_{\delta 13}$ and $TE_{\delta 15}$ ($0 < \delta \leq 1$) modes. The gain is found to be increasing in higher modes. This is also evident from the findings that gain of RDRA starts decreasing or reducing even in higher modes, when magnetic dipoles start overlapping. This overlapping of dipoles can be seen when the wavelength used is very small.

Top loading excited both even and odd modes. Simulations have shown that 1st and 3rd resonances got shifted toward 2nd resonance, when the space 's' between top and bottom RDRAs varies. Merging of neighboring resonance modes could be done using this method. This is an excellent phenomenon, which can be used for bandwidth enhancement. This merging of bands helped to increase the antenna bandwidth. Thus, existing patch antenna gain and bandwidth can also be increased by using the concept of higher-order modes. Blocking or shifting of any modes has become possible in RDRA.

E and H fields perturbations in RDRA can be introduced by carrying out small changes in the structure, or this can be obtained by input excitation currents. This perturbation gets converted into eigenvector or eigenvalues. The perturbations are proportional to eigenvector and resonant mode. The number of modes is directly related to number of lobes occuring in radiation patterns. There are two ways in which the number of modes can be increased in RDRA: One is by increasing the RDRA dimensions a, b, and d, and the other is by increasing the input excitation frequency.

7.5 Simulated HFSS Results

From Fig. 7.10, it is evident that single resonant mode, as one half-wave variation take place in x-direction.

The gain of the antenna got enhanced due to increase of RDRA height. TE_{111} mode resonant frequency is 10 GHz, TE_{113} mode frequency is 12 GHz and TE_{115} frequency is 15 GHz. Figure 7.12 shows the magnetic and electric field distributions inside RDRA. The return loss and radiation pattern are shown in that figure.

It has been seen from Fig. 7.13 that gain has been enhanced. Figure 7.14 shows TE_{115} mode, which has higher resonant frequency. Figure 7.15 shows gain versus frequency plot. Figure 7.16 depicts the spacing between short magnetic dipoles.

The above results obtained from the analysis of RDRA carried out revealed that higher-order modes offer high gain until dipole overlapping does not take place. The decrease in gain due to overlapping of short magnetic dipoles takes place. This will happen when there is a less spacing between two short magnetic dipoles. Hence, minimum spacing between short magnetic dipoles must be equal to 0.4λ. If the spacing between short magnetic dipoles is less than this limit, then the gain will be reduced. This is depicted by simulations in Fig. 7.14. TE_{113} gain has been reduced even at TE_{115} as shown in Fig. 7.18. Now, if we obtain TE_{115} mode with increase in RDRA height, then more gain can be obtained. This is the reason why gain at TE_{115} is less than TE_{113} as shown in Figs. 7.14 and 7.15.

$$\text{At } h = \lambda/2$$

When h = 15 mm, three modes got generated, i.e., TE_{113}, TE_{115}, and TE_{117} corresponding to 10, 12, and 15 GHz, respectively. It is clear that gain has been decreased at higher modes due to the reason explained earlier that spacing between short dipoles placed at the center of the field is less than 0.4 λ. At frequency 10 GHz inside RDRA, there is proper spacing between these dipoles; hence, the gain is maximum. Various excited modes are shown in Figs. 7.16, 7.17, 7.18, 7.19, 7.20, 7.21, 7.22 and 7.23 are excited, i.e., resonant modes in RDRA. Their results have also been taken on S_{11} plots.

$$h = \lambda$$

Here, the height of RDRA has been chosen as h = 30 mm, and mode was operating at 10.5 GHz. The highest gain was due to the same concept of spacing of short magnetic dipoles. From Figs. 7.17, 7.18, 7.19 and 7.20, very important fact is noticed that when spacing between short magnetic dipoles was reduced, then the order of mode becomes high, while the power of main lobe was distributed to side lobes. Thus, the gain of the antenna was reduced at higher mode if the spacing is less than 0.4 λ.

In the above figures, generation of higher modes, limitation, and their effect on antenna gain have been clearly shown.

7.6 Modes at Varying Heights of RDRA

The comparison between three RDRA's of different heights have been made. It was noticed that RDRA having less height and operating at lower mode offers less gain but higher bandwidth. On increasing the height of RDRA, the gain of the antenna is found to be higher along with directivity, thus narrowing the beam width. Figure 7.24 shows that gain is increasing, when the height of RDRA is increased.

7.7 Distortions Due to Overlap of Dipole Moment

RDRA of height $\lambda/3$, $\lambda/2$, and λ operating around 11–15 GHz consists of TE_{111}, TE_{113}, and TE_{115} modes. This fulfills the requirement of separation of magnetic fields by spacing 0.4 λ. But when RDRA with same height operates at higher frequency, then the spacing between dipoles is reduced. The power of main lobe is distributed to the side lobes, which creates the loss of antenna power and gain. Hence, any desired resonant modes inside the device can be excited for desired radiation pattern at known resonant frequency. The higher modes amplitude coefficients equation has been developed. Modes can be used to visualize corresponding radiation pattern and polarization of the antenna. Modes give physical insight into eigenvalue for determining resonant frequency and feeding point for 50 Ω impedance.

7.8 Prototype and Anechoic Chamber Experimentations

Prototype RDRA was made, and it was tested inside the anechoic chamber using VNA. Results for radiation pattern and other antenna parameters have been taken and are shown in Figs. 7.26, 7.27, 7.28, 7.29, 7.30, 7.31 and 7.32. Each figure is captioned below for the results.

It was seen in RDRA of particular height, more number of higher-order mode can be excited by applying another excitation on the top loaded RDRA as shown in Fig 7.3. The reduced spacing '*s*' between top and bottom RDRAs, merged even modes, thus increased bandwidth of the antenna. The RDRA under top loading converted few odd modes to nearest even mode. Thus, both even and odd modes were made available due to top loading. Thus, spacing '*s*' seems to control bandwidth of RDRA. High gain, miniaturization, high bandwidth, directive antenna can be designed by having proper control or maneuvering resonant modes.

The design of this antenna offers wide scope of achieving wide bandwidth along with high gain. The application of this antenna includes satellite tracking, air traffic control Wi-fi, Wi-max, and mobile communication.

By developing control on modes, we can control beam width of antenna and can restrict the reception of signal to a particular area and hence it can be used for military applications. Presently, we face the problem of TV signal reception during rainy season, due to the absorption of signal by rain drops due to signal being single polarized either TE or TM. This can be minimized by application of dual polarized or hybrid antenna. The other application could be miniaturization of antenna. By keeping the dimensions of antenna fixed, the mode of antenna can be changed by changing the permittivity of RDRA and thereby changing the frequency. To automate the mode generation, microcontroller-based lookup table can generate possible combinations of bandwidth, gain, and frequency.

7.9 Adjacent Modes Combination for Broadband Applications

The merging of adjacent bands or neighboring modes of RDRA can be used for enhancement of the bandwidth. By varying the aspect ratio, three resonant bands can be obtained for useful operation as shown in Fig. 7.33.

The dimensions of RDRA are given in Table 7.2. Figure 7.33b shows the return loss of the antenna with three bands resonating at 2.89 GHz at dominant mode, i.e., TE_{111} mode and at 3.61 GHz for TE_{121} mode and at 4.6 GHz for TE_{131}. Figure 7.34 shows *H* and *E* fields distributions inside RDRA. The direction of the electric field is indicated by arrow.

Figure 7.35 depicted that the lower gain at lower mode and high gain at higher modes.

7.10 Effect of Air Gap Between RDRA and Ground Plane

The effect of the air gap between ground plane and RDRA is shown in Fig. 7.36. Table 7.3 shows the variation in the resonant bands of the antenna. Effect of introducing the gap between RDRA and ground plane is depicted in Fig. 7.36. Results obtained by simulation along with the results obtained by calculations clearly indicate the effect of air gap. The modes are spreading as the frequency of the modes is shifted in forward direction with respect to increase in the gap.

Tables 7.1, 7.2, 7.3, 7.4, 7.5 and 7.6 are the results tabulated for various simulations.

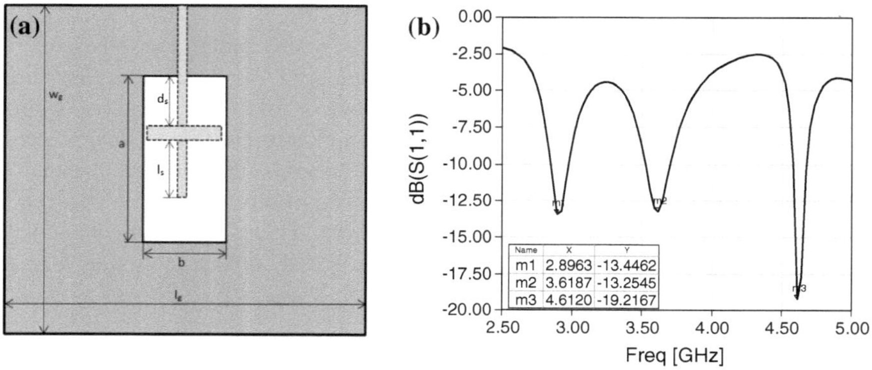

Fig. 7.33 **a** Top view of the DR. **b** Return loss

Table 7.2 Simulated and calculated resonant frequencies of the modes in DR

Mode/gap	Results	$G = 0$	$G = 0.01$	$G = 0.02$
TE_{111}	Simulated	2.89	2.83	2.88
	Calculated	2.84	2.85	2.87
TE_{112}	Simulated	3.67	3.83	Omitted
	Calculated	3.69	3.71	3.74
TE_{113}	Simulated	4.64	4.72	4.81
	Calculated	4.62	4.69	4.73

7.11 Effect of Asymmetrical Wells Inside RDRA

When two asymmetrical wells are inserted, then these modes start to merge together and all the bands are shifted. This is shown by return loss graph in Fig. 7.38. Comparison between asymmetrical wells and without asymmetrical wells has been made. This is to note that by adding wells, higher frequency bands get shifted more as compared to lower frequency bands as shown in Fig. 7.40. Dimensions of the structure are shown in Table 7.4. The results were taken for various values of width of well as 0.5, 1.1, 2.0 mm respectively are shown in Fig. 7.38.

7.12 Effect of Moat Insertion Inside RDRA

In this way of merging of modes takes place when air gap is inserted in the structure of RDRA as shown in Fig. 7.41 then we get broader bandwidth. Plot for reflection coefficient in Fig. 3.9 shows clearly the effect of moat in the structure. All the dimensions of the structure are shown in Table 7.3.

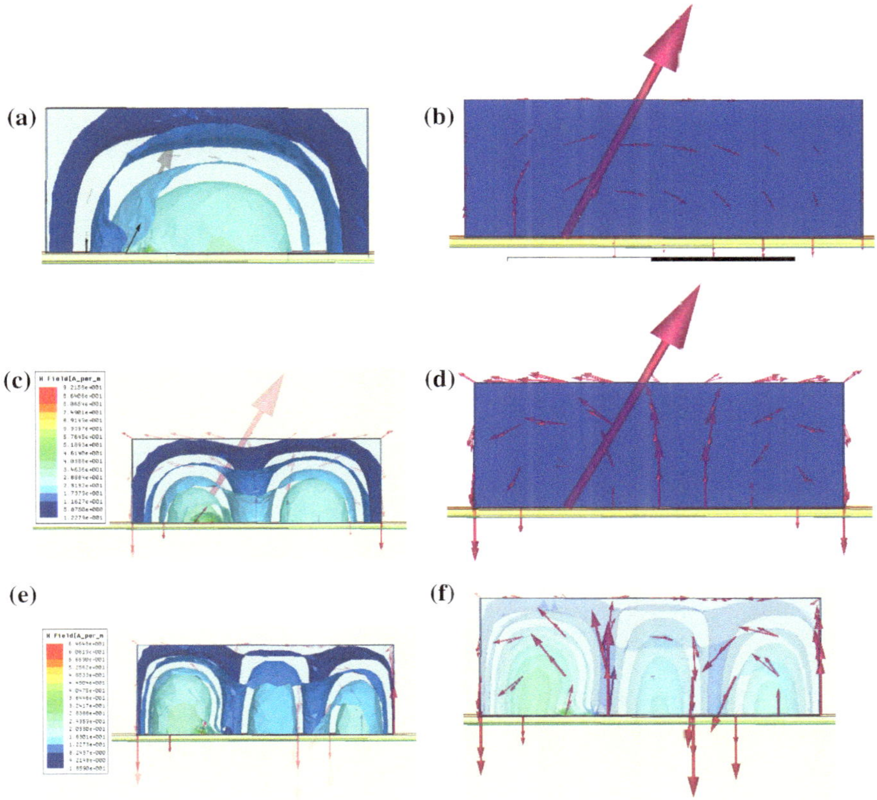

Fig. 7.34 Magnetic and electric field distributions inside DR. **a** *H* field TE_{111} mode, **b** *E* field TE_{111} mode, **c** *H* field TE_{112} mode, **d** *E* field TE_{112} mode, **e** *H* field TE_{113} mode, **f** *E* field TE_{113} mode

7.13 Effect of *a*/*b* and *d*/*b* Aspect Ratio

The effect of length and width of RDRA is such that if we increase the dimensions then there can be large number of modes generated. The effect of *a*/*b* and *d*/*b* ratio has been speculated in the manner such that when the ratio *a*/*b* is increased, the modes come closer to each other and merged, and when the ratio *d*/*b* is increased, resonant frequencies of all modes are diverged. Further, if the ratio *d*/*b* is reduced, then the modes are merged. So here, we increased the *a*/*b* ratio and reduced the *d*/*b* ratio, and then we pointed aristocratically that intensified changes like modes have been merged to increase the bandwidth of the device. The important thing to note is that mode TE_{112} has been merged, and there are the resonant frequencies of modes TE_{111} and TE_{113} only. Figure 7.31 reflects the effect of the overall process.

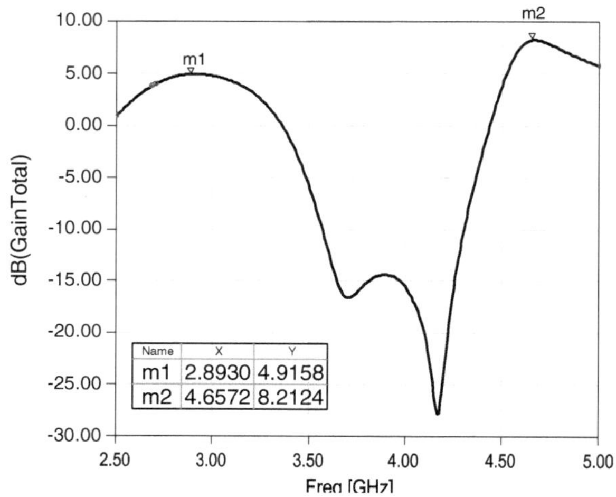

Fig. 7.35 Showing the gain of the antenna for frequency range

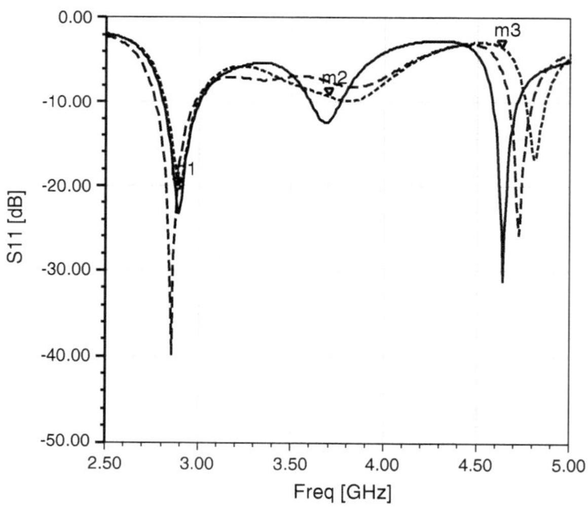

Fig. 7.36 Effect of air gap between RDRA and ground plane

Table 7.3 Modes and their resonant frequencies

Mode	Resonant frequency (GHz)	Gain (dB)
TE_{111}	4.56	5.2
TE_{112}	4.96	–
TE_{113}	5.56	8.78

Figures 7.35, 7.36, 7.37, 7.38, 7.39, 7.40, 7.41, 7.42, 7.43 and 7.44 presented effect on change in aspect ratios. Tables 7.3 and 7.4 indicated design parameters. Tables 7.5 and 7.6 show resonant modes.

Table 7.4 Dimensions of all the structures

Dimension				
Parameter/structure	Aperture-coupled DR	DR with two asymmetrical wells	DR after increasing in *a*/*b* and decreasing in *d*/*b* ratio	DR with moat
Ground plane and substrate (mm) and permittivity	70 × 70 and FR_4 epoxy (4,4)	70 × 70 and FR_4 epoxy (4,4)	70 × 70 and FR_4 epoxy (4,4)	70 × 70 and FR_4 epoxy (4,4)
DR (mm)	28 × 9 × 10	28 × 9 × 10	30 × 19 × 4	30 × 19 × 4 Inner dimensions: 16.5 × 10 × 4 Moat gap (mm): Gi = 0.1, G_2 = 5.3, G_3 = G_4 = 0.2
Microstrip (mm)	35 × 1.15	38 × 1.15	37 × 1.15	40 × 1.15
Slot (mm)	10 × 2	10 × 2	7 × 1	13.5 × 2
D_s (mm)	8	7	12	13
L_s (mm)	6	6	4	5

Table 7.5 Comparison among all the structures

Comparison between all the structures				
Parameter/structure	Aperture-coupled DR	DR with two Asymmetrical Wells	DR after Increasing in *a*/*b* and decreasing in *d*/*b* ratio	DR with Moat
Modes and frequency (GHz)	TE_{111}(2.92), TE_{112}(3.70), TE_{113}(4.64)	TE_{111}(2.92), TE_{112}(3.70), TE_{113}(4.64)	Modes are merged with each other, gain is enhanced, and the bandwidth is increased	Modes are merged with each other, gain is enhanced, and the bandwidth is increased drastically
Gain (dB)	1.7	2.1	3	4.8
Bandwidth (GHz)	Less	Less	Bandwidth is enhanced by merging the modes	Bandwidth is enhanced and is larger, and modes have the closer resonant frequency to each other

Table 7.6 RDRA simulated resonant modes' parameters

Height h (mm)	Excitation at top of the DR	Generated Mode	Frequency f (GHz)	Gain (dB)	Bandwidth (GHz)
5	No	$TE^x_{\delta 11}$	11.3	5.6	10.1–12.0
	Yes	$TE^x_{\delta 11}$	11.8	4.8	9.5–17.7
		$TE^x_{\delta 12}$	15	4.5	
10	No	$TE^x_{\delta 11}$	10	5	9.25–12.1
		–	–	5	12.81–14.85
		$TE^x_{\delta 15}$	13.7	7	16.5–17.1
	Yes	$TE^x_{\delta 14}$	13	6	9.4–16.8
15	No	$TE^x_{\delta 13}$	10	9.1	9.48–11.4
		$TE^x_{\delta 17}$	16	5.6	15.8–16.4
		–	–	6	18.4–18.6
	Yes	$TE^x_{\delta 14}$	11.7	5	9.3–17.6
		$TE^x_{\delta 16}$	13.7	4.3	
		$TE^x_{\delta 18}$	16.7	4.5	
30	No	$TE^x_{\delta,1,5}$	9.8	8.2	9.30–11.85
		$TE^x_{\delta,1,11}$	12.56	6	12.45–12.70
		$TE^x_{\delta,1,15}$	15.95	5.8	15.70–16.20
		$TE^x_{\delta,1,17}$	17	5.9	16.90–17.25
		–	–	6	18.20–18.50
	Yes	$TE^x_{\delta,1,6}$	10.83	5	9.35–17.1
		$TE^x_{\delta,1,12}$	13.58	5.1	
		$TE^x_{\delta,1,14}$	14.88	4.7	
		$TE^x_{\delta,1,16}$	15.7	5.1	

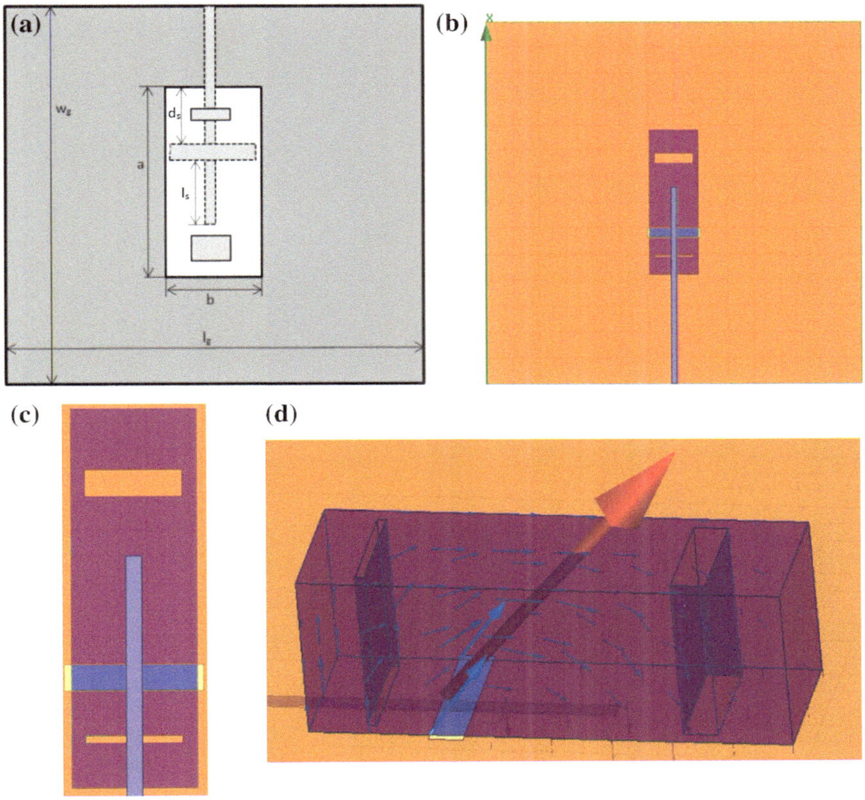

Fig. 7.37 Structure after insertion of two asymmetrical wells

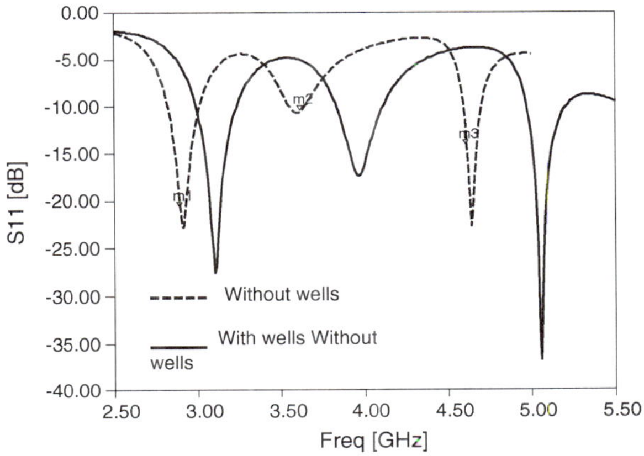

Fig. 7.38 Showing the effect of addition of wells

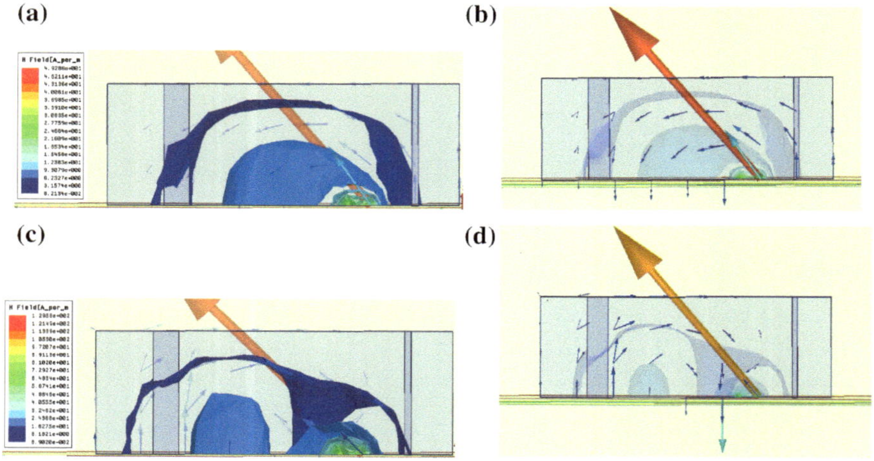

Fig. 7.39 Merging of second band in first and field distributions. **a** H_x field at $f = 2.9$ GHz, **b** E field at $f = 2.9$ GHz, **c** H_x field at 3.6 GHz, **d** E field at 3.6 GHz

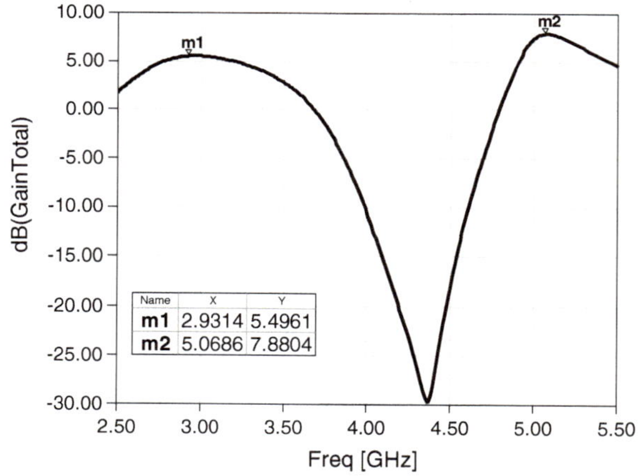

Fig. 7.40 Gain after insertion of wells

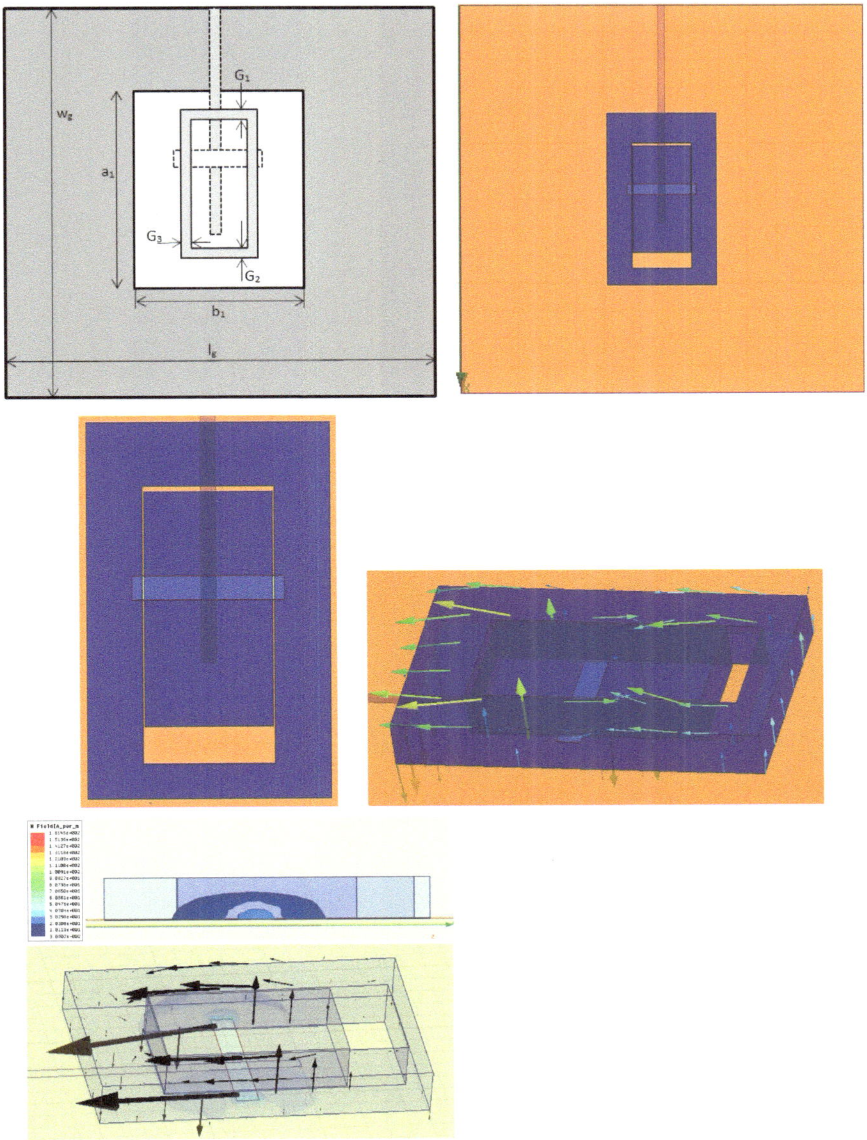

Fig. 7.41 Structure of the antenna after insertion of moat inside RDRA and field distribution at frequency 4.56 GHz

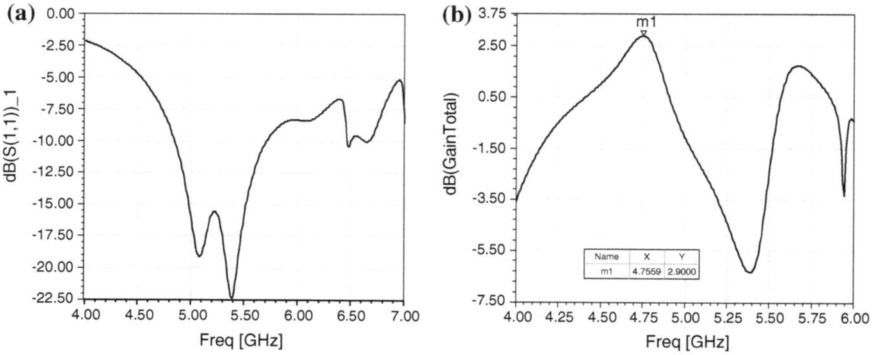

Fig. 7.42 **a** Frequency response showing return loss. **b** Gain of the antenna over frequency

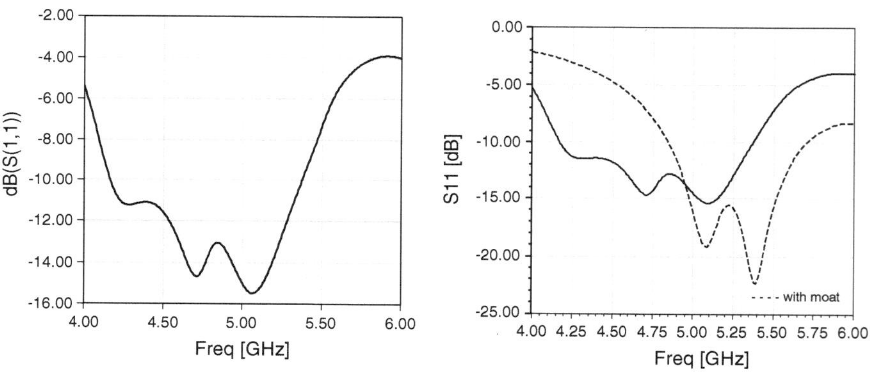

Fig. 7.43 Return loss and comparison with the structure with moat

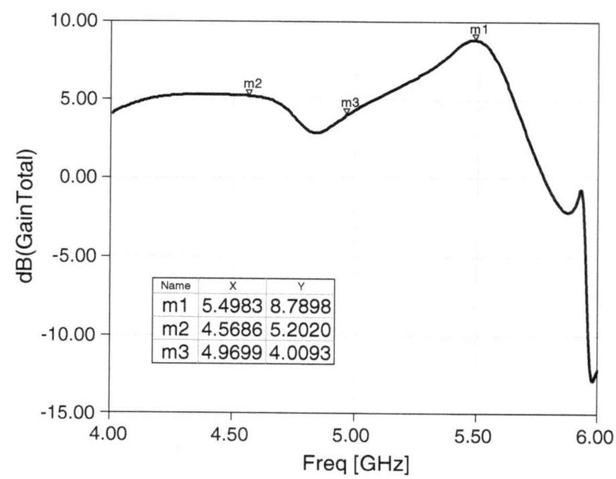

Fig. 7.44 Gain after increasing *a*/*b* ratio and decreasing *d*/*b* ratio

It is found that merging of modes can be very advantageous for broadband applications. The aspect ratio plays very important role in this phenomenon. Excitation applied at top of the RDRA converted odd modes into nearby even modes.

Reference

1. Mongia RK, lttipiboon A, Cuhaci M, Roscoe D (1994) Radiation Q-facter of rectangular dielectric resonator antennas theory and experiment. In: IEEE international symposium on Antennas Propagation Society, Seattle, WA, pp 764–767, June 1994

Chapter 8
RDRA Angular Excitation Mathematical Model and Resonant Modes

Abstract This chapter narrates angular excitation mathematical modeling of RDRA. The shift in radiation pattern and resonant modes have been realized based of angular shift in input. Slot is the source of input voltage to RDRA. Slot size and orientation effects loading of RDRA. The resonant characteristics of a RDRA are dependent on shape, DRA volume and excitation. The excitation current can be defined in terms of magnetic vector potential "*A*" based on applied current densities "*J*". This "*A*" can be expressed in terms of *E* and *H* fields or as "*S*" Poynting vector.

Keywords Slot · Angular variation · Change in radiation pattern · Resonating modes · Power flux · HFSS · VNA · Hardware model · Anechoic chamber

8.1 Introduction

Slot is the source of input to RDRA. Slot size and orientation is responsible for loading of RDRA. The angular orientation of slot has been investigated in this chapter with simulations and experimentation. The resonant characteristics of a RDRA depend upon the shape and size of the (volume) dielectric material along with feeding style. It is to be appreciated that in a RDRA, it is the dielectric material that resonates when excited by the feed. This phenomenon takes place due to displacement currents generated in the dielectric material. The excitation current can be defined in terms of magnetic vector potential "*A*" based on the current densities "*J*" inside the resonator, at any far-field point. This "*A*" can be expressed in terms of *E* and *H* fields. Later, this is expressed as "*S*" Poynting vector. Now the flux described can be treated with boundary conditions to find Radiated power P_{rad} into space. Figure 8.1 presented RDRA excited at slot angle. Figures 8.1 and 8.2 are HFSS model of RDRA. In Fig. 8.3, slot is shifted with certain amount of angle. If two slots are placed at 90°, circular polarization will take place. If one slot area is larger than the other, then LHCP (left-hand circular polarization) or RHCP (right-hand circular polarization) will take place. Figure 8.4 RDRA is excited at 45° angle. Figures 8.5, 8.6 and 8.7 presented radiation pattern at slot angles. Using two

R.S. Yaduvanshi and H. Parthasarathy, *Rectangular Dielectric Resonator Antennas*, DOI 10.1007/978-81-322-2500-3_8

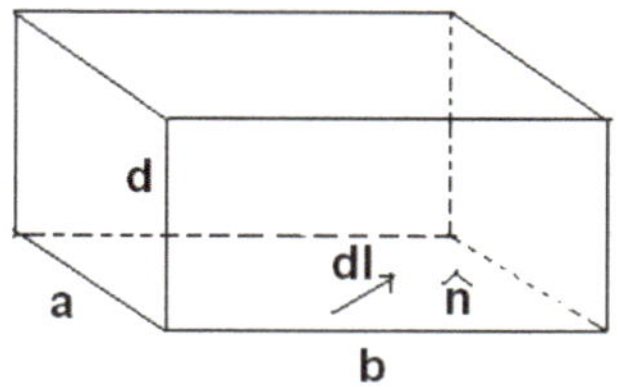

Fig. 8.1 RDRA with slot at an angle $(\o_0, \phi_0)$, a, b, and d are dimensions

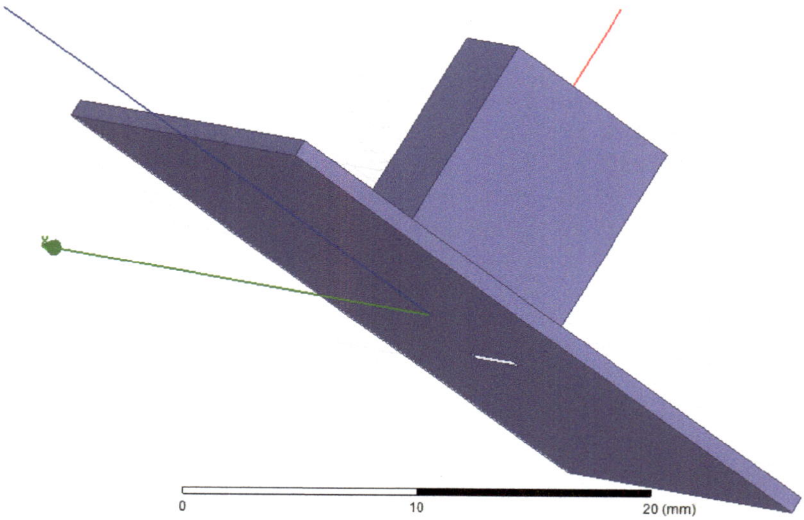

Fig. 8.2 Let rectangular DRA excited by slot at an angle (θ_i, ϕ_i)

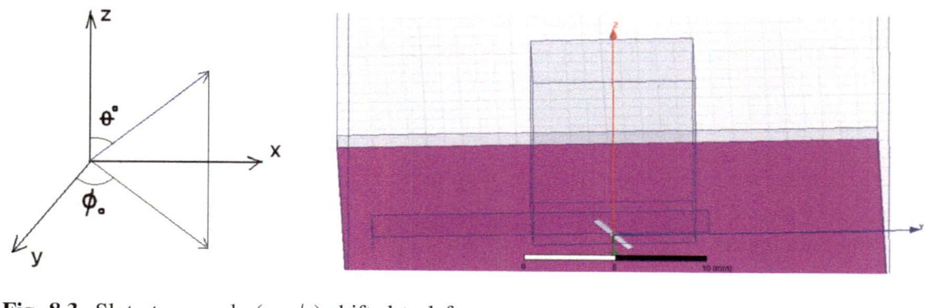

Fig. 8.3 Slot at an angle $(\o_i, \phi_i)$ shifted to left

cross slots circular polarization can be integrated. If two slot of different lengths are used then due to differential signal LHCP and RHCP can be generated. This indicates that a mechnism for polarization cantrol can become possible if these slots are arranged in a particular manner.

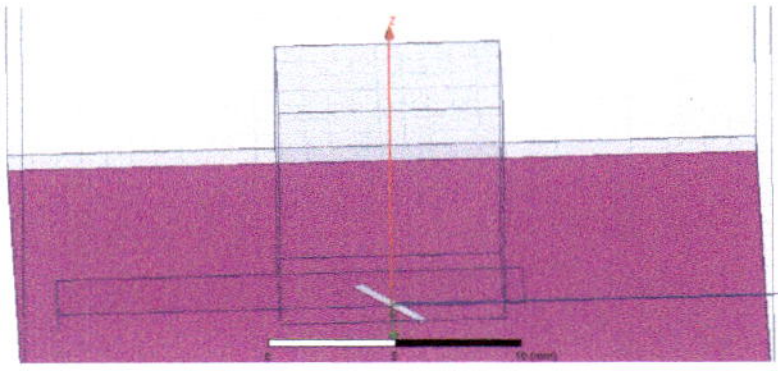

Fig. 8.4 RDRA angular excitation left side (30°–40°)

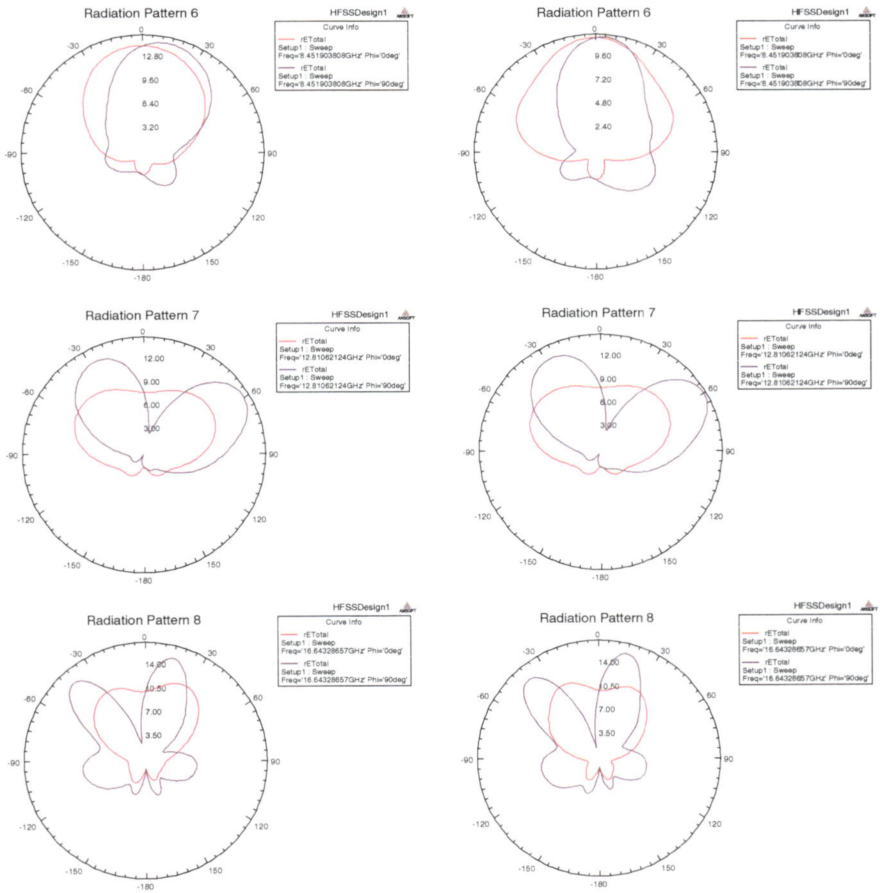

Fig. 8.5 Radiation pattern at an angle (30°–40° to the left)

The detailed description of radiation phenomenon is given below. The resonator RDRA radiates from the fringing fields. The resonator acts as tuned sequential RLC circuits having different values of LC or resonant cavity with an electric field perpendicular to the resonator, that is, along the *Z*-direction. The magnetic field has

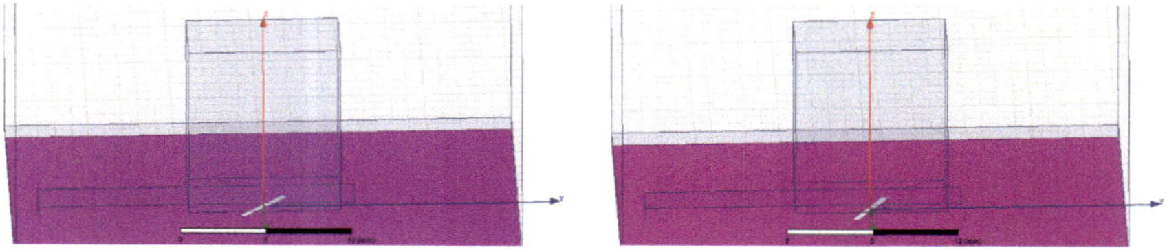

Fig. 8.6 Angular excitations in RDRA to the right side

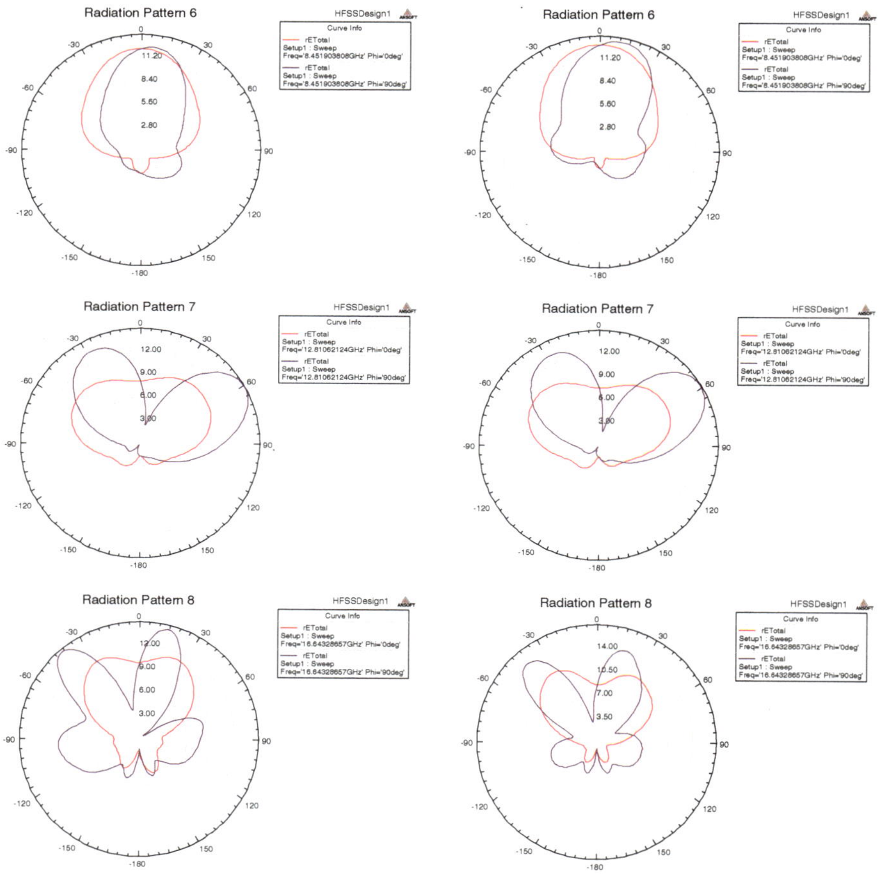

Fig. 8.7 Radiation pattern when slot is shifted to the right

vanishing tangential components at the four side walls. The four extended edge surfaces around RDRA serve as the effective radiating apertures. These fringing fields extend over a small distance around the side walls and can be replicated as fields E_x that are tangential to the substrate surface. The only tangential aperture field on these walls is $E = E_z$, because the tangential magnetic fields vanish by the boundary conditions. The ground plane can be eliminated using the image theory, resulting in doubling the aperture magnetic currents, that is, $J = n \times E$. Hence, the effective tangential fields can be expressed in terms of the field E_z. Now, radiated power pattern can be compared with the modes generated inside the resonator. The surface current density can be the main source of *E–H* fields pattern when applied with boundary conditions inside the resonator. This can be correlated with far-field pattern. The physics of this radiation is based on the fringing effect due to dipole moments. First derivative is velocity fields, and then, the second derivative on dipole moments can be termed as acceleration, which is main source of radiations. Hence, steering of the resonant modes mainly depends on excitation. E_z, H_z, or both E_z and H_z fields at any instant of time can define TM, TE and HEM modes.

8.2 Angular Shift in Excitation

Let aperture-coupled microstrip with slot and stub (feed) is situated in *xy* plane of RDRA at bottom part and slot placed at an angle (θ_i, ϕ_i) as shown in Fig. 8.1. The resonator modes and radiation pattern generated have been investigated as follows:

1. H_z, E_z fields are longitudinal. These have been expressed in terms of orthonormality with signals $u_{mnp}(x, y, z)$ and $v_{mnp}(x, y, z)$ at frequency ω_{mnp} based on the Maxwell's equations with given boundary conditions of RDRA.
2. At $z = 0$; surface (x, y) excitation is applied with slot and surface current density $\{J_{sx}(x, y, t), J_{sy}(x, y, t)\}$ is developed into RDRA.
3. The surface electric current density is equated with generated magnetic fields into RDRA:

$$\{J_s\ (x,\ y,\ \delta) = \left(J_{sx}, J_{sy}\right) = (\hat{n} \times H) = \left(-H_y,\ H_x\right)\};$$

   at $z = 0$; amplitude coefficients are obtained on expansion of H_z, E_z in terms of C_{mnp} and D_{mnp}.
4. Equating tangential component of E_z at boundary, i.e., $\underline{E}_y|_{z=0}$ to zero, the amplitude coefficients D_{mnp} for H_z and C_{mnp} of E_z are expressed.
5. *Feed position in xy plane can be defined as follows*:

$$(x_0, y_0)(\phi_0, \theta_0)$$

6. $f(x, y) = \left\{\begin{matrix} 1 & x_0 \leq x \leq \delta l(length) \\ & |y| \leq W\ (\text{width}) \\ 0 & \text{otherwise} \end{matrix}\right\}$
7. Excitation current in time domain can be expressed as:

$$J_s(x, y, z, t)$$

$so, J_s = \frac{I(mnp)}{\omega} e^{j\eta(mnp)}, (\eta_x, \eta_y, o) - (\cos\phi_o \; \sin\phi_o, 0)\theta(x) = \begin{Bmatrix} 1 & \eta > 0 \\ 0 & \eta < 0 \end{Bmatrix}$; where η is the angle of variation in excitation.

Here, we apply excitation through slot d*l* at some specific angle. Later, shift in the position of slot is provided. Change in radiation pattern or resonant modes is investigated with mathematical equation, simulations and experimentations on RDRA.

$$\frac{\mu I \overrightarrow{\mathrm{d}l} e^{-jkr}}{4\pi r} = \vec{A};\quad \text{where } A \text{ is magnetic vector potential} \tag{8.1}$$

$$A = V \cdot \int E\,\mathrm{d}l$$

$$E = -j\omega \cdot \vec{A}$$

Radiated power

$$\frac{|E|^2}{2\eta} = \frac{\omega^2 |\vec{A}|^2}{2\eta}, \qquad \sqrt{\frac{\mu}{\epsilon}} = \eta = \text{impedance.} \tag{8.2}$$

$$\underline{\vec{A}} = \frac{\mu}{4\pi} \int_{\text{Volume}} \frac{J(\underline{r}',\omega) e^{-jk|\underline{r}-\underline{r}'|}}{|\underline{r} - \underline{r}'|} d^3 r';\ \text{at source.} \tag{8.3}$$

We know that radiation pattern can be defined by $E_\theta,\ E_\phi$

$$E_\theta = -j\omega A_\theta \ \text{ and } \ A_\theta = \hat{\theta} \cdot A$$

Antenna current density can be expressed as follows:

$$J \cdot (r',\ \omega) = \sum_{mnp} J_s[mnp, r'] \quad e^{j\omega(mnp)t} \tag{8.4}$$

The magnetic vector potential in terms of J can be written as follows:

$$\begin{aligned} \underline{A} &= \frac{\mu}{4\pi} \sum_{mnp} \int \frac{\underline{J}_s[mnp,\ \underline{r}'] e^{j\omega(mnp)\left(t - \frac{|\underline{r}-\underline{r}'|}{c}\right)}}{|\underline{r} - \underline{r}'|} \mathrm{d}s(\underline{r}'); \quad \mathrm{d}s \text{ is surface} \\ &= \frac{\mu}{4\pi} \frac{e^{jkn}}{|\underline{r} - \underline{r}'|} \sum_{mnp} \int_s \underline{J}_s[mnp, \underline{r}'] e^{j\omega(mnp)\hat{\underline{r}}\cdot\underline{r}'} \mathrm{d}s(\underline{r}') \end{aligned} \tag{8.5}$$

$$E_\theta,\ E_\phi,\ H_\phi = E_\theta/\eta,\ H_\theta = -{E_\phi}/{\eta}.$$

Radiated power can be given as follows:

$$P_{\text{rad}} = \frac{1}{2\eta}\left(|E_\theta|^2 + |E_\phi|^2\right) \tag{8.6}$$

$$\begin{cases} \hat{\theta} = \hat{x}\cos\varphi\cos\theta + \hat{y}\sin\varphi\cos\theta - \hat{z}\sin\theta \\ \hat{\phi} = -\hat{X}\sin\varphi + \hat{Y}\cos\varphi \end{cases}$$

$$\begin{aligned} E_\theta = &\frac{\mu}{4\pi r^2}\text{Re}\sum_{mnp}\int_s \{J_{sx}[mnp,\ \underline{r}']\cos\varphi\cos\theta \\ &+J_{sy[mnp,\ \underline{r}']}\sin\phi\cos\theta - J_{sz[mnp,\ \underline{r}']}\sin\theta\Big\} \\ &\exp\left(j\omega\frac{(mnp)}{c}\right)(X'\cos\phi\sin\theta + Y'\sin\phi\sin\theta + Z'\cos\theta)\text{d}s(\underline{r}')]\ e^{j\omega(mnp)t} \end{aligned} \tag{8.7}$$

$$\begin{aligned} E_\varphi = &\text{Re}\sum_{mnp}\int\Big\{-J_{sx}(mnp,\underline{r}')\sin\phi + J_{sy[mnp,\ \underline{r}']}\cos\phi\Big\} \\ &e^{j\frac{\omega(mnp)}{c}}(X'\cos\varphi\sin\theta + Y'\sin\phi\sin\theta + z'\cos\theta)\text{d}s(\underline{r}')e^{j\omega(mnp)t}\text{d}s(\underline{r}') \end{aligned} \tag{8.8}$$

Radiated power P_{rad} can thus be defined as:

$$P_x[\hat{r}|\ mnp] = \int_s J_{sx}(mnp,\ \underline{r}')e^{j\frac{\omega(mnp)\hat{r}\cdot\underline{r}'}{c}}\text{d}s(\underline{r}') \tag{8.9a}$$

$$P_y[\hat{r}|\ mnp] = \int_s J_{sy}(mnp,\ \underline{r}')e^{j\frac{\omega(mnp)\hat{\underline{r}}.\underline{r}'}{c}}\text{d}s(\underline{r}') \tag{8.9b}$$

$$P_z[\hat{r}|\ mnp] = \int_s J_{sz}(mnp,\ \underline{r}')e^{j\frac{\omega(mnp)\hat{\underline{r}}.\underline{r}'}{c}}\text{d}s(\underline{r}') \tag{8.9c}$$

$$\hat{r}(\theta,\phi) = \hat{X}\cos\phi\ \sin\theta + \hat{Y}\sin\phi\ \sin\theta + \hat{Z}\cos\theta.$$

Let

$s = mnp$ for convenience, then

$$\begin{aligned} E_\theta &= \text{Re}\sum_s\Big\{P_x[\hat{r}|s]\cdot\cos\phi\cos\theta + P_y[\hat{r}|s]\sin\phi\cos\theta - P_z[\hat{r}|s]\sin\theta\}e^{j\omega(s)t}\Big\} \\ &= \text{Re}\sum_s E_{s\theta}e^{j\omega(s)t} \end{aligned} \tag{8.10}$$

where $s = (mnp) = \begin{bmatrix} 000 \\ 001 \\ 010 \end{bmatrix}$ and so on till $s = [111]$. Similarly

$$E_{\phi} = \text{Re}\sum_{s} (-P_x[\hat{r}|s] \sin\varphi + P_y[\hat{r}|s] \cos\phi) e^{j\omega(s)t} \tag{8.11}$$

Hence, Radiation Pattern of RDRA: Power flux per unit solid angle will describe the pattern. Power radiation pattern can be defined as follows:

$$\begin{aligned}
\frac{|E_{\theta}|^2 + |E_{\phi}|^2}{2_{\eta}} &= \frac{1}{2}\left\{\left(\sum_{s} E_{s\theta} e^{j\omega(s)t} + E_{s\phi} e^{j\omega(s)t}\right)\right\} \\
&\quad \times \frac{1}{2}\left\{\left(\sum_{s} H_{s\theta} e^{j\omega(s)t} + \sum_{s} H_{s\phi} e^{j\omega(s)t}\right)\right\} \\
&= \frac{1}{4}\left(\sum_{s} E_s \times H_m^* e^{j(\omega_s - \omega_m)t} + \sum_{s} E_s^* \times H_m e^{j(\omega_m - \omega_s)t}\right) \\
&= \frac{1}{4}\sum_{s}\left[E_{s\theta} \times H_{s\text{ø}}^* + E_{s\text{ø}}^* \times H_{s\theta}\right] \\
&= \frac{1}{2}\text{Re}\sum_{s}\left(E_{s\theta} \times H_{s\text{ø}}^*\right) \\
&= \frac{1}{2}\left(E_{s\theta}\hat{\theta} + E_{\text{ø}}\hat{\text{ø}}\right) \times \left(\frac{E_{s\theta}^*}{\eta}\hat{\varphi} - \frac{E_{s\text{ø}}}{\eta}\hat{\theta}\right) \\
&= \sum_{s} \frac{|E_{s\theta}|^2}{2\eta}\hat{r} + \frac{|E_{s\text{ø}}|^2}{2\eta}\hat{r}
\end{aligned} \tag{8.12}$$

Poynting vector

$$\begin{aligned}
S \cdot \hat{r} = &\frac{1}{2\eta}\sum \Big[\omega(s)^2 |P_x(\hat{r}|s) \cos\phi \cos\theta + P_y(\hat{r}|s) \sin\varphi \cos\theta - P_z(\hat{r}|s) \sin\theta|^2 \\
&+ \omega(s)^2 |P_x(\hat{r}|s) \sin\varphi - P_y(\hat{r}|s) \cos\varphi|^2\Big]
\end{aligned} \tag{8.13}$$

Radiated power per unit solid angle or Poynting vector

$$\begin{aligned}
S \cdot \hat{r}(r, \theta, \phi) = &\frac{1}{2\eta}\sum_{mnp} \omega(mnp)^2 \{|P_x(\theta, \phi|mnp) \cos\phi \cos\theta \\
&+ P_y(\theta, \phi|mnp) \sin\phi \cos\theta - P_z(\theta, \phi|mnp) \sin\theta|^2 \\
&+ |P_x(\theta, \phi|mnp) \sin\phi \pm P_y(\theta, \phi|mnp) \cos\phi|^2\}
\end{aligned} \tag{8.14}$$

8.3 Radiation Pattern Based on Angle $(\varnothing_0, \phi_0)$ Variation in xy Plane

$$\text{Let } \tilde{D}_{(mnp)} = \frac{\pi p}{d}\frac{\hat{D}_{mnp}}{h_{mn}^2}\ e^{j\psi mnp}$$

$$\tilde{C}_{mnp} = \frac{\epsilon}{h_{mn}^2} e^{j\left(\phi(mnp)-\pi/2\right)}\hat{C}_{mnp}$$

$$H_\perp = \left[\sum_{mnp} \text{Re}\ \{\tilde{D}_{mnp}e^{j\omega(mnp)t}\}\nabla_\perp \tilde{u}_{mnp}(x,y,z)\right]\sum_{mnp} \text{Re}\ \left\{\tilde{C}_{mnp}e^{j\omega(mnp)t}\right\}\nabla_\perp \tilde{u}_{mnp}(x,y,z) \tag{8.15}$$

Probe orientation

$$\hat{n}(\theta_0,\ \phi_0) = \hat{x}\ \cos\phi_0\ \sin\theta_0 + \hat{y}\ \sin\phi_0\ \sin\theta_0 + \hat{z}\ \cos\phi_0$$

$$\hat{n}\times H = J_s$$

$$h_{mn}^2 = \gamma^2 + k^2 = n_x\hat{x} + n_y\hat{y} + n_z\hat{z}$$

$$\gamma = \pm\frac{jp\pi}{d};\ \text{for all wave guide and}$$

$$\text{let}\quad \gamma = \frac{\mathrm{d}}{\partial z},\quad j\omega = \frac{\mathrm{d}}{\mathrm{d}t};\ \text{for all cavity resonator}$$

$$\delta l\ \cos\theta_0 = \delta;\ \text{Probe length}$$

Matrix-based computations are as follows:

$$\begin{Bmatrix} E_z & E_\perp \\ H_z & H_\perp \end{Bmatrix}$$

$$\begin{aligned} H_x &= J_{sy} \\ H_y &= -J_{sx} \\ H_\perp &= H_x\hat{x} + H_y\hat{y} \end{aligned}$$

$$E_z = \sum_{mnp} \text{Re}\left[C_{mnp}u_{mnp}(x,y,z)\right]\exp(j\omega_{mnp}t) \tag{8.16}$$

$$H_\perp = \frac{-\gamma}{h_{mnp}^2}\quad \nabla_\perp H_z + \frac{j\omega\epsilon}{h_{mnp}^2}\nabla_\perp E_z\times\hat{z} \tag{8.17}$$

$$H_z = u_{mn}(x,y)\exp\left(\frac{j\pi pz}{d}\right) \tag{8.18}$$

$$H_z = u_{mn}(x,y)C(mnp)e^{j\left(\omega(mnp)t-\frac{jp\pi z}{d}\right)}$$

$$H_z = \mathrm{Re}\left\{C_{mnp}u_{mn}(x,y)e^{\pm j\left(\omega(mnp)t-\frac{jp\pi z}{d}\right)}\right\}$$

$$\begin{aligned} H_z =& \mathrm{Re}\left\{C_1(mnp)u_{mn}(x,y)e^{j(\omega(mnp)t-\frac{jp\pi z}{d})}\right\} \\ &+ \mathrm{Re}\left\{C_2(mnp)u_{mn}(x,y)e^{-j(\omega(mnp)t+\frac{jp\pi z}{d})}\right\} \end{aligned} \tag{8.19}$$

$$H_\perp = \left[\frac{1}{h_{mn}^2}\frac{\mathrm{d}}{\mathrm{d}z}\nabla_\perp H_z(x,y,z,t)\right] + \left[\frac{\epsilon}{h_{mn}^2}\frac{\partial}{\partial t}\nabla_\perp E_z(x,y,z,t)\right] \tag{8.20}$$

Hence

$$E_z(x,yz,t) = \sum_{mnp} C_{mnp}u_{mnp}(x,y,z)\cos(\omega(mnp)t+\varphi(mnp)) \tag{8.21a}$$

$$H_z(x,yz,t) = \sum_{mnp} D_{mnp}u_{mnp}(x,y,z)\cos(\omega(mnp)t+\psi(mnp)) \tag{8.21b}$$

8.4 Replacing Probe with Slot of Finite Dimensions (Ls, Ws) at an Angle (θ_0, ϕ_0)

We replace excitation probe to slot u_{mnp} and v_{mnp} by $\tilde{u}_{mnp}$ and $\tilde{v}_{mnp}$

$$\tilde{u}_{mnp}(x,y,z) = \frac{2\sqrt{2}}{\sqrt{abd}}\sin\left(\frac{m\pi x}{a}\right)\sin\left(\frac{n\pi y}{b}\right)\sin\left(\frac{\rho\pi z}{d}\right) \tag{8.22a}$$

$$\tilde{v}_{mnp}(x,y,z) = \frac{2\sqrt{2}}{\sqrt{abd}}\cos\left(\frac{m\pi x}{a}\right)\cos\left(\frac{n\pi y}{b}\right)\cos\left(\frac{\rho\pi z}{d}\right) \tag{8.22b}$$

$$J_s(x,y,\delta) = \frac{I\left[_{mnp}\right]e^{\eta'\omega(mnp)t}}{\omega};\ \text{where } \eta' \text{ is an angle}$$

$$\begin{aligned} H_\perp\big|_{z=\delta} =& \sum_{mnp}\mathrm{Re}\left\{\tilde{D}_{mnp}e^{j\omega(mnp)t}\right\}\nabla_\perp\tilde{v}_{mnp}(x,y,z) \\ &- \sum_{mnp}\mathrm{Re}\left\{\tilde{C}_{mnp}e^{j\omega(mnp)t}\right\}\nabla_\perp\tilde{u}_{mnp}(x,y,z) \end{aligned} \tag{8.23}$$

$$\hat{n} \times H$$

E_x, E_y, H_x, H_y are the fields in terms of surface current density due to applied probe current at an angle η′; J_{sx}, J_{sy} can be expressed as current density using Fourier coefficients; C_{mnp} and D_{mnp} H_yH_x fields can be computed from E_xE_y fields; *propagation terms* $h_{mn}^2 = \gamma^2 + k^2$

$$H_\perp = \frac{I_{mnp}}{\omega_{(\text{width})}} e^{j\eta(mnp)} \{\hat{x} - (\sin(\phi_0)\ \cos(\phi_0)\hat{y})\} g(x, y)$$

$$H_\perp = \sum_{mnp} \text{Re}\{\tilde{D}_{mnp} e^{j\omega(mnp)t} \frac{2\sqrt{2}}{\sqrt{abd}} \nabla_\perp \left\{\cos\left(\frac{m\pi x}{a}\right) \cos\left(\frac{n\pi y}{b}\right)\right\}. \tag{8.24}$$

$$\begin{aligned} H_\perp|_{z=0} &= H_X(x, y, 0)\hat{x} + H_Y(x, y, 0)\hat{y} \\ &= \left\{J_{sy}(x, y, t) - J_{sx}(x, y, t)\right\} \end{aligned}$$

$$J_{sy}(x, y, t) = \sum \text{Re}\left\{\tilde{D}_{mnp} e^{j\omega(mnp)t}\right\} \left(\frac{m\pi}{a}\right) \frac{2\sqrt{2}}{\sqrt{abd}} \sin\left(\frac{m\pi x}{a}\right)\ \cos\left(\frac{n\pi y}{b}\right)$$

$$\begin{aligned} &\int_0^a \int_0^b J_{sy}(x, y, t) \frac{2}{\sqrt{ab}} \sin\left(\frac{m\pi x}{a}\right) \cos\left(\frac{n\pi y}{b}\right) \text{d}x\text{d}y \\ &= \text{Re}\left\{\tilde{D}_{mnp} e^{j\omega(mnp)t}\right\} \left(\frac{m\pi}{a}\right) \sqrt{\frac{2}{d}} \sin\left(\frac{m\pi x}{a}\right) \end{aligned}$$

Hence,

$$\begin{aligned} J_{sy}(x, y, t) &= \text{Re}\left\{f_y(x, y) e^{j\omega(mnp)t}\right\} \\ &= \int_0^a \int_0^b f_y(x, y) \frac{2}{\sqrt{ab}} \sqrt{\frac{2}{d}}, \left(\frac{m\pi}{a}\right) \tilde{D}_{mnp} \sin\left(\frac{m\pi x}{a}\right) \cos\left(\frac{n\pi y}{b}\right) \text{d}x\text{d}y \end{aligned} \tag{8.25}$$

Now

$$\tilde{D}_{mnp} = \sqrt{\frac{2}{d}} \frac{a}{m\pi} \int_0^a \int_0^b f_y(x, y) \frac{2}{\sqrt{ab}} \sin\left(\frac{m\pi x}{a}\right) \cos\left(\frac{n\pi y}{b}\right) \text{d}x\text{d}y \tag{8.26}$$

$$H_y = -J_{sx}$$

$$-J_{sx}(x,y,t) = H_y|_{z=0}$$
$$= \sum_{mnp} \mathrm{Re}\left\{\tilde{D}_{mnp} e^{j\omega(mnp)t}\right\}\left(-\frac{n\pi}{b}\right)\frac{2\sqrt{2}}{\sqrt{abd}}\cos\left(\frac{m\pi x}{a}\right)\cos\left(\frac{n\pi y}{b}\right) \tag{8.27}$$

$$J_{sx}(x,y,t) = \mathrm{Re}\left\{f_x(x,y)e^{+j\omega(mnp)t}\right\}$$

$$\tilde{D}_{mnp} = \frac{2}{ab}\sqrt{\frac{2}{d}}\frac{b}{n\pi}\int_0^a\int_0^b f_x(x,y)\cos\left(\frac{m\pi x}{a}\right)\cos\left(\frac{n\pi y}{b}\right)\mathrm{d}x\mathrm{d}y$$

$$J_{sx}(x,y,t) = \sum_{mnp} \mathrm{Re}\{f_x\{x,y|mnp\}e^{j\omega(mnp)t} \tag{8.28}$$

Similarly, compute $\tilde{D}_{mnp}$:

$$J_{sy}(x,y,t) = \sum_{mnp} \mathrm{Re}\{f_y\{x,y|mnp\}e^{j\omega(mnp)t} \tag{8.29}$$

8.5 HFSS Computed Radiation Pattern with Shifted (θ_i, ϕ_i) Slot Positions

Angular excitation at 45° and 30° left side.

Results of angular excitation on radiation pattern have been evaluated on HFSS and shown in Fig. 8.5.

Angular shifts in excitation at 45° and 30° right side and radiation pattern are shown in Fig. 8.7 and summarized are placed in Table 8.1.

Results of Radiation pattern when angular excitation is given to the right side have been shown in Fig. 8.7.

Table 8.1 Summarized results

Frequency in GHz	S_{11} in dB	S_{11} in dB slot position at 45° left	S_{11} in dB slot position 45° right	S_{11} in dB cross-slot	Gain in dB	Gain in dB at left 45°	Gain in dB right 45°	Gain (cross-slot) in dB
8.45	−11.2	−11.6	−10.8	−11.5	1.5	3.75	2.7	1.5
12.81	−13.3	−12.4	−13.4	−12.2	1.1	1.0	1.1	1.2
16.64	−14.7	−18.1	−18.0	−15.9	1.5	1.8	1.6	3.0

Table 8.1 described results of antenna parameters in tabular form. It has been observed that angular variation in excitation has direct impact on the radiation pattern as well as number of modes generated. This has been verified by the plots given above. We have taken measurements of radiation pattern on varying slot of feed at 30° and 45° to left and right from its original position. These plots have been verified at two different frequencies. This completes the solution.

8.6 Experimentations

Figures 8.8, 8.9, 8.10, 8.11, 8.12, 8.13, 8.14, 8.15 and 8.16 present the experimental results of RDRA. Their significance is placed below each figure. The RDRA made from acrylic glass sheets having dimensions of 9, 6 and 3 cm. The silicon oil having $\varepsilon = 2.2$ was used as RDRA dielectric material. The resonant frequency of RDRA was measured to 4.55 GHz. The measurements were taken at various angular positions of the slot. Aperture-coupled feed RDRA is shown in Fig. 8.1. The feed position was shifted to investigate RDRA S_{11} using VNA 40 GHz. The results are shown in Figs. 8.2, 8.3, 8.4 and 8.5.

Fig. 8.8 RDRA under measurements with VNA

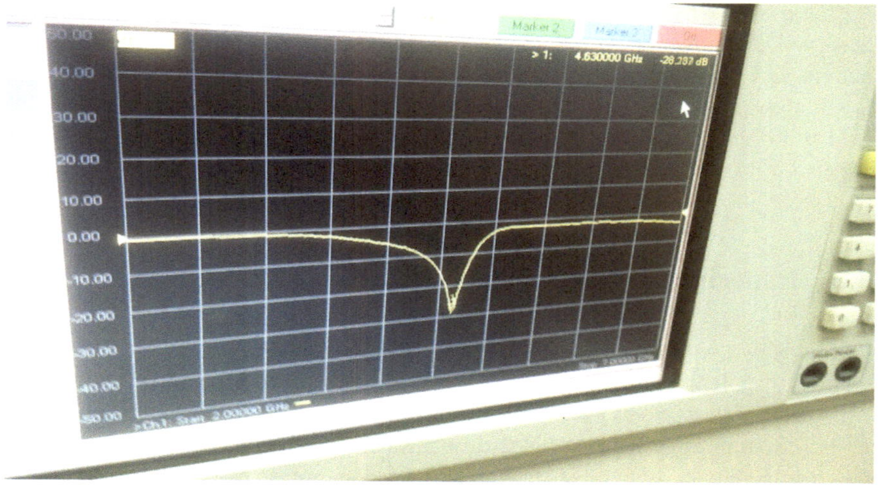

Fig. 8.9 −28.23 dB measured S_{11} of RDRA at 4.63 GHz

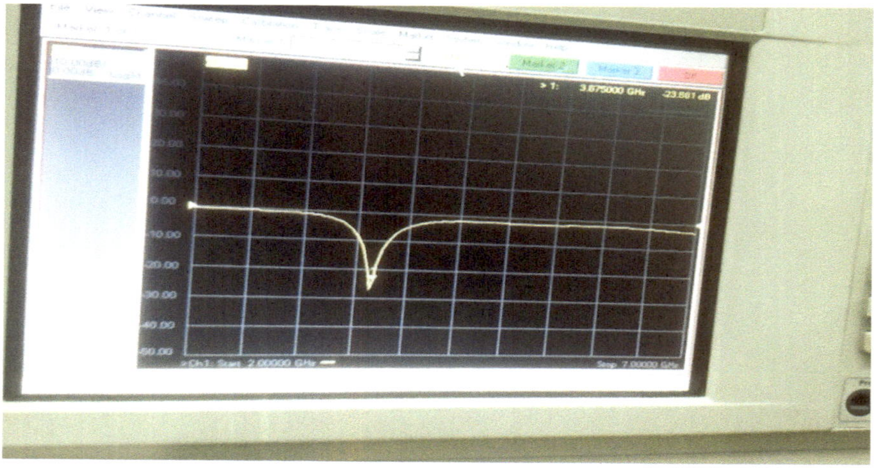

Fig. 8.10 S_{11} RDRA with shifted slot resonant frequency 3.67

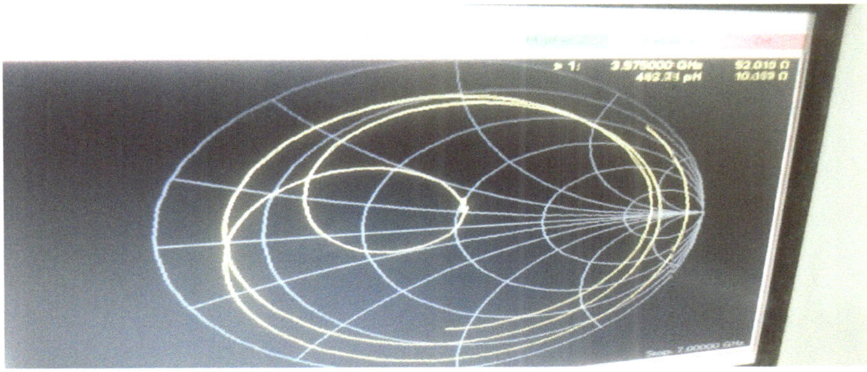

Fig. 8.11 Smith chart showing proper Z_{11} of RDRA

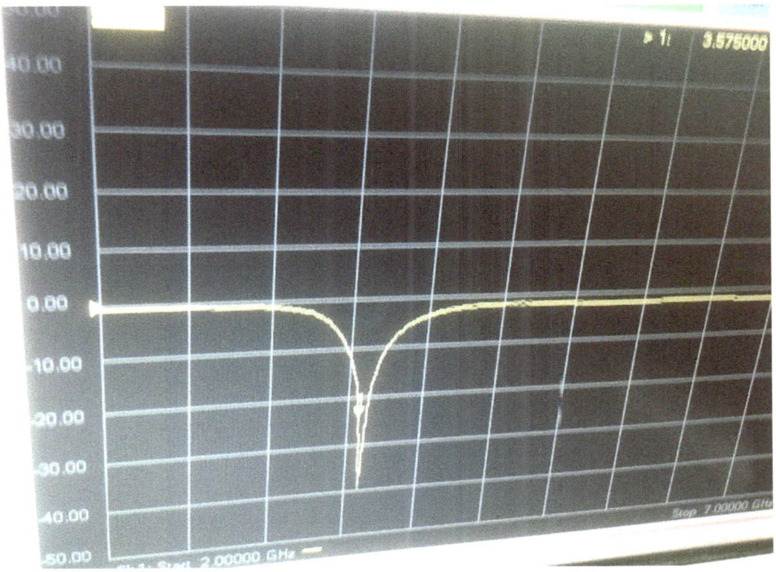

Fig. 8.12 S_{11} at shifted slot frequency 3.57 GHz

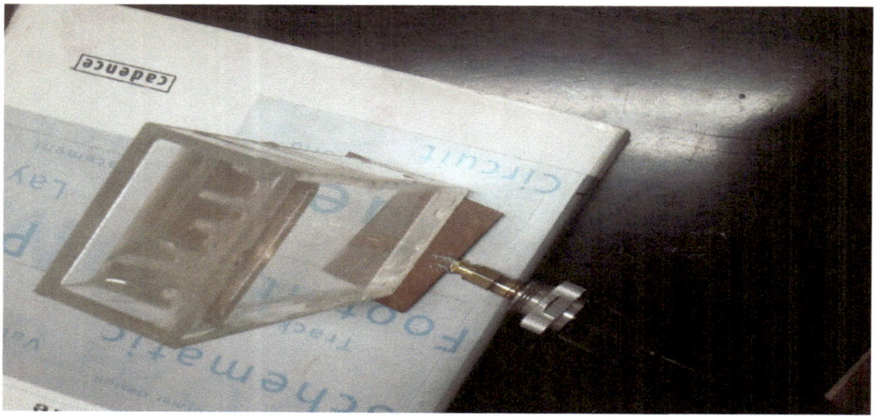

Fig. 8.13 Measurements of RDRA with aspect ratio changed

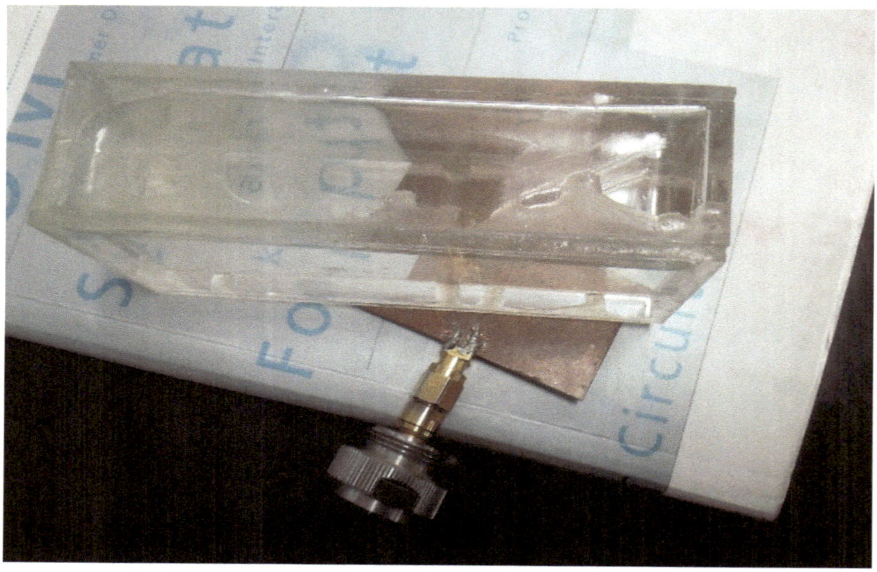

Fig. 8.14 RDRA aspect ratio changed

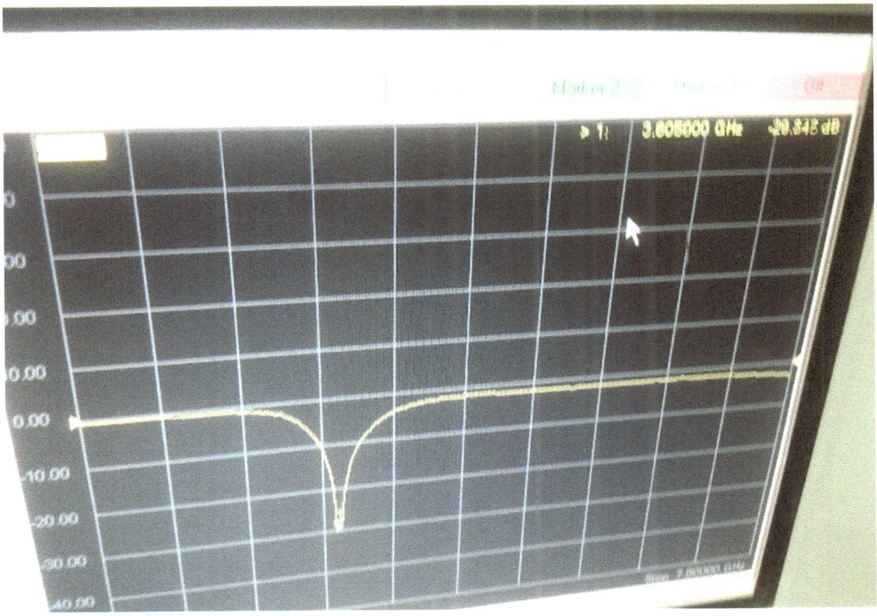

Fig. 8.15 Shifted frequency observed (3.60 GHz)

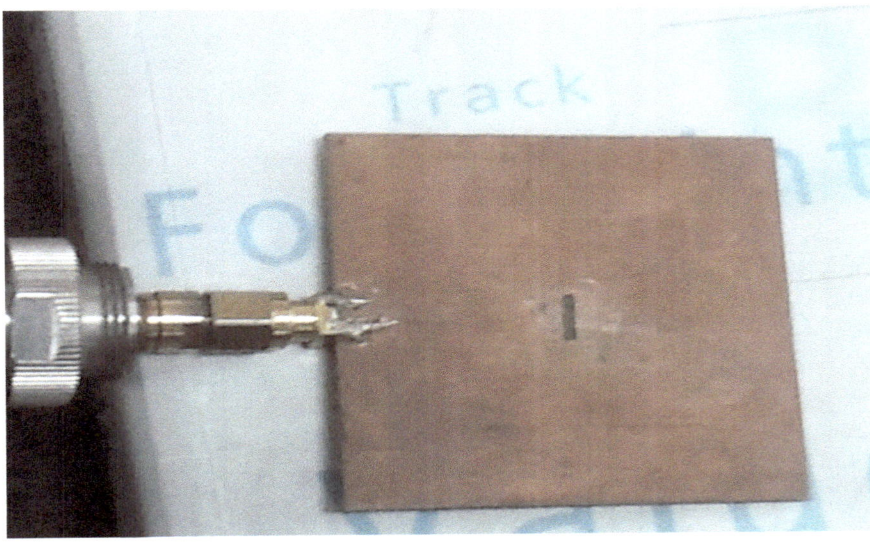

Fig. 8.16 Aperture-coupled feed showing slot

The results obtained with VNA have clearly shown shift in resonant frequency due to feed orientation. It indicated that resonant modes are changing based on the slot orientation. Hence, it is clearly evident that radiation pattern can be steered with slot position in RDRA. If these results can be placed in look up table, then microcontroller-based orientation can result into automated antenna. This automated antenna can be very useful for military applications. These cross slot can be arranged in such a manner that circular polarization becomes possible in RDRA. Also by varying lengths of cross slots, left hand or right hand polarization can be achieved. The circular ploarization makes the signal robust and help to reduce electromagnetic polution.

Chapter 9
Sensitivity Analysis of Rectangular DRA

Abstract Sensitivity analysis of rectangular DRA depending on dielectric material, and *a*, *b*, and *d* dimensions. These dimensions decides resonant frequency of RDRA. The resonant modes are formed when realized with excitation. The resonant frequency solution is worked with MATLAB and HFSS software. When these dimensions are changed, resonant frequency of RDRA also changes. Variance method has been tried out to evaluate error.

Keywords Isolated RDRA · Ground plane RDRA · Resonant frequency · Sensitivity analysis · Variance · Error minimization

Rectangular DRAs of dielectric material having *a*, *b*, and *d* dimensional length have been analyzed for frequency and resonant modes. RDRA is shown in Fig. 9.1. These have been solved based on MATLAB and HFSS. Figure 9.2 presented rectangular DRA with *a*, *b*, and *d* dimensions. Table 9.1 has shown RDRA dimensions and their corresponding resonant frequencies. Figure 9.3 indicated resonant modes with RDRA height. Plot of frequency versus length "*a*" variation is shown in Figs. 9.4, 9.5 and 9.6. HFSS simulated modes in RDRA with S_{11} parameters are shown in Figs. 9.8 and 9.9 (Fig. 9.7).

$\delta a, \delta b, \delta d$ are (small change in length) random variables, and computed functions are $f(\delta_{mnp})$ and ω_{mnp}. The variance functions are $\sigma_a, \sigma_b, \sigma_d$. These are mainly dependent on *a*, *b*, and *d*. Taylor's expansion is restricted to second-order variable. Hence, $\delta a, \delta b, \delta d$ are mapped in terms of $\sigma_a, \sigma_b, \sigma_d$ using diagonal matrix. C_{mnp}, D_{mnp}, are amplitude coefficients which depend on the RDRA *a*, *b*, or *d*.

Frequency relationship can be determined based on *a*, *b* and *d* length variation as given below:

$$\frac{\delta}{\delta d}, \frac{\delta}{\delta b}, \frac{\delta \omega(mnp|a, b, d)}{\delta a}$$

R.S. Yaduvanshi and H. Parthasarathy, *Rectangular Dielectric Resonator Antennas*, DOI 10.1007/978-81-322-2500-3_9

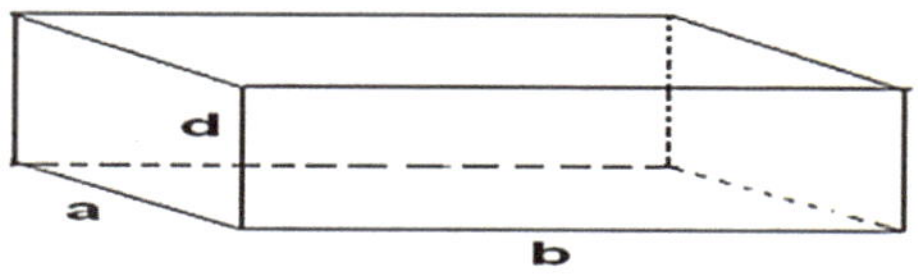

Fig. 9.1 Isolated RDRA

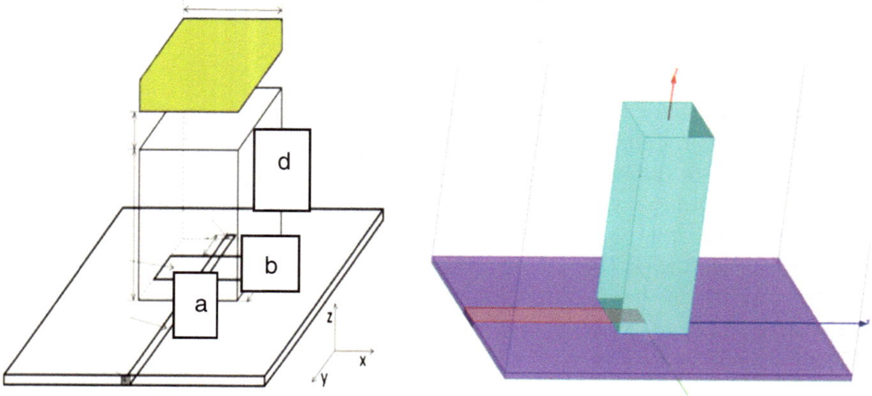

Fig. 9.2 Rectangular DRA with *a*, *b*, and *d* dimensions

Table 9.1 RDRA dimensions

	x (mm)	y (mm)	z (mm)	ε_r	Material used
RDRA	$a = 7$ $a = 6$ $a = 5$	$b = 7$ $b = 6$ $b = 5$	$d = 10$ $d = 15$ $d = 30$	10	Sapphire
Substrate	20	30	0.5	3.38	$Arlon_{25}N$(tm)
Ground plane	20	30	–	–	–
Microstrip feed line	15	1.11	–	–	–

DRA dimensions (mm)			Resonant frequencies (GHz) simulated
a	b	d	
7	7	10	$f_1 = 13.46$
6	6	15	$f_2 = 13.85$
5	5	30	$f_3 = 14.21$

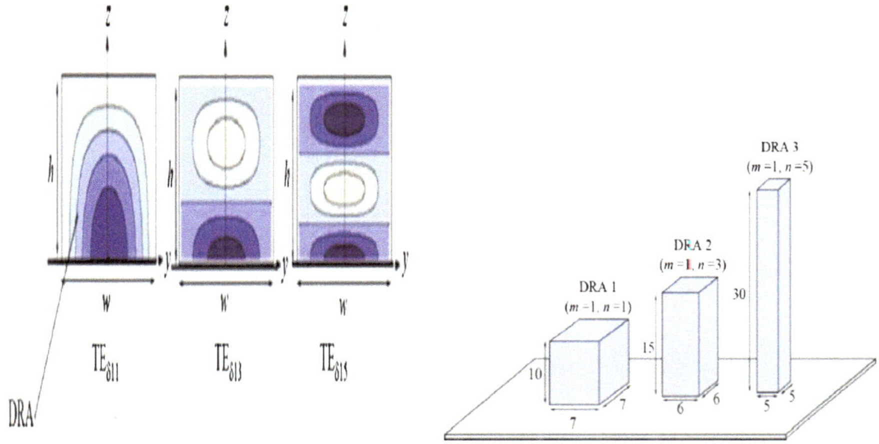

Fig. 9.3 Resonant mode and RDRA height relationship

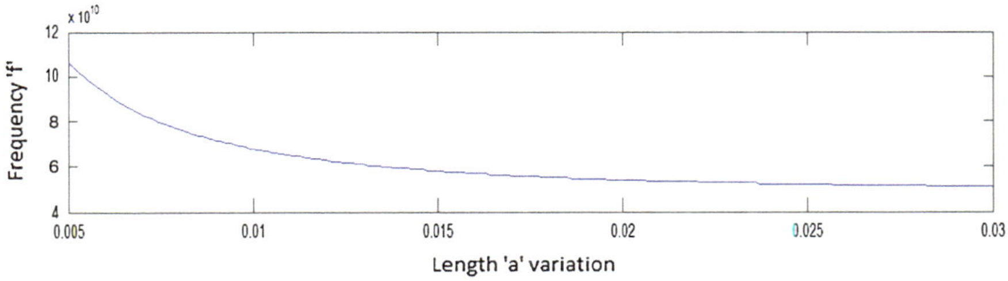

Fig. 9.4 Plot of frequency versus length "a" variation

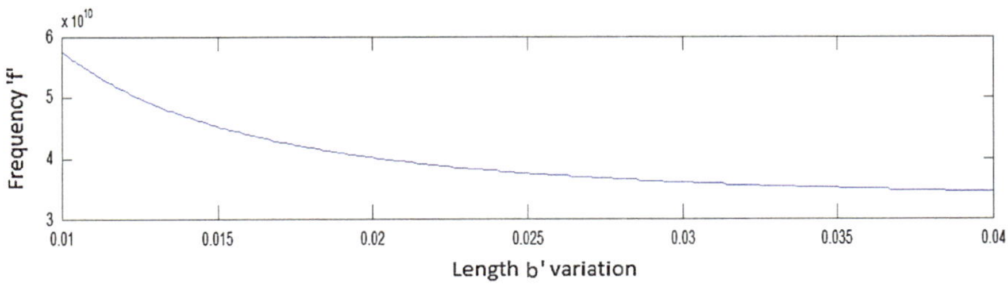

Fig. 9.5 Plot of frequency versus length "b" variation

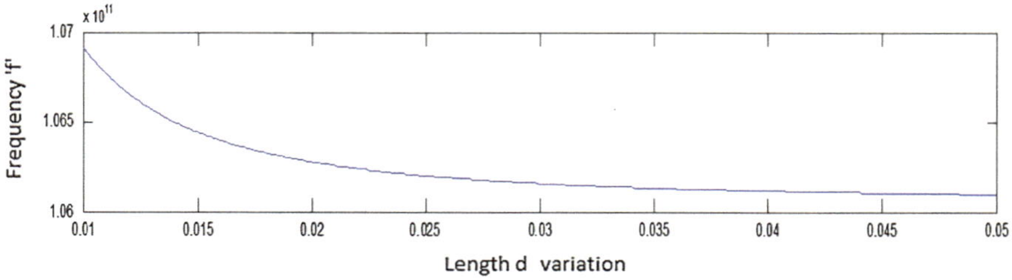

Fig. 9.6 Plot of frequency versus length "*d*" variation

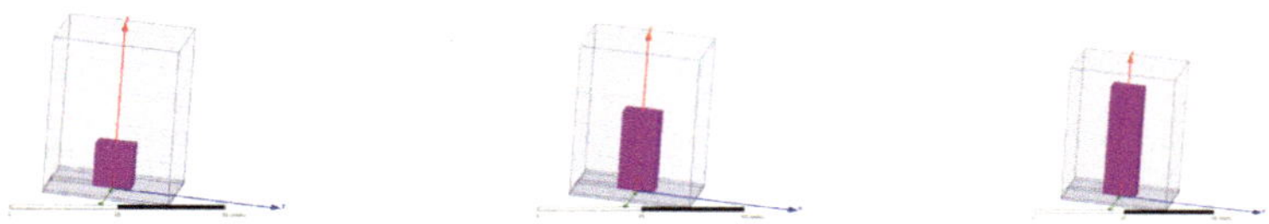

Fig. 9.7 RDRA having three different heights for increasing resonant modes

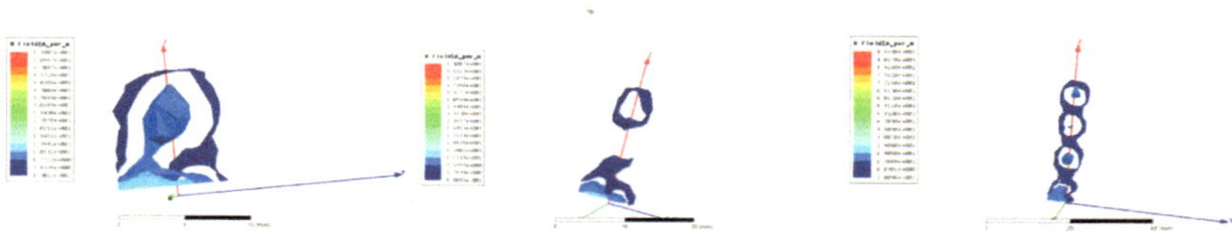

Fig. 9.8 Modes' pattern

H_z and E_z fields are expressed based on principle of orthogonality as:

$$E_z(x,y,z,t) = \sum_{mnp} \mathrm{Re}\left(e^{j\omega(mnp)t} C(mnp)\right) u_{mnp}(x,y,z) \tag{9.1a}$$

$$H_z(x,y,z,t) = \sum_{mnp} \mathrm{Re}\left(e^{j\omega(mnp)t} C(mnp)\right) v_{mnp}(x,y,z) \tag{9.1b}$$

At $z = 0$, E_z field

$$E_z(t,x,y,0) = \sum_{mnp} \mathrm{Re}\left(e^{j\omega(mnp)t} C(mnp)\right) u_{mn}(xy)\sqrt{\frac{2}{d}} \tag{9.2}$$

and

$$H_x, H_y = -J_{sy}, J_{sx}$$

C_{mnp}, D_{mnp} are amplitude coefficients.

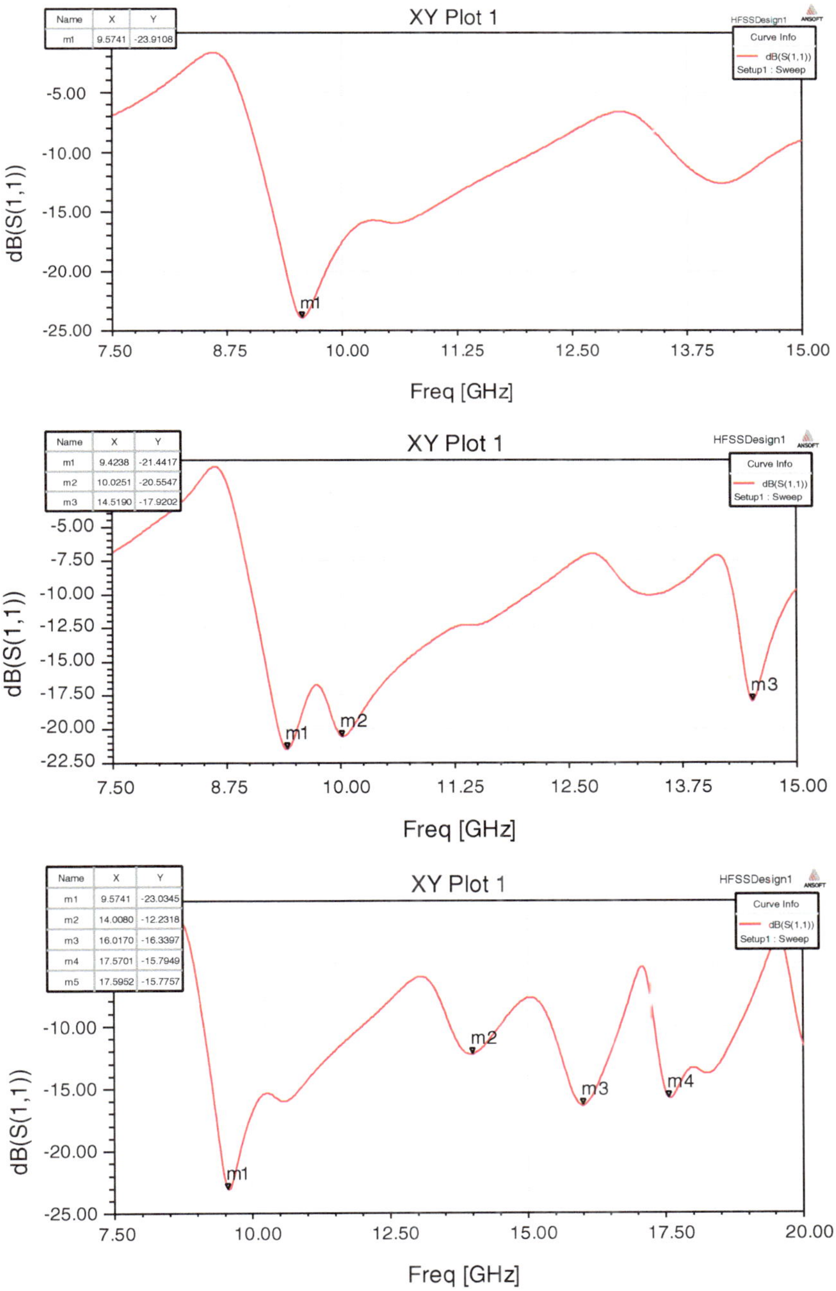

Fig. 9.9 Resonant frequency based on RDRA height

We have computed

$$C_{mnp} = \frac{2\sqrt{2}}{\sqrt{abd}} \int J_{sx}(X, Y) \sin\left(\frac{m\pi x}{a}\right) \sin\left(\frac{n\pi y}{b}\right) \mathrm{d}x\mathrm{d}y \tag{9.3}$$

Now if

$$\begin{aligned} a &\to a + \delta a \quad (a \text{ is increased to } a + \delta a) \\ b &\to b + \delta b \quad (b \text{ is increased to } b + \delta b) \\ d &\to d + \delta d \quad (d \text{ is increased to } d + \delta d) \end{aligned}$$

We need to compute

$$C_{(mnp|a + \delta a, b + \delta b, d + \delta d)}$$

and similarly

$$\omega_{(mnp|a + \delta a, b + \delta b, d + \delta d)}$$

This can be approximated by mean variance method and Taylor's expansion

N normal distribution with mean zero
σ variance function

To compute error

$$\begin{bmatrix} \delta a \\ \delta b \\ \delta d \end{bmatrix} \sim N\left(0, \begin{bmatrix} \sigma_a^2 & 0 & 0 \\ 0 & \sigma_b^2 & 0 \\ 0 & 0 & \sigma_d^2 \end{bmatrix}\right)$$

(By Taylor's expansion)

$$\begin{aligned} C_{(mnp|a + \delta a, b + \delta b, d + \delta d)} = C_{(mnp|a,b,d)} &+ \frac{\delta C(mnp)}{\delta a}\delta a + \frac{\delta C(mnp)}{\delta b}\delta b + \frac{\delta C(mnp)}{\delta d}\delta d \\ &+ \frac{1}{2}\left(\frac{\delta^2 C(mnp)}{\delta a^2}\delta a^2 + \frac{\delta^2 C(mnp)}{\delta b^2}\delta b^2 + \frac{\delta^2 C(mnp)}{\delta d^2}\delta d^2 \right. \\ &\left. + 2\frac{\delta^2 C(mnp)}{\delta a \delta b}\delta a \delta b + 2\frac{\delta^2 C(mnp)}{\delta a \delta d}\delta a \delta d + 2\frac{\delta^2 C(mnp)}{\delta b \delta d}\delta b \delta b\right) \end{aligned} \tag{9.4}$$

Hence, variance or error can be written as

$$\begin{aligned} \langle C_{(mnp|a + \delta a, b + \delta b, d + \delta d)} - C^2_{(mnp|a, b, d)} \rangle &= \left|\frac{\delta C(mnp)}{\delta a}\right|^2 \left\langle |\delta a|^2 \right\rangle \\ &+ \left|\frac{\delta C(mnp)}{\delta b}\right|^2 \left\langle |\delta b|^2 \right\rangle + \left|\frac{\delta C(mnp)}{\delta d}\right|^2 \left\langle |\delta d|^2 \right\rangle \\ &= \sigma_a^2 \left|\frac{\delta C(mnp)}{\delta a}\right|^2 + \sigma_b^2 \left|\frac{\delta C(mnp)}{\delta b}\right|^2 + \sigma_d^2 \left|\frac{\delta C(mnp)}{\delta d}\right|^2 \end{aligned} \tag{9.5}$$

Similarly, we can compute:

$$\left\langle \omega_{(mnp|a+\delta a,b+\delta b,d+\delta d)} - \omega_{(mnp|a,b,d)} \right\rangle$$

Error value

$$\omega_{(mnp)} + \delta\omega_{(mnp)} - \left\langle |\delta\omega|^2 \right\rangle$$

Error or variance:

$$= \left|\frac{\delta\omega(mnp)}{\delta a}\right|^2 \sigma_a^2 + \left|\frac{\delta\omega(mnp)}{\delta b}\right|^2 \sigma_b^2 + \left|\frac{\delta\omega(mnp)}{\delta d}\right|^2 \sigma_d^2$$

$$\omega = \pi\sqrt{\frac{m^2}{a^2} + \frac{n^2}{b^2} + \frac{p^2}{d^2}}$$

The change in frequency based on RDRA change in dimension in x, y, and z direction is given below:

$$\frac{\partial\omega}{\partial a} = \frac{\frac{-m^2\pi}{a^3}}{\sqrt{\frac{m^2}{a^2} + \frac{n^2}{b^2} + \frac{p^2}{d^2}}} \tag{9.6a}$$

$$\frac{\partial\omega}{\partial b} = \frac{\frac{-n^2\pi}{b^3}}{\sqrt{\frac{m^2}{a^2} + \frac{n^2}{b^2} + \frac{p^2}{d^2}}} \tag{9.6b}$$

$$\frac{\partial\omega}{\partial d} = \frac{\frac{-p^2\pi}{d^3}}{\sqrt{\frac{m^2}{a^2} + \frac{n^2}{b^2} + \frac{p^2}{d^2}}} \tag{9.6c}$$

This gives the complete solution of RDRA sensitivity analysis. The higher-order modes of a rectangular DRA were used to produce radiation patterns with enhanced gain. The advantage of this approach is for enhancing gain. The maximum achievable gain on mode $m = 1$, $n = 7$ to increase Directivity to 13.7 dBi. Such DRA designed at 11 GHz with height 35 mm, this investigation focused on rectangular DRAs, for excitation of the appropriate higher-order modes in RDRAs.

9.1 MATLAB Simulation

Matlab Program for Sensitivity Analysis

```
clear all;

clc;

close all;

c=3*10^8;

m=1;

n=1;

p=1;

E=10;


a=5*10^-3:.1*10^-3:30*10^-3;

b=10*10^-3;

d=15*10^-3;

for i=1:length(a)

f(i)=c/(2*pi)*sqrt(E)*sqrt((m*pi/a(i))^2+(n*pi/b)^2+(pp*pi/(2*d)^2));

end

a1=15*10^-3;

b1=10*10^-3:.1*10^-3:40*10^-3;

d1=20*10^-3;
```

```
for k=1:length(b1)

f1(k)=c/(2*pi)*sqrt(E)*sqrt((m*pi/a1)^2+(n*pi/b1(k))^2+(pp*pi/(2*d1)^2));

end



a2=10*10^-3;

b2=5*10^-3;

d2=10*10^-3:.1*10^-3:50*10^-3;

for t=1:length(d2)

f2(t)=c/(2*pi)*sqrt(E)*sqrt((m*pi/a2)^2+(n*pi/b2)^2+(pp*pi/(2*d2(t))^2));

end



subplot(3,1,1);plot(a,f);title('plot a vs f when a is varying');

subplot(3,1,2);plot(b1,f1);title('plot b vs f when b is varying');

subplot(3,1,3);plot(d2,f2);title('plot d vs f when d is varying');
```

9.2 HFSS Simulations

Now using HFSS software, we shall verify.

This gives the complete solution for RDRA sensitivity analysis.

It has been observed that resonant modes have been increasing based on increase in dipole moment, i.e., modes are proportional to the height of RDRA.

9.2.1 HFSS Result

See below Figs. 9.10, 9.11, 9.12, 9.13 and Table 9.2.

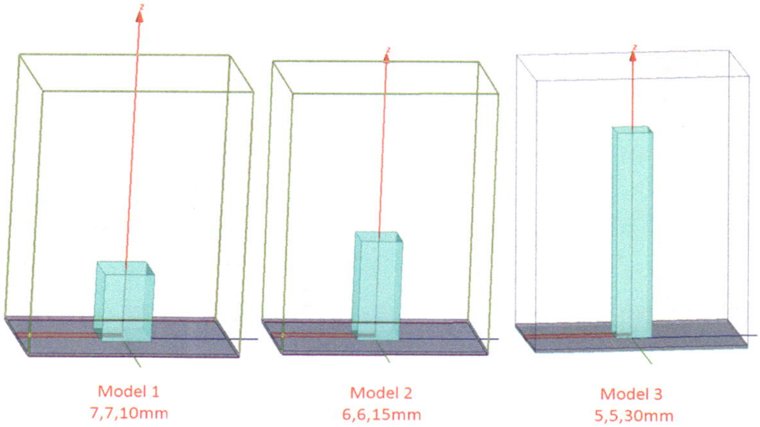

Fig. 9.10 HFSS models of RDRA

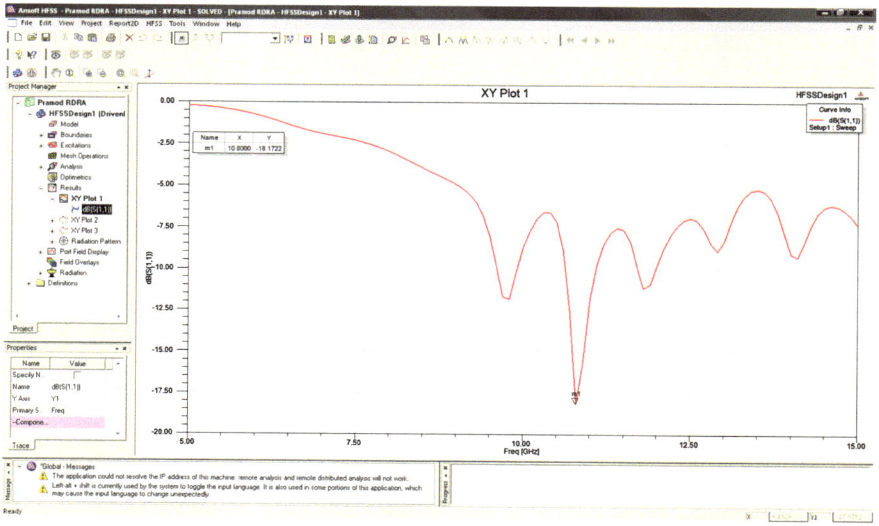

Fig. 9.11 Return loss versus frequency with dimensions $a = 5$ mm, $b = 5$ mm, $d = 30$ mm shows Return loss 18 dBi at $f = 10.95$ GHz

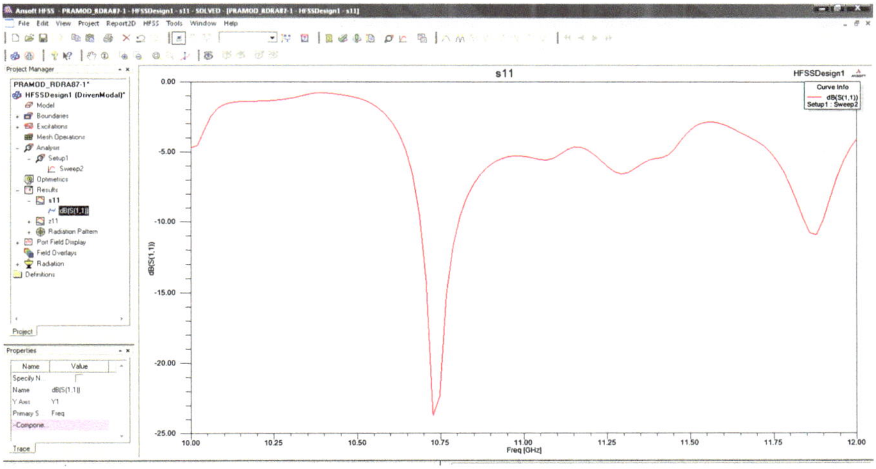

Fig. 9.12 Return loss versus frequency with dimensions a = 6 mm, b = 6 mm, d = 15 mm shows return loss 24 dBi at f = 10.95 GHz

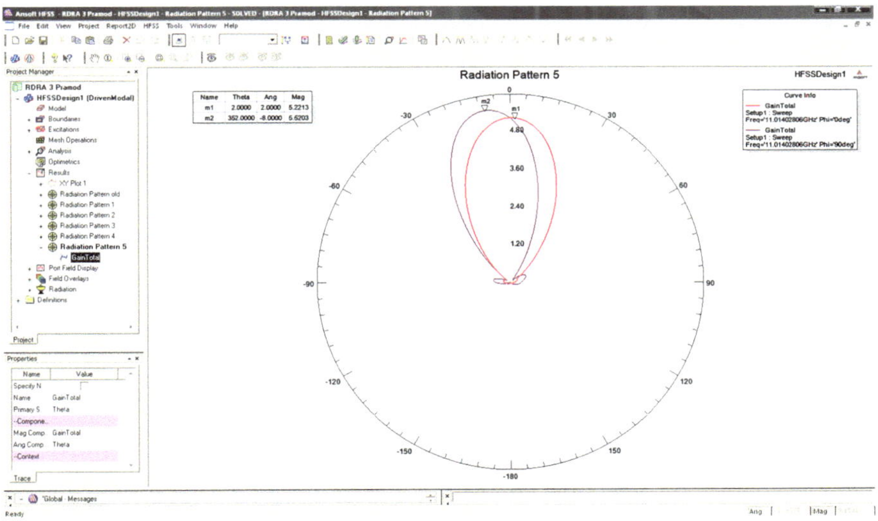

Fig. 9.13 Gain = 5.5 at f = 11 GHz for DRA 1

Table 9.2 Dimensions table

	x (mm)	y (mm)	z (mm)	ε_r	Material used
RDRA	a = 7 a = 6 a = 5	b = 7 b = 6 b = 5	d = 10 d = 15 d = 30	10	TMM10i
Substrate	20	30	0.5	3.38	$Arlon_{25}N(tm)$
Ground plane	20	30	–	–	–
Microstrip feed line	19.2	1.1672	–	–	–
Lumped element	1.1672	0.5	–	–	–

9.3 Radiation Pattern

See below Figs. 9.14 and 9.15.

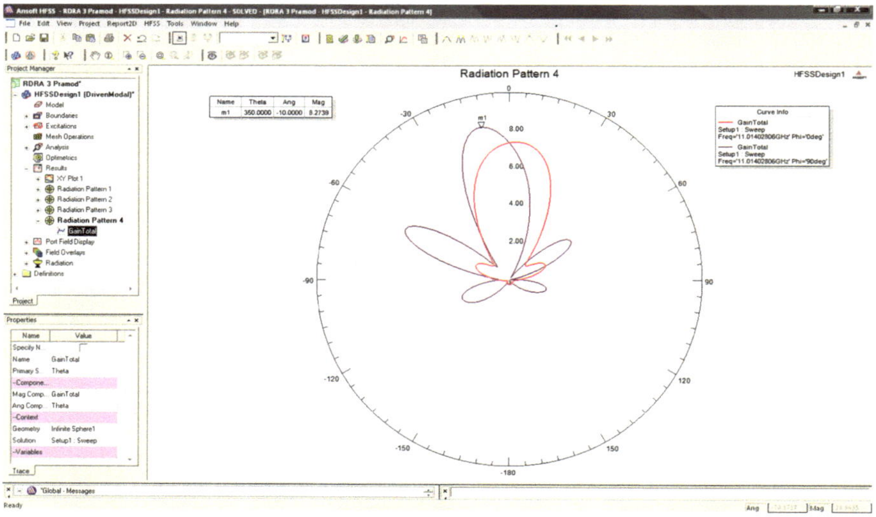

Fig. 9.14 Gain = 8.2 at f = 11 GHz for DRA 2

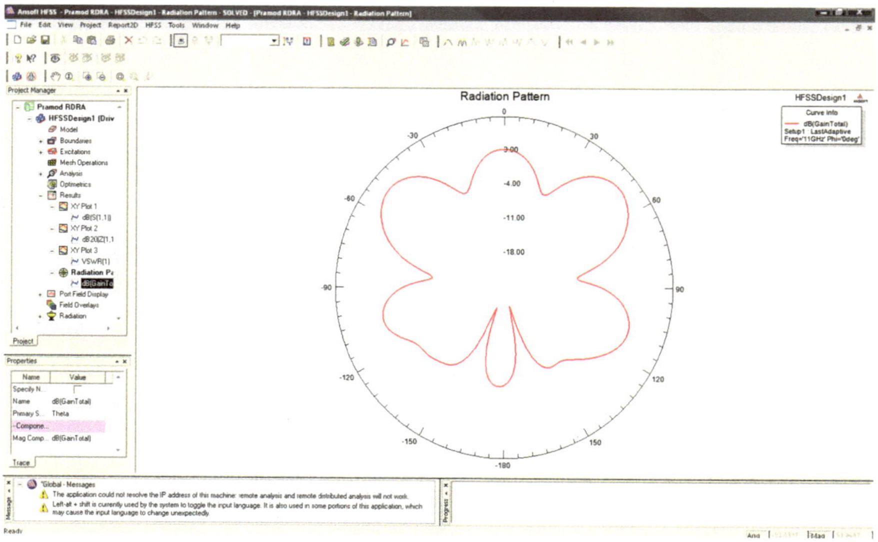

Fig. 9.15 Gain = 9.5 at f = 11 GHz for DRA 3

Chapter 10
Hybrid Modes in RDRA

Abstract In this chapter, new kind of resonant mode, i.e., hybrid mode in RDRA (rectangular dielectric resonator antenna), is described using mathematical modeling. RDRA is excited by inserting RF feed probe or microstripline having finite dimensions, carrying electric and magnetic currents at a given frequency. The charge conservation equations then imply the presence of electric charge densities and magnetic charge densities within the resonator. From Maxwell's equations, we derive vector Helmholtz equations for the electromagnetic fields. The vector sources provide electric charge which gets converted into magnetic charge. In one of the models, sidewalls of the resonator are perfect magnetic conductors, and top and bottom surfaces are perfect electric conductors. Thus, the boundary conditions on the fields are such that the tangential components of the magnetic field vanish on sidewalls and the normal components of the magnetic field vanish at top and bottom surfaces. The normal components of the electric field vanish on sidewalls. H_z can therefore be expanded as linear combinations of sin functions in xy direction along with z-component of the source. For the H_z, Helmholtz equation can be expanded in terms of sin functions (assuming that these sources vanish on boundary), with z-dependent coefficients.

Keywords Hybrid modes · Mathematical model · Normal component · Tangential component · Conservation equation · Magnetic energy · Electrical energy · Field diversity · Fourier basis function

10.1 Introduction

RDRA (rectangular dielectric resonator antenna) is excited by inserting RF feed probe or microstripline having finite dimensions, carrying electric and magnetic currents at a given frequency. The charge conservation equations then imply the presence of electric charge densities and magnetic charge densities within the resonator at that particular frequency. To completely solve for the fields with the

R.S. Yaduvanshi and H. Parthasarathy, *Rectangular Dielectric Resonator Antennas*, DOI 10.1007/978-81-322-2500-3_10

resonator, we therefore set up four Maxwell's equations taking into account magnetic and electric currents and charge densities. From these equations, we derive vector Helmholtz equations for the electromagnetic fields with vector sources determined from gradient and curl of the electric charge, magnetic charge, and current densities. The sidewalls of the resonator are perfect magnetic conductors, and top and bottom surfaces are perfect electric conductors. Thus, the boundary conditions on the fields are such that the tangential components of the magnetic field vanish on sidewalls and the normal components of the magnetic field vanish at top and bottom surfaces. The normal components of the electric field vanish on sidewalls. H_z can therefore be expanded as linear combinations of sin functions in xy direction along with z-component of the source. For the H_z, Helmholtz equation can be expanded in terms of sin functions (assuming that these sources vanish on boundary), with z-dependent coefficients.

On substitution of these expressions into the Helmholtz equation for $H_{z,}$ source then gives us a second-order linear differential equation for the coefficient functions of z in H_z with a source term. This is solved, and the solution consists of a superposition of a source (particular solution or inhomogeneous solution) term and a homogeneous term (i.e., general solution of the homogeneous part). In hybrid modes, total solution is developed, i.e., homogeneous and inhomogeneous. Two constants in the homogeneous part are determined by applying the vanishing boundary conditions on H_z at top and bottom surfaces, i.e., at $z = 0$, d. Likewise applying boundary conditions on the vanishing of the normal component of the E field on the sidewalls, expressions are determined for H_x and H_y. Then, resonance is seen, i.e., the electromagnetic field inside the resonator is proportional to $\frac{1}{\delta}$, where δ is the frequency perturbations determined from Dirac delta functions. This completely solves the problem of RDRA modes.

Hybrid modes can be generated by superposition of TE and TM modes inside RDRA. In this case of RDRA, hybrid modes have been generated by using a probe of finite dimension (d) is inserted into z direction and excitation of this probe (d length) current is given to rectangular copper plane (x, y) as shown in Fig. 10.1 The current density can be determined based on KAM (Kolmogorov–Arnold–Moser) time-averaging method and using δ—Dirac delta function. The principle of orthogonality is finally applied to determine C_{mnp} and D_{mnp} amplitude coefficients of hybrid modes with fields in homogeneous and particular case. Here, particular case will have inhomogeneous medium with source applied. H_z and E_z fields have been computed simultaneously to generate hybrid modes and their coefficients.

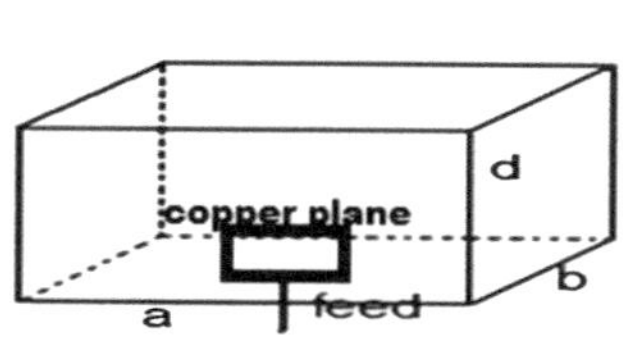

Fig. 10.1 RDRA with copper x, y rectangular plane and feed d length

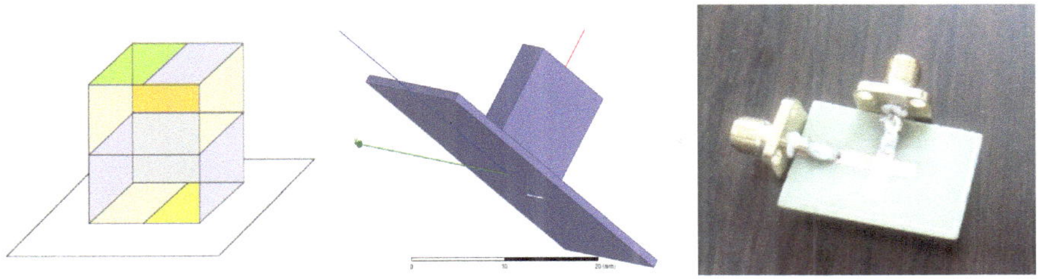

Fig. 10.2 Rectangular DRA

Figure 10.1 indicates RDRA for generating hybrid modes. Figure 10.2 shows wall configuration of rectangular DRA required for hybrid modes. The hybrid modes offer high efficiency and polarization diversity. Frequency bandwidth can be conserved by using polarization diversity. Maxwell's equation is applied to find solution of RDRA. The eigen functions are obtained by solving Helmholtz equation. The transverse components of E_x, E_y, H_x, and H_y have been expressed in terms of longitudinal components E_z and H_z. The RDRA has been excited with a RF feed probe of finite radius and small length inserted through ground plane into RDRA along z-axis. Surface current density on walls of resonator is produced due to excitation given at feed point $\left(\frac{a}{2}, \frac{b}{2}, z\right)$ of rectangular resonator. The azimuthal component of magnetic fields inside the resonator is introduced, which is also equivalent to z-component of surface current density. The modal longitudinal coefficients are E_z and H_z. The radiation pattern or power distribution among these different eigen modes is controlled by current distribution inside the resonator. The inner product or reaction term of eigen function will be equal to corresponding eigen mode. It is because magnetic currents are equal to electric currents in an antenna, due to orthonormality principle or conservation of energy methods. Some of these power coefficients can be made zero by canceling a particular resonant mode or blocking a particular eigen function. This is possible for TE and TM modes. The same frequency is introduced inside the guide with phase opposite to each other. Extracting a particular resonant mode is also possible if surface current density of that mode is made large enough by input excitation. When we apply input excitation frequency matching to the desired mode, weighted magnitude of that particular mode coefficient becomes large and corresponding mode gets excited in the RDRA. The mode merging can also be made possible by introducing shift in more than two modes toward a common desired point. Equal weightage of TE and TM at same frequency with opposite phase can cancel the mode. Higher-order modes can also be generated in RDRA as shown in Fig. 10.1. Higher modes can provide higher gain and high directivity to prevent EM pollution in microwave devices.

Figure 10.3 shows circular polarization of EM waves. The transverse components E_x, E_y, H_x, H_y are the components determined in terms of longitudinal components E_z and H_z. These transverse fields satisfy Helmholtz equations, are expressible in terms of $u(mnp)\ e^{j\omega(mnp)t}$ and $v(mnp)\ e^{j\omega(mnp)t}$ i.e., fourier basis functions. E_x, E_y, H_x, H_y fields are also expressible in terms partial derivatives of $u(mnp)$, $v(mnp)$ and

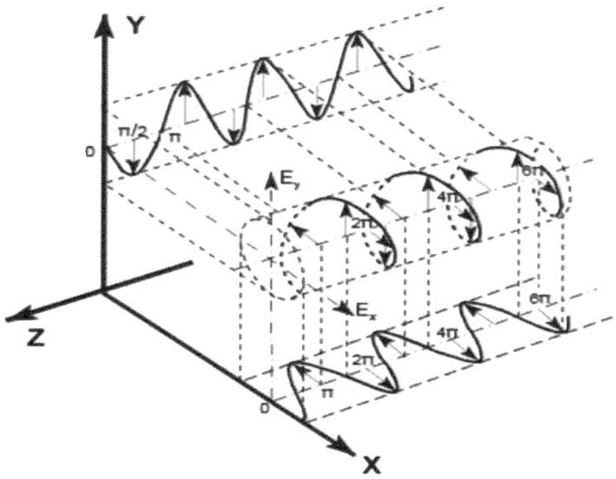

Fig. 10.3 Circular polarization

hence if $C(mnp)$, $D(mnp)$ denotes the linear combinational coefficients of $u(mnp)$, $v(mnp)$ for E_z, H_z, then the same coefficients appear in E_x, E_y, H_x, H_y. These amplitude coefficients $C(mnp)$, $D(mnp)$ can be determined by matching H_x, H_y at $z = 0$, to the surface current density of RDRA when feed is at $z = 0$. If the surface excitation at $z = 0$ has frequency component other than $\omega(mnp)$, say ω, then the field amplitude components corresponding to this excitation are determined from given below terms. The source frequency, which is other than $\omega(mnp)$, introduces decay in the resonator due to finite conductivity of the medium. Using the KAM theory of averaging, the resonator extracts out only $\omega(mnp)$ frequencies with amplitude. Equivalently, if source contents are switched on for a finite duration and then switched off, the only the dominant $\omega(mnp)$ frequencies will be present in the resonator. This situation is analogous to connecting a voltage source to an LC oscillator for a finite duration and then switching it off. If, however, the source is not switched off, then the other (non-dominant frequencies) will also be present and these can be computed based on above-mentioned methods. The composite structure having combination of PMC and PEC walls can generate hybrid modes (HEM). The HEM can be further classified HE as odd hybrid modes and EH as even hybrid modes. The applications for higher modes generation, mode shifting, mode merging, and mode control can be made in antenna design. They can impart wide design space in the field of antenna. These designs can be used in beam control and regulation.

10.2 Mathematical Model

Maxwell's equations:

For magnetic fields:

$$(\nabla^2 + k^2)\begin{pmatrix} H_x \\ H_y \\ H_z \end{pmatrix} = \begin{pmatrix} (\nabla \times J)_x(x, y, z, \omega) \\ (\nabla \times J)_y(x, y, z, \omega) \\ (\nabla \times J)_z(x, y, z, \omega) \end{pmatrix}; \tag{10.1}$$

Similarly, for electric fields:

J	current density
ψ_n	function
a_n, b_n	amplitude coefficients
u_{mnp}, v_{mnp}	Fourier basis function
hmn	cut off frequency

$$\begin{pmatrix} E_x \\ E_y \\ E_z \end{pmatrix} = \begin{bmatrix} \sum a_n\psi_n + \sum b_n\varphi_n \\ \sum a_n\psi'_n + \sum b_n\varphi'_n \\ \sum a_n\psi''_n + \sum b_n\varphi''_n \end{bmatrix} = \sum a_n \begin{bmatrix} \psi_n \\ \psi'_n \\ \psi''_n \end{bmatrix} + \sum b_n \begin{bmatrix} \varphi_n \\ \varphi'_n \\ \varphi''_n \end{bmatrix}$$
$$\begin{pmatrix} E_x \\ E_y \\ E_z \end{pmatrix} = \frac{1}{h(mn)^2}\frac{\partial^2 u_{mnp}}{\partial z \partial x} C_{mnp}; \tag{10.2}$$

$$\underline{E} = \begin{pmatrix} \sum\limits_{mnp} \frac{C_{mnp}}{h(mn)^2}\frac{\partial^2 u_{mnp}}{\partial x \partial z} - \sum \frac{\mu D_{mnp}}{h(mn)^2} j\omega(m,n,p)\frac{\partial^2 v_{mnp}}{\partial y \partial z} \\ \sum\limits_{mnp} \frac{1}{h(mn)^2} C_{mnp}\frac{\partial^2 u_{mnp}}{\partial y \partial z} + \sum \frac{\mu}{h(mn)^2} j\omega(m,n,p) D_{mnp}\frac{\partial^2 v_{mnp}}{\partial x \partial z} \\ C_{mnp}\, u_{mnp} \end{pmatrix} e^{j\omega(mnp)t}; \tag{10.3}$$

$$\begin{aligned} \underline{E}(x,y,z,t) &= \text{electric field component} \\ &= \sum_{mnp} C(mnp) \begin{pmatrix} \psi^{\mathrm{E}}_{mnp_x}(x,y,z) \\ \psi^{\mathrm{E}}_{mnp_y}(x,y,z) \\ \psi^{\mathrm{E}}_{mnp_z}(x,y,z) \end{pmatrix} e^{j\omega(mnp)t} \\ &\quad + \sum_{mnp} D(mnp) \begin{pmatrix} \varphi^{\mathrm{E}}_{mnp_x}(x,y,z) \\ \varphi^{\mathrm{E}}_{mnp_y}(x,y,z) \\ \varphi^{\mathrm{E}}_{mnp_z}(x,y,z) \end{pmatrix} e^{j\omega(mnp)t}; \end{aligned} \tag{10.4}$$

with duality:

$$\begin{aligned} \underline{H}(x,y,z,t) &= \text{magnetic field component} \\ &= \sum_{mnp} C(mnp) \begin{pmatrix} \psi^{\mathrm{H}}_{mnp_x}(x,y,z) \\ \psi^{\mathrm{H}}_{mnp_y}(x,y,z) \\ \psi^{\mathrm{H}}_{mnp_z}(x,y,z) \end{pmatrix} e^{j\omega(mnp)t} \\ &\quad + \sum_{mnp} D(mnp) \begin{pmatrix} \phi^{\mathrm{H}}_{mnp_x}(x,y,z) \\ \phi^{\mathrm{H}}_{mnp_y}(x,y,z) \\ \phi^{\mathrm{H}}_{mnp_z}(x,y,z) \end{pmatrix} e^{j\omega(mnp)t}; \end{aligned} \tag{10.5}$$

$$\hat{n} \times H = J_s = \text{ surface current density on walls;}$$
$$H_{sx}\ (x, y, 0) = J_{sy}\ \ (x, y)$$
$$H_{sy}\ (x, y, 0) = -J_{sx}\ \ (x, y).$$

Hence,

$$\begin{aligned} J_{sy}(x, y, \delta t) &= \text{current density into RDRA} \\ &= \sum_{mnp} C(mnp)\psi^{\mathrm{H}}_{mnp_x}(x, y, 0)e^{j\omega(mnp)t} \\ &\quad + \sum_{mnp} D(mnp)\varphi^{\mathrm{H}}_{mnp_x}(x, y, 0)e^{j\omega(mnp)t}, \end{aligned} \tag{10.6}$$

$$\begin{aligned} -J_{sx}(x, y, dt) &= \sum_{mnp} C(mnp)\psi^{\mathrm{H}}_{mnp_y}(x, y, 0)e^{j\omega(mnp)t} \\ &\quad + \sum_{mnp} D(mnp)\varphi^{\mathrm{H}}_{mnp_y}(x, y, 0)e^{j\omega(mnp)t}. \end{aligned} \tag{10.7}$$

If d is the length of probe inserted into RDRA, $\psi^{\mathrm{H}},\ \psi^{\mathrm{E}}, \phi^{\mathrm{H}}, \phi^{\mathrm{E}}$ equations, we get from Linear combinations of sine and cosine terms given below:

$$\psi^{\mathrm{H}}_{mnp_x}(x, y, 0) \propto \cos \cdot \sin \tag{a}$$

$$\psi^{\mathrm{H}}_{mnp_y}(x, y, 0) \propto \sin \cdot \cos \tag{b}$$

$$\varphi^{\mathrm{H}}_{mnp_x}(x, y, 0) \propto \sin \cdot \cos \tag{c}$$

$$\varphi^{\mathrm{H}}_{mnp_y}(x, y, 0) \propto \cos \cdot \sin \tag{d}$$

Amplitude coefficients can be determined from principle of orthonormality: $C(mnp)\left\langle \psi^{\mathrm{H}}_{mnp_y}(x, y, 0), \psi^{\mathrm{H}}_{mnp_y} \right\rangle$ inner product or reaction terms can be written as follows:

$$\begin{aligned} &C(mnp)\left\langle \phi^{\mathrm{H}}_{mnp_x}, \psi^{\mathrm{H}}_{mnp_x} \right\rangle + D(mnp)\left\langle \phi^{\mathrm{H}}_{mnp_x}, \psi^{\mathrm{H}}_{mnp_x} \right\rangle \\ &= \lim_{T\to\infty} \frac{1}{2T} \int_{-T}^{+T} \left\langle J_{sy}(x, y, t),\ \psi^{\mathrm{H}}_{mnp_x}(x, y, 0) \right\rangle e^{-j\omega(mnp)t} \mathrm{d}t; \end{aligned} \tag{10.8}$$

and

$$C(mnp)\left\langle \phi^{\mathrm{H}}_{mnp_y}, \psi^{\mathrm{H}}_{mnp_y} \right\rangle + D(mnp)\left\langle \phi^{\mathrm{H}}_{mnp_y}, \psi^{\mathrm{H}}_{mnp_y} \right\rangle$$
$$= \lim_{T\to\infty} \frac{1}{2T} \int_{-T}^{+T} \langle J_{sx}(x,y,t), \psi^{\mathrm{H}}_{mnp_y}(x,y,0) \rangle \, e^{-j\omega(mnp)t} \mathrm{d}t. \tag{10.9}$$

These are the solutions of amplitude coefficients $D(mnp)$ and $C(mnp)$ using time-averaging KAM method (Kolmogorov–Arnold–Moser).

These are the solutions of hybrid modes.

Helmholtz Equations

$(\nabla^2 - \frac{1}{c^2}\frac{\partial}{\partial t^2})\psi(x,y,z,t) = 0$; Helmholtz equation in time domain
$(\nabla^2 + \frac{\omega^2}{c^2})\hat{\psi}(x,y,z,\omega) = 0$; Helmholtz equation in frequency domain
$\hat{\psi}(x,y,z,\omega) \;=\; X(x)Y(y)Z(z)$; separation of variables will be used

Solutions for RDRA on application of boundary conditions are as follows:

$(\nabla^2 + k^2)E_z = 0;$	TM mode
$(\nabla^2 + k^2)H_z = 0;$	TE mode
$H_z = 0;$	on all walls
$E_x, E_y = 0;$	$z = 0, d;$
$E_x = 0;$	$x = 0, a;$
$E_y = 0;$	$x = 0, b;$

$$\left[\underline{E}_\perp = -\frac{1}{h^2}\frac{\partial}{\partial z}\underline{\nabla}_\perp E_z - \frac{\mu}{h^2}\frac{\partial}{\partial t}\underline{\nabla}_\perp H_z \times \hat{z}\right]; \quad \text{Hybrid mode electric fields;} \tag{10.10}$$

$$\left[\underline{H}_\perp = \frac{1}{h^2}, \frac{\partial}{\partial z}\underline{\nabla}_\perp H_z - \frac{\mu}{h^2}\frac{\partial}{\partial t}\underline{\nabla}_\perp E_z \times \hat{z}\right]; \quad \text{Hybrid mode electric fields.} \tag{10.11}$$

Transverse components are as follows:

$$E_x = -\frac{1}{h^2}\frac{\partial^2 E_z}{\partial x \partial t} - \frac{\mu}{h^2}\frac{\partial^2 H_z}{\partial y \partial t};$$
$$E_y = -\frac{1}{h^2}\frac{\partial^2 E_z}{\partial y \partial z} + \frac{\mu}{h^2}\frac{\partial^2 H_z}{\partial x \partial t};$$

Boundary conditions are as follows:

$$\frac{\partial^2 E_z}{\partial x \partial z} = 0, \quad x = 0, a;$$
$$\frac{\partial E_z}{\partial x} = 0, \quad x = 0, a;$$
$$\frac{\partial E_z}{\partial y} = 0, \quad y = 0, b.$$

There are three types of resonant modes at any known frequency $\omega(mnp)$:

$$\text{TE, TM, HEM modes} \rightarrow \omega(mnp)(\text{resonant mode}).$$

$$\frac{\partial^2 E_z}{\partial x \partial z} = 0, \quad z = 0, d;$$
$$\frac{\partial E_z}{\partial z} = 0, \quad z = 0, d.$$

For homogeneous medium without source terms:

$$E_z = \sum C(mnp) e^{j\omega(mnp)t} \, u(mnp)(x, y, z, t);$$
$$H_z = \sum D(mnp) e^{j\omega(mnp)t} \, v(mnp)(x, y, z, t).$$

10.3 Modes in Homogeneous Medium with Source Terms

For homogeneous medium case

$$\underline{E}^{(\text{hom})}(x, y, z, t) = \sum C(mnp) e^{j\omega(mnp)t} \underline{\psi}^{\text{E}}_{mnp}(\underline{r}) + \sum D(mnp) e^{j\omega(mnp)t} \underline{\psi}^{\text{E}}_{mnp}(\underline{r}); \tag{10.12}$$

where $r = x, y, z$

$$\underline{H}^{(\text{hom})}(x, y, z, t) = \sum C(mnp) e^{j\omega(mnp)t} \, \underline{\psi}^{\text{H}}_{mnp}(\underline{r}) + \sum D(mnp) e^{j\omega(mnp)t} \, \underline{\psi}^{\text{H}}_{mnp}(\underline{r}); \tag{10.13}$$

$$J_s(x, y) = J_{sx}(x, y)\hat{x} + J_{sy}(x, y)\hat{y}$$
$$J \rightarrow J_s(x, y)\delta(z)$$
$$J_s(x, y, \omega)$$

Maxwell's equation:

$$\begin{aligned} \nabla \times H &= J + j\omega\epsilon\underline{E} \\ \nabla \times E &= -j\omega\mu\underline{H} \end{aligned}$$

E or H can be calculated

$$\begin{aligned} &-\nabla^2 H = \nabla \times J + \omega^2\mu\epsilon\underline{H} \\ &(\nabla^2 + k^2)\underline{H} = -\underline{\nabla} \times \underline{J} \text{ (with source)}; \\ &\underline{H} = H_{\text{hom}} + H_{\text{part}} \\ &J_s(x, y, z) = \underline{J}_{sy}(x, y)(\delta(z) - d) = -J_{y,z} \\ &(\nabla \times J) = J_{z,y} - J_{y,z} \\ &(\nabla^2 + k^2)H_x = -J_{sy}(x, y)(\delta'(z) - d) \\ &(\nabla^2 + k^2)H_y = -J_{sx}(x, y)(\delta'(z) - d) \\ &(\nabla^2 + k^2)H_z = (J_{syx} - J_{sxy})(\delta'(z) - d) \end{aligned}$$

Hence, we can compute

$$\begin{aligned} &(H_x, H_y, H_z)(x, y, z, \omega) \\ &v(mnp) = \sin(x)\sin(y)\sin(z) \\ &u(mnp) = \cos(x)\cos(y)\cos(z) \end{aligned}$$

$$\begin{aligned} H_z &= \text{from } \underline{E}_\perp, \underline{H}_\perp \text{ equations:} \\ &= \sum D^{\text{source}}(m, n, p, \omega)v_{mnp}(r) + C^{\text{source}}(m, n, p, \omega)u_{mnp}(r); \end{aligned} \tag{10.14}$$

$$(\nabla^2 + k^2)H_z = \sum_{mnp} \left(\frac{-\omega(mnp)^2}{c^2} + \frac{\omega^2}{c^2} \right) \cdot Cz^s(m, n, p, \omega)v_{mnp(r)}. \tag{10.15}$$

Hybrid modes can be generated by introducing non-resonant terms. Set infinity magnitude of coefficients for non-resonant frequency, and ω and $\{\omega(mnp)\}$. ω are non-resonant terms.

10.4 Current Density in RDRA

$$J_{sx}(x, y) = \sum_{mnp} \omega^2 - \omega(mnp)^2 C^s(mnp, \omega)v(mnp)(x, y, d);$$

ω is non-resonant frequency, and $\omega(mnp)$ is resonant frequency

$$= c^2(J_{sy,x}(x,y) - J_{sx,y}(x,y)$$

Hence, integrating and multiplying $(\delta(z) - d)$

$$\begin{aligned}&(\omega^2 - \omega(mnp)^2 C^{\text{source}}(mnp,\omega)\\&= \int_a^0 \int_b^0 c^2\ (J_{sy,x}(x,y,\omega) - J_{sx,y}(x,y,\omega))v(mnp)(x,y,d)\mathrm{d}x\,\mathrm{d}y;\end{aligned} \tag{10.16}$$

$$C_z^s(mnp\omega) = \frac{1}{\left(\omega^2 - \omega(mnp)^2\right)} \int (J_{sy,x}(x,y,\omega) - J_{sx,y}(x,y,\omega))v(mnp)(x,y,d)\mathrm{d}x\ \mathrm{d}y. \tag{10.17}$$

Similarly, we can compute $C_x^{(s)} C_y^{(s)}$ terms which are the desired solutions of hybrid resonant modes.

10.5 *E* and *H* Fields

E, *H*, fields:

$$\begin{aligned}\nabla \times E &= -j\omega\mu H\\ \nabla \times H &= j\omega\epsilon E + J\\ -\nabla^2 E &= -j\omega\mu(j\omega\epsilon)\\ E &= \frac{\nabla \times H - J}{J\omega\epsilon}\end{aligned}$$

hence, $C_z^s(mnp) = \frac{c^2}{(\omega^2 - \omega^2(mnp))} \int J_{sy,x}(x,y,\omega) - J_{sy,y}(x,y,\omega)v_{mnp}(x,y,d)\ \mathrm{d}x\,\mathrm{d}y$ is the solution of hybrid modes. Similarly, other hybrid mode coefficients can be worked out.

$$\begin{matrix} C_x^s(mnp) & d_x^s(mnp)\\ C_y^s(mnp) & d_y^s(mnp)\\ C_z^s(mnp) & d_z^s(mnp)\end{matrix}$$

Thus, complete solution of hybrid modes has been obtained.

10.6 Mathematical Modeling of Hybrid Modes

$$n \times H = J_s \quad \text{on each wall}$$
$$n \times E = M_s \quad \text{on each wall}$$

The mathematical derivation of Hybrid modes is given below. This is purely based on solution of Maxwell's equation.

First, we develop solution of rectangular waveguide and switch to resonator. The waveguide solution is very simple. Figure 10.4 indicates field configuration inside RDRA. These waveguide equations will have both the fields H_z and E_z as given below:

$$(\nabla_{\perp} - \gamma\hat{z}) \times (E_z\hat{z} + E_{\perp}) = -j\omega\mu(H_z\hat{z} + H_{\perp}) \tag{10.18}$$

$$(\nabla_{\perp} - \gamma\hat{z}) \times (H_z\hat{z} + H_{\perp}) = -j\omega\epsilon(E_z\hat{z} + E_{\perp}) \tag{10.19}$$

$$\nabla_{\perp}E_z \times \hat{z} - \gamma\hat{z} \times E_{\perp} = -j\omega\mu H_{\perp} \tag{10.20}$$

$$\nabla_{\perp}H_z \times \hat{z} - \gamma\hat{z} \times H_{\perp} = -j\omega\epsilon E_{\perp} \tag{10.21}$$

$$\nabla_{\perp}E_z + \gamma E_{\perp} = -j\omega\mu\hat{z} \times H_{\perp} \tag{10.22}$$

$$\nabla_{\perp}E_z + \gamma E_{\perp} = \frac{-j\omega\mu}{\gamma}(\nabla_{\perp}H_z \times \hat{z} - j\omega\epsilon E_{\perp}) \tag{10.23}$$

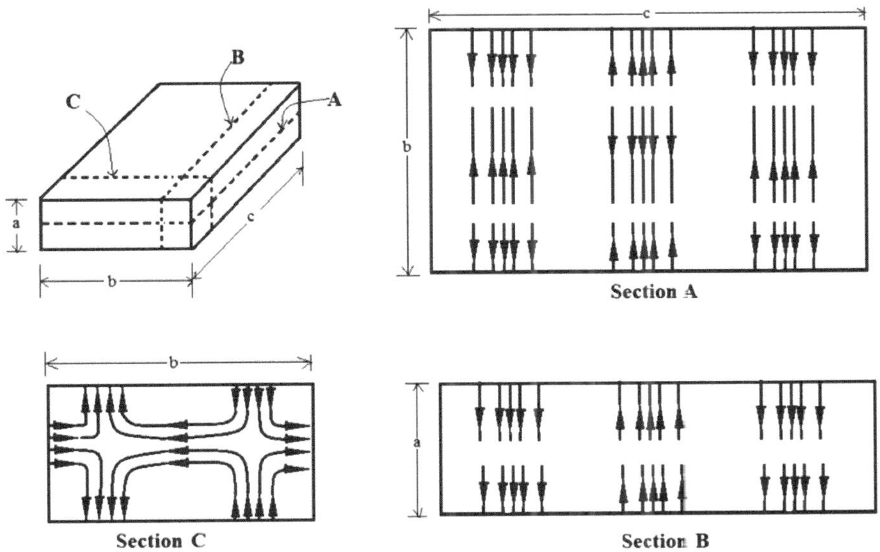

Fig. 10.4 Resonating modes in RDRA

$$\nabla_{\perp} E_z + \frac{j\omega\mu}{\gamma} \nabla_{\perp} H_z \times \hat{z} = -\left(\frac{\omega^2 \mu\epsilon}{\gamma} + \gamma\right) E_{\perp}$$
$$\gamma^2 + \omega^2 \mu\epsilon = h^2$$

Waveguide equations based on Helmholtz equations are as follows:

$$E_{\perp} = \frac{-\gamma}{h^2} \nabla_{\perp} E_z - \frac{j\omega\mu}{h^2} \nabla_{\perp} H_z \times \hat{z} \tag{10.24}$$

$$H_{\perp} = \frac{-\gamma}{h^2} \nabla_{\perp} H_z + \frac{j\omega\mu}{h^2} \nabla_{\perp} E_z \times \hat{z} \tag{10.25}$$

Resonator equations are obtained by simply replacing $[\gamma]$ by $\left[-\frac{\mathrm{d}}{\mathrm{d}z}\right]$:

$$E_{\perp} = -\frac{1}{h^2} \frac{\mathrm{d}}{\mathrm{d}z} \nabla_{\perp} E_z - \frac{\mu}{h^2} \frac{\mathrm{d}}{\mathrm{d}t} \nabla_{\perp} H_z \times \hat{z} \tag{10.26}$$

$$H_{\perp} = \frac{1}{h^2} \frac{\mathrm{d}}{\mathrm{d}z} \nabla_{\perp} H_z + \frac{\mu}{h^2} \frac{\mathrm{d}}{\mathrm{d}t} \nabla_{\perp} E_z \times \hat{z} \tag{10.27}$$

$$\left(\nabla^2 + h^2\right) \begin{pmatrix} E_z \\ H_z \end{pmatrix} = 0; \quad \text{Helmholtz equation}$$

Boundary conditions in RDRA have been taken as, top and bottom walls of resonator are PEC other four sides walls are PMC.

$$\begin{array}{ll} H_z = 0; & \text{at}\, x = 0,\, a,\ \text{and}\ y = 0, d;\ z = 0,\, d; \\ E_x = E_y = 0; & \text{at}\ z = 0,\, d; \\ H_x = 0; & \text{at}\ y = 0,\, b; \\ H_y = 0; & \text{at}\ x = 0,\, a. \end{array}$$

Standard fields are as follows:

$$H_z = \sum \mathrm{Re}\left(D(mnp) e^{j\omega(mnp)t} v_{mnp}(\underline{r})\right) \tag{10.28}$$

$$E_z = \sum \mathrm{Re}\left(C(mnp) e^{j\omega(mnp)t} u_{mnp}(\underline{r})\right) \tag{10.29}$$

Standard orthogonal fields are as follows:

$$v_{mnp} = \frac{2\sqrt{2}}{\sqrt{abd}} \sin\left(\frac{m\pi x}{a}\right) \sin\left(\frac{m\pi y}{b}\right) \sin\left(\frac{m\pi z}{d}\right) \tag{10.30}$$

$$u_{mnp} = \frac{2\sqrt{2}}{\sqrt{abd}} \cos\left(\frac{m\pi x}{a}\right) \cos\left(\frac{m\pi y}{b}\right) \cos\left(\frac{m\pi z}{d}\right) \tag{10.31}$$

Equations (10.30) and (10.31) have been obtained from expansion of Helmholtz equation by separation of variables method:

$$(\nabla^2 + h^2)H_z = 0$$

Hence, $h^2 = h_{mn}^2 = \pi^2\left(\frac{m^2}{a^2} + \frac{n^2}{b^2}\right)$; this gives the resonant frequency of RDRA.

Tensor product of linear combination can appear as given below:

$$H_z = \mathcal{L}\Big\{\cos\left(\frac{m\pi x}{a}\right) \cos\left(\frac{n\pi x}{a}\right), \cos\left(\frac{n\pi x}{a}\right) \sin\left(\frac{m\pi x}{a}\right), \sin\left(\frac{m\pi x}{a}\right) \cos\left(\frac{n\pi x}{a}\right), \sin\left(\frac{n\pi x}{a}\right) \sin\left(\frac{m\pi x}{a}\right)\Big\}$$

where $\mathcal{L}$ denotes linear combinations. It turns out that depending on the nature of wall or surface (PEC or PMC), four possible linear combinations can appear ($\cos \otimes \sin$, $\sin \otimes \cos$, and $\sin \otimes \sin$, $\cos \otimes \cos$). Also,

$$\omega^2 \mu\epsilon + \gamma^2 = h_{mn}^2$$

Hence,

$$H_z = 0; \quad \text{when}\, x = 0,\ \cos\ \text{terms are ruled out from}\ x.$$
$$H_z = 0; \quad \text{when}\, y = 0,\ \text{against cos terms are ruled out from}\, y.$$
$$H_z = \sin\left(\frac{m\pi x}{a}\right) \sin\left(\frac{n\pi y}{b}\right)(C_1 e^{\gamma_{mnz}} + C_2 e^{-\gamma_{mnz}})$$
$$H_z = 0; \quad \text{when}\, z = 0, d$$
$$(C_1 + C_2) = 0$$
$$(e^{\gamma_{mnd}} - e^{-\gamma_{mnd}}) = 0, \beta$$
$$C_1 = C_2; \quad \sin(\gamma_{mnd}) = 0$$
$$\gamma_{mn} = j\beta_{mn}$$
$$\beta_{mnd} = \pi p$$

Hence,

$$\beta_{mn} = \frac{p\pi}{d}$$

$$H_z = \sin\left(\frac{m\pi x}{a}\right) \sin\left(\frac{n\pi y}{b}\right) \sin\left(\frac{p\pi z}{d}\right) \tag{10.32}$$

$\omega^2\mu\epsilon - \left(\frac{p\pi}{d}\right)^2 = h_{mn}^2$; hence, resonant frequency can be determined as follows:

$$\omega\sqrt{\mu\epsilon} = \pi\sqrt{\frac{m^2}{a^2} + \frac{n^2}{b^2} + \frac{p^2}{d^2}};$$

Here, we note that resonant frequency in hybrid mode is same for TE and TM modes.

Now

$$\begin{aligned} H_x &= \frac{1}{h^2}\frac{\partial}{\partial z}\frac{\partial H_z}{\partial x} + \frac{\mu}{\mathrm{h}^2}\frac{\partial}{\partial t}\frac{\partial E_z}{\partial y}; \\ H_x &= 0; \quad \text{at}\, y = 0, b; \\ \frac{\delta H_x}{\delta x} &= 0; \quad \text{at}\, y = 0, \mathrm{b}; \\ \frac{\delta E_z}{\delta y} &= 0; \quad \text{at}\, y = 0, b; \\ E_x &= 0; \\ E_y &= 0, \quad z = 0, d; \end{aligned}$$

Hence,

$$E_x = \frac{1}{h^2}\frac{\partial}{\partial z}\frac{\partial}{\partial x}E_z - \frac{\mu}{h^2}\frac{\partial}{\partial t}\frac{\partial H_z}{\partial y};$$

$$\begin{aligned} \frac{\partial^2 E_z}{\partial x \partial z} &= 0 \quad \text{at} \quad z = 0, d \\ \frac{\partial E_z}{\partial t} &= 0 \quad \text{at} \quad z = d \end{aligned}$$

z-dependence of E_z is $\cos\left(\frac{\pi p z}{d}\right)$, $E_x = 0$; when $x = 0, a$;

$$E_z = \cos\left(\frac{m\pi x}{a}\right)\cos\left(\frac{n\pi y}{b}\right)\cos\left(\frac{p\pi z}{d}\right) \tag{10.33}$$

This is the way of getting E_z and H_z longitudinal components by the method of separation of variables.

10.7 General Solution of Hybrid Modes (HEM)

The investigations are based on first applying waveguide theory, it models regarding the electromagnetic fields to vary with z-axis, i.e., these are exploited into the Maxwell curl equations, then manipulating them to express the transverse components of the fields in terms of partial derivatives of the longitudinal components of the fields w.r.t. x and y axis (i.e., the transverse coordinates). Waveguide models of four different rectangular DRAs with specified boundary conditions filled with homogeneous material having linear permittivity have been mathematically developed and realized to determine TE and TM modes' propagating fields. These have resulted in different sine–cosine combinations. Propagation of these fields have been split as inside and outside of the RDRA with an interfacing surface having two different permittivity. The solution is transcendental equation which purely characterizes rectangular DRA resonant frequency and propagating fields. The amplitude coefficient of these fields C_{mnp} and D_{mnp} inside the DRA can be determined by comparing time-averaged magnetic energies equal to time-averaged electrical energies by KAM method based on principle of orthonormality. The transverse components E_x, E_y, H_x, H_y are the components determined in terms of longitudinal components E_z, H_z. These transverse fields satisfy Helmholtz equations, are expressible in terms of $u(mnp)\ e^{j\omega(mnp)t}$ and $v(mnp)\ e^{j\omega(mnp)t}$ Fourier basis function. E_x, E_y, H_x, H_y fields are also expressible in terms partial derivatives of $u(mnp)$, $v(mnp)$, and hence if $C(mnp)$, $D(mnp)$ denotes the linear combinational coefficients of $u(mnp)$, $v(mnp)$ for E_z, H_z, then the same coefficients appear in E_x, E_y, H_x, H_y. These coefficients $\mathrm{C}(mnp)$, $D(mnp)$ can be determined by matching H_x, H_y at $z = 0$ to the surface current density of RDRA, when feed is at $z = 0$. If the surface excitation at $z = 0$ has frequency component other than $\omega(mnp)$, say ω, then the field amplitude components corresponding to this excitation are determined. Both the fields E_z and H_z will remain excited at any instant of time in resonator, and then, these modes can be termed as hybrid modes. Our solution is developed based on homogeneous medium in the resonator.

$$E_z = \sum \mathrm{Re}\left\{C(mnp)e^{j\omega(mnp)t}\right\}u_{mnp}(\underline{r}) \tag{10.34}$$

$$H_z = \sum \mathrm{Re}\ \left\{D(mnp)e^{j\omega(mnp)t}\right\}v_{mnp}(\underline{r}) \tag{10.35}$$

$$\begin{aligned} E_{\perp} = \sum \frac{1}{h_{mn}^2} \mathrm{Re}\Bigg[&\left\{C_{mnp}e^{j\omega(mnp)t}\right\}\frac{\partial}{\partial z}\nabla_{\perp}u_{mnp}(\underline{r}) \\ &- \frac{\mu}{h_{mn}^2}\mathrm{Re}\left\{j\omega(mnp)D(mnp)e^{j\omega(mnp)t}\right\}\left(\nabla_{\perp}v_{mnp}(\underline{r}) \times \hat{z}\right)\Bigg] \end{aligned} \tag{10.36}$$

$$E_{\perp} = \hat{x}\,E_x + \hat{y}\,E_y;$$
$$H_{\perp} = \hat{x}\,H_x + \hat{y}\,H_y;$$
$$\nabla_{\perp} = \hat{x}\frac{\partial}{\partial x} + \hat{y}\frac{\partial}{\partial y};$$

$$\underline{E} = \begin{pmatrix} E_x \\ E_y \\ E_z \end{pmatrix} = \sum C_{mnp} e^{j\omega(mnp)t}\,\underline{\Psi}^{E}_{mnp}(\underline{r}) + \sum D(mnp) e^{j\omega(mnp)t}\,\underline{\phi}^{E}_{mnp}(\underline{r}) \quad (10.37)$$

$$\underline{H} = \begin{pmatrix} E_x \\ E_y \\ E_z \end{pmatrix} = \sum \left[C_{mnp} e^{j\omega(mnp)t}\,\underline{\Psi}^{H}_{mnp}(\underline{r}) + D(mnp) e^{j\omega(mnp)t}\,\underline{\phi}^{H}_{mnp}(\underline{r})\right] \quad (10.38)$$

$$\underline{\Psi}^{\mathrm{E}}_{mnp}(\underline{r}) = \begin{pmatrix} \frac{1}{h^2_{mn}} & \frac{\partial^2}{\partial x \partial z} u_{mnp} \\ \frac{1}{h^2_{mn}} & \frac{\partial^2}{\partial x \partial z} u_{mnp} \\ & u_{mnp} \end{pmatrix}$$

where $\underline{\Psi}^{\mathrm{E}}_{mnp}$ is the electric field vector coming from the z-components of electric field, i.e., TM mode.

And $\underline{\phi}^{\mathrm{E}}_{mnp}$ is the electric field vector coming from the z-components of magnetic field, i.e., TE mode.

Similarly, magnetic field vectors $\underline{\phi}^{\mathrm{H}}_{mnp}(\underline{r})$ and $\underline{\Psi}^{\mathrm{H}}_{mnp}(\underline{r})$ can be solved.

$$\underline{\phi}^{\mathrm{H}}_{mnp}(\underline{r}) = \begin{pmatrix} \frac{1}{h^2_{mn}} & \frac{\partial^2}{\partial x \partial z} v_{mnp} \\ \frac{1}{h^2_{mn}} & \frac{\partial^2}{\partial x \partial z} v_{mnp} \\ & v_{mnp} \end{pmatrix}$$

$$\underline{\phi}^{\mathrm{E}}_{mnp}(\underline{r}) = \begin{pmatrix} \frac{-j\mu\omega(mnp)}{h^2_{mn}} & \frac{\partial}{\partial y} v_{mnp} \\ \frac{j\mu\omega(mnp)}{h^2_{mn}} & \frac{\partial}{\partial x} v_{mnp} \\ & 0 \end{pmatrix}$$

and,

$$\underline{\Psi}^{\mathrm{H}}_{mnp}(\underline{r}) = \begin{pmatrix} \frac{j\mu\omega(mnp)}{h^2_{(mn)}} \frac{\partial}{\partial y} u_{mnp} \\ \frac{-j\mu\omega(mnp)}{h^2_{(mn)}} \frac{\partial}{\partial x} u_{mnp} \\ 0 \end{pmatrix}$$

Hence, general hybrid equations can be written as follows:

$$\underline{E} = \sum \left[Cmnp\underline{\Psi}^{\mathrm{E}}_{mnp}(\underline{r}) + D(mnp)\underline{\phi}^{\mathrm{E}}_{mnp}(\underline{r}) \right] e^{-j\omega(mnp)t} \tag{10.39}$$

$$\underline{H} = \sum \left[C_{mnp}\underline{\Psi}^{\mathrm{H}}_{mnp}(\underline{r}) + D(mnp)\underline{\phi}^{\mathrm{H}}_{mnp}(\underline{r}) \right] e^{-j\omega(mnp)t} \tag{10.40}$$

Solution of the RDRA can be developed by using these above two equations. For this, we insert a probe of δ length having R radius into rectangular DRA. This is pointing toward z-axis.

Figure 10.5 gives a clear picture of RDRA with feed associated, and Fig. 10.6 shows the structure of cylindrical probe.

$$x = \frac{a}{2} + R\cos\phi;$$
$$y = \frac{b}{2} + R\sin\phi;$$
$$z = 0.$$

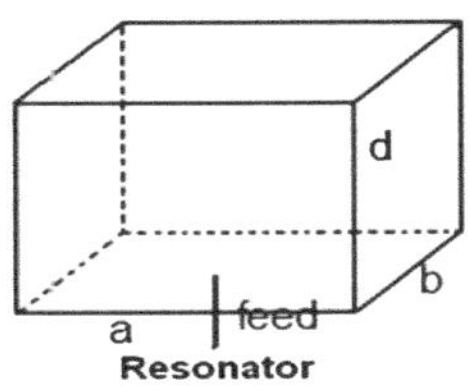

Fig. 10.5 RDRA with feed

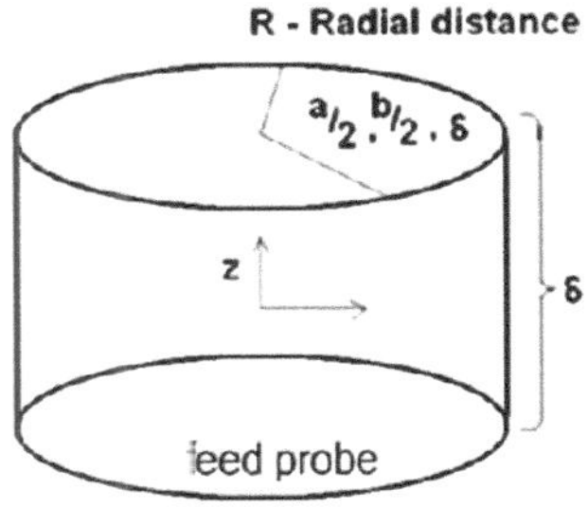

Fig. 10.6 Feed probe

It is expressed based on Cartesian to cylindrical coordinates.
$\hat{n} \times \underline{H} = \underline{J}_s$: This is based on boundary conditions inside the RDRA.

$$\hat{\rho}\left(H_z\hat{z} + H_\phi\hat{\phi}\right) = \underline{J}_s$$
$$H_\phi\hat{z} - H_z\hat{\phi} = \underline{J}_s$$

and

$$H_\phi\left(\frac{a}{2} + R\cos\phi,\ \frac{b}{2} + R\sin\phi, z\right) = J_{sz}(\phi, z) \quad (0<z<\delta,\ 0<\phi<2\pi)$$
$$H_z\left(\frac{a}{2} + R\cos\phi, \frac{b}{2} + R\sin\phi, z\right) = J_s\phi(\phi, z)$$

$$-J_s\phi(\phi, z, t) = \sum D(mnp)v_{mnp}\left(\frac{a}{2} + R\cos\phi, \frac{b}{2} + R\sin\phi, z\right)e^{j\omega(mnp)t}$$

$$\begin{aligned} H_\phi\left(\frac{a}{2} + R\cos\phi, \frac{b}{2} + R\sin\phi, z\right) &= -H_x\sin\phi + H_y\cos\phi \\ &= J_{sz}(\phi, z) \quad (0<z<\delta,\ 0<\phi<2\pi) \\ &= -J_{s\phi}(\phi, z) \end{aligned}$$

$$\begin{aligned} J_{sz}(\phi, z, t) = \sin\phi\sum\Bigg[& -c_{mnp}\frac{j\mu\omega(mnp)}{h_{mn}^2}\frac{\partial v_{mnp}}{\partial y}\left(\frac{a}{2} + R\cos\phi, \frac{b}{2} + R\sin\phi, z\right) \\ & -\frac{\mathrm{d}_{(mnp)}}{h_{mn}^2}\frac{\partial^2 v_{mnp}}{\partial z\partial x}\left(\frac{a}{2} + R\cos\phi, \frac{b}{2} + R\sin\phi, z\right)\Bigg]e^{+j\omega(mnp)t} \\ + \cos\phi\sum\Bigg[& C_{mnp}\left(-\frac{j\mu\omega(mnp)}{h_{mn}^2}\right)\frac{\partial}{\partial x}u_{mnp}\left(\frac{a}{2} + R\cos\phi, \frac{b}{2} + R\sin\phi, z\right) \\ & +\frac{D_{(mnp)}}{h_{mn}^2}\frac{\partial^2 v_{mnp}}{\partial z\partial x}\left(\frac{a}{2} + R\cos\phi, \frac{b}{2} + R\sin\phi, z\right)\Bigg]e^{+j\omega(mnp)t}; \end{aligned} \tag{10.41}$$

$$-J_{s\phi}(\phi, z, t) = \sum_{mnp} C_{mnp}e^{j\omega(mnp)t}X_{mnp}(\phi, z);$$

where,

$$X_{mnp}(\phi, z) = v_{mnp}\left(\frac{a}{2} + R\cos\phi, \frac{b}{2} + R\sin\phi, z\right); \tag{10.42}$$

$$J_{sz}(\phi, z, t) = \sum_{mnp} C_{mnp}\eta^{(1)}_{mnp}(\phi, z) + D_{mnp}\eta^{(2)}_{mnp}(\phi, z)e^{j\omega(mnp)t}$$

where,

$$\eta^{(1)}_{mnp}(\phi, z) = \frac{-j\omega(mnp)t\,\mu\sin\phi}{h^2_{mn}}\frac{\partial}{\partial y}u_{mnp}\left(\frac{a}{2} + R\cos\phi, \frac{b}{2} + R\sin\phi, z\right)$$

$$\eta^{(2)}_{mnp}(\phi, z) = \frac{-\sin\phi}{h^2_{mn}}\frac{\partial^2 v_{mnp}}{\partial x\partial z}\left(\frac{a}{2} + R\cos\phi, \frac{b}{2} + R\sin\phi, z\right)e^{-j\omega(mnp)t}\widetilde{X}_{mnp}(\phi, z)$$

$$C_{mnp} = \lim_{T\to\infty} -\frac{1}{2T}\int J_{s\phi}(\phi, z, t)\tilde{X}_{mnp}(\phi, z)e^{-j\omega(mnp)t}\,\mathrm{d}t\,\mathrm{d}\phi\,\mathrm{d}z \tag{10.43}$$

$$|t| < T$$
$$0 < \phi < 2\pi$$
$$0 < z < \delta$$

$$\int \left|\widetilde{X}_{mnp}(\phi, z)\right|^2 \mathrm{d}\phi\,\mathrm{d}z$$

$$C_{mnp}\int\left[\left|\eta^{(1)}_{mnp}(\phi, z)\right|^2 \mathrm{d}\phi\,\mathrm{d}z + D_{mnp}\int\overline{\eta^{(1)}_{mnp}(\phi, z)}\,\eta^{(2)}_{mnp}(\phi, z)\right]\mathrm{d}\phi\,\mathrm{d}z$$
$$= \lim_{T\to\infty}\frac{1}{2T}\int\overline{\eta^{(1)}_{mnp}(\phi, z)}J_{sz}(\phi, t)e^{-j\omega(mnp)t}\,\mathrm{d}\phi\,\mathrm{d}z\,\mathrm{d}t \tag{10.44}$$
$$|t| \le T$$

$$C_{mnp}\int\left[\eta^{(1)}_{mnp}(\phi,\ z)\,\eta^{(2)}_{mnp}(\phi,\ z)\,\mathrm{d}\phi\,\mathrm{d}z\,\mathrm{d}t + D_{(mnp)}\int\left|\eta^{(2)}_{mnp}(\phi,\ z)\right|^2\mathrm{d}\phi\,\mathrm{d}z\right]$$
$$= \lim_{T\to\infty}\frac{1}{2T}\int J_{sz}(\phi, z, t)\,\eta^{(2)}_{mnp}(\phi, z)e^{-j\omega(mnp)t}\mathrm{d}\phi\,\mathrm{d}z\,\mathrm{d}t$$
$$|t| < T \tag{10.45}$$

If we keep $J_{s\phi} = 0$, from Eqs. (10.23) and (10.24), we get $C(mnp)$ and $D(mnp)$.

The study of electric and magnetic fields for maxima and minima inside RDRA introduces us to define mode number. By applying perturbations, higher modes can be excited. The increase in the electrical length of the antenna on higher-order mode causes higher antenna gain. Short and open boundaries are the basis of modes. The half-wavelength resonant modes with odd numbers only will be excited when ground plane is used as even modes get short-circuited due to ground plane. The polarization of even and odd modes is opposite. The higher modes will have higher resonant frequency. A number of higher modes also modify the radiation patterns, i.e., the modes' number will be equal to number of lobes in final radiation pattern. Highly directive patterns can be obtained at higher modes. Bandwidth of higher-order modes will be decreased. HEM 1, 3, 5, 7... are odd modes that can be written as HE. Similarly HEM 2, 4, 6, 8,... are even modes or EH mode. The care must be taken to select this hybrid number n because it has a direct relationship with the radiation pattern of far fields or beam shape. Gain of antenna can also decrease abruptly due to dispersion at higher modes. This is introduced when the dipole moment starts overlapping. Based on various solutions, hybrid modes can be memorized for any particular mode with desired radiation patterns. Automated applications using microcontroller can generate lookup table for desired radiation patterns or beam pattern for any desired frequency as well as gain. Thus, hybrid modes can be used for automated RDRA reconfigurability.

10.8 HFSS Results

Figures 10.7, 10.8, 10.9, 10.10, 10.11, 10.12 and 10.13 are results of simulated and experimentations of RDRA. Type of result is mentioned in each picture.

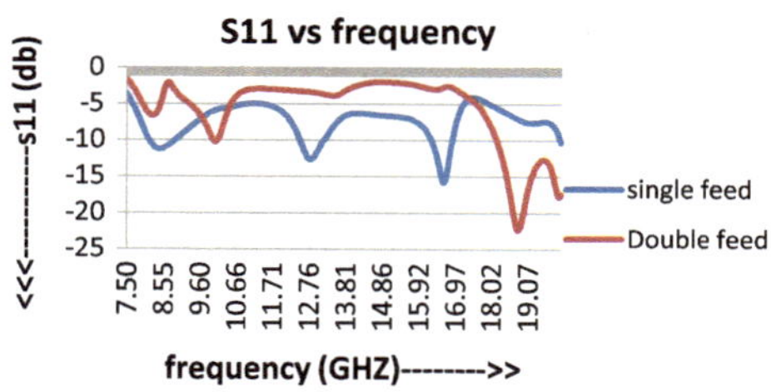

Fig. 10.7 The excitation is given by TE and TM modes at the same time

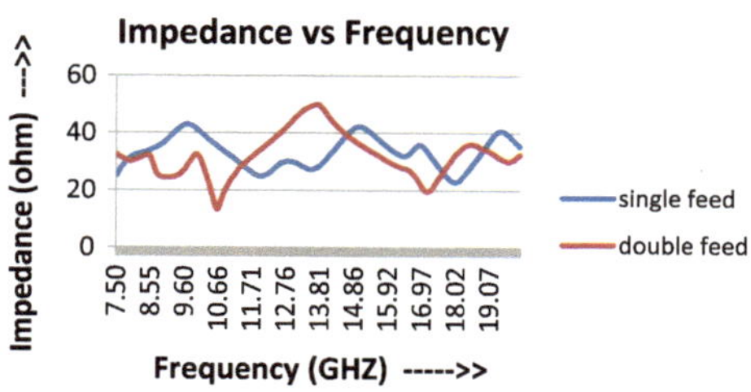

Fig. 10.8 The excitation is given by TE and TM modes at the same time

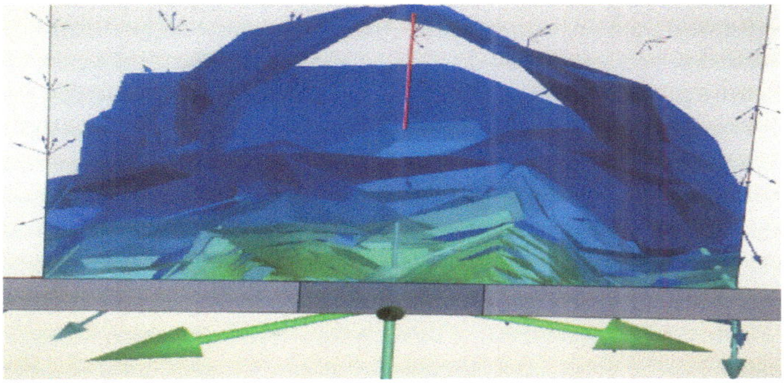

Fig. 10.9 The HE fields in RDRA

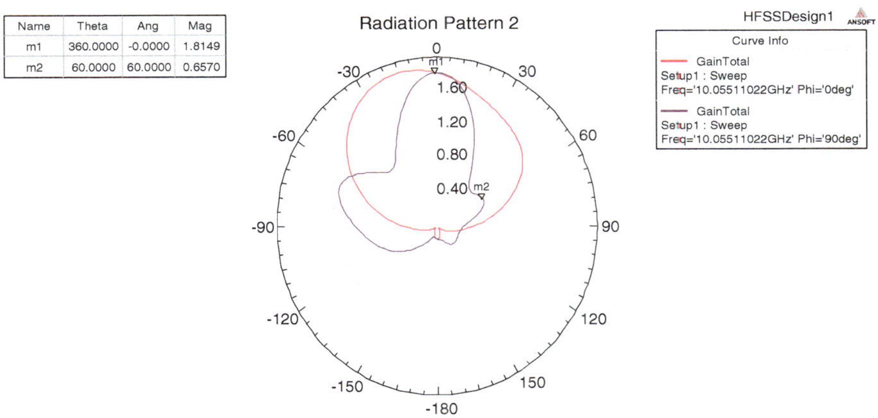

Fig. 10.10 Orthogonal polarization due to *two feeds*

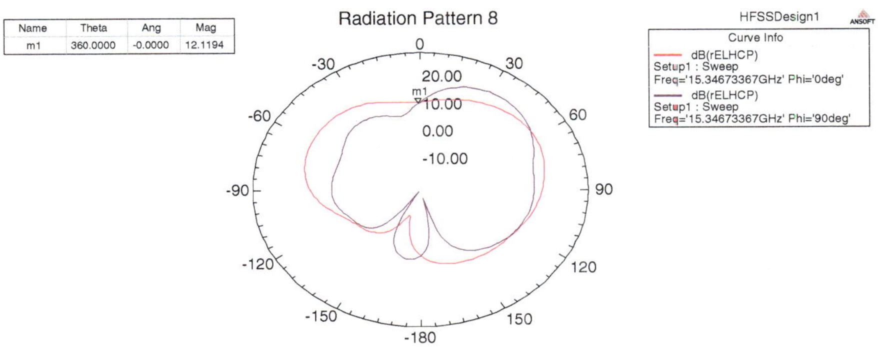

Fig. 10.11 Radiation pattern

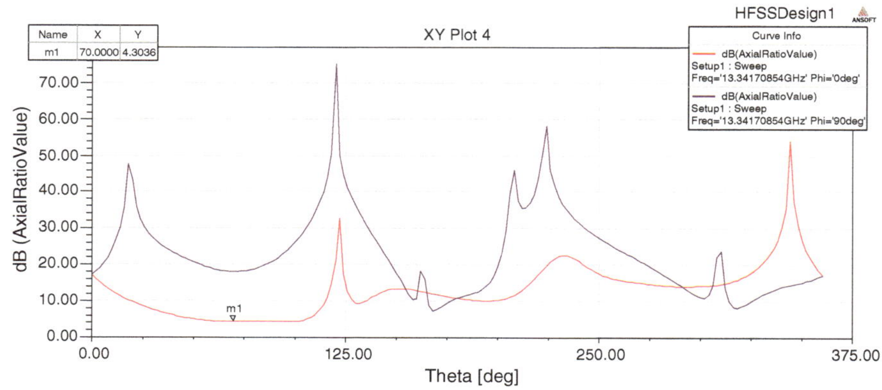

Fig. 10.12 Plot of axial ratio indicating polarization

10.9 Prototype RDRA Results

See Fig. 10.13.

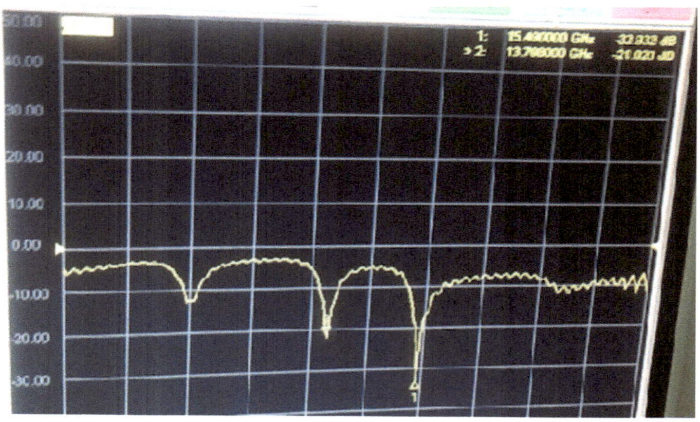

Fig. 10.13 Prototype *double-feed* RDRA S_{11} measurement

Chapter 11
Inhomogeneous Permittivity, Permeability, and Conductivity Solution in Rectangular DRA

Abstract Fields solution for inhomogeneous permittivity, permeability, and conductivity in rectangular DRA was found. The solution is very complex. This is based on the solution of rectangular waveguide filled with inhomogeneous permittivity, permeability, and conductivity material. These amplitudes are called "the waveguide modes" and are of the form cos–sin. Depending on the nature of wall surfaces (PEC or PMC), four possible linear combinations can appear (cos–sin, sin–cos, sin–sin, and cos–cos). The discrete modes enable us to visualize the resonator as collection of L, C oscillators with different *L*, *C* values.

Keywords Inhomogeneous · Permittivity · Permeability · Linear combinations · Sequential RLC circuits · Discrete modes · Complex solution

11.1 Introduction

Solution of rectangular waveguide with inhomogeneous permittivity, permeability, and conductivity of the medium was found. These amplitudes are called "the waveguide modes" and are of the form cos–sin and sin–cos which denotes linear components. It turns out that depending on the nature of wall surfaces (PEC or PMC), four possible linear combinations can appear (cos–sin, sin–cos, sin–sin, and cos–cos). In a rectangular DRA, we've got to applying in additional boundary conditions on top and bottom surfaces to be the linear combinations as compared to waveguide. They have two possible linear combinations of sin cos. Thus, the possible frequencies obtained by solving them comes out to be an equivalent, but computationally simpler way to pass on from waveguide physics to resonator physics is to just replace by—in all the waveguide formulae that express the tangential field components in terms of the longitudinal components. This is done after solving the full 3-D Helmholtz equations using separation of variable in x, y, z. The discrete modes enable us to visualize the resonator as collection of L, C oscillators with different L, C values. The outcome of all these analyses enables us to write down the fields inside the resonator, as superposition of four or three vector-valued basis functions.

R.S. Yaduvanshi and H. Parthasarathy, *Rectangular Dielectric Resonator Antennas*, DOI 10.1007/978-81-322-2500-3_11

11.2 Mathematical Model

$$\epsilon(x,y) = \epsilon_0(1+\delta\cdot\ \chi_e(x,y)) \tag{11.1a}$$

$$\mu(x,y) = \mu_0(1+\delta\cdot\chi_m(x,y)) \tag{11.1b}$$

$$\begin{aligned} \underline{E}(\underline{r}) &= \underline{E}(x,y)\exp(-\gamma z) \\ \underline{H}(\underline{r}) &= \underline{H}(x,y)\exp(-\gamma z) \\ \text{Curl } \underline{E}(\underline{r}) &= -j\omega\mu(x,y)\underline{H}(\underline{r}) \end{aligned} \tag{11.1c}$$

$$\begin{aligned} \text{Curl } \underline{H}(\underline{r}) &= (\sigma + j\omega\epsilon_1)(x,y)\underline{E}(\underline{r}) \\ &= j\omega\epsilon_1(x,y)\left(1-\frac{j\sigma(x,y)}{\omega\epsilon_1(x,y)}\right)\underline{E}(\underline{r}) = j\omega\epsilon(x,y)\underline{E}(\underline{r}) \end{aligned} \tag{11.1d}$$

We define

$$\epsilon(x,y) = \epsilon_1(x,y)\left(1-\frac{j\sigma(x,y)}{\omega\epsilon_1(x,y)}\right) = \epsilon_0(1+\delta\cdot\chi_e(x,y))$$

Note that χ_e and χ_m are the frequency dependent complex functions of $(\omega,\ x,\ y)$ and by substituting into the Maxwell equations, it gives the following equations:

$$E_{z,y} + \gamma\, E_y = -j\omega\mu H_x, \tag{11.3a}$$

$$-\gamma E_x - E_{z,x} = -j\omega\mu H_y, \tag{11.3b}$$

$$E_{y,x} - E_{x,y} = -j\omega\mu H_z, \tag{11.3c}$$

$$H_{z,y} + \gamma H_y = j\omega\epsilon E_x, \tag{11.4a}$$

$$\gamma H_x + H_{z,x} = -j\omega\epsilon E_y, \tag{11.4b}$$

$$H_{y,x} - H_{x,y} = j\omega\epsilon E_z \tag{11.4c}$$

where $E_x, E_y, E_z, H_x, H_y, H_z$ are the functions (complex) of X, Y, and frequency ω only. We can arrange the equations [i.e., (11.3a), (11.3b), (11.4a) and (11.4b)] as follows:

$$\begin{pmatrix} \gamma & j\omega\mu \\ j\omega\epsilon & \gamma \end{pmatrix}\begin{pmatrix} E_y \\ H_x \end{pmatrix} = \begin{pmatrix} -E_{z,y} \\ -H_{z,x} \end{pmatrix},$$

$$\begin{pmatrix} j\omega\epsilon & -\gamma \\ \gamma & -j\omega\mu \end{pmatrix}\begin{pmatrix} E_x \\ H_y \end{pmatrix} = \begin{pmatrix} H_{z,y} \\ -E_{z,x} \end{pmatrix}.$$

Thus,

$$\begin{pmatrix} E_y \\ H_x \end{pmatrix} = \frac{1}{h^2}\begin{pmatrix} \gamma & -j\omega\mu \\ -j\omega\epsilon & \gamma \end{pmatrix}\begin{pmatrix} -E_{z,y} \\ -H_{z,x} \end{pmatrix},$$

$$\begin{pmatrix} E_x \\ H_y \end{pmatrix} = \frac{1}{h^2}\begin{pmatrix} -j\omega\mu & \gamma \\ -\gamma & j\omega\epsilon \end{pmatrix}\begin{pmatrix} -E_{z,y} \\ -H_{z,x} \end{pmatrix},$$

$$\begin{aligned} h^2 = h^2(x,y) &= \gamma^2 + \omega^2\epsilon\mu \\ &= \gamma^2 + \omega^2\epsilon_0\mu_0(1+\delta\chi_e)(1+\delta\chi_m) \\ &= h_0^2 + k^2\delta(\chi_e+\chi_m) + k^2\delta^2\chi_e\chi_m \end{aligned}$$

where

$$h_0^2 = \gamma^2 + \omega^2\epsilon_0\mu_0 = \gamma^2 + k^2, \quad k^2 = \omega^2\epsilon_0\mu_0$$

Then,

$$\begin{aligned} E_x &= -\frac{1}{h^2}\left(\gamma E_{z,x} + j\omega\mu H_{z,y}\right), \\ E_y &= \frac{1}{h^2}\left(-\gamma E_{z,y} + j\omega\mu H_{z,x}\right), \\ H_x &= \frac{1}{h^2}\left(-\gamma H_{z,x} + j\omega\epsilon E_{z,y}\right), \\ H_y &= -\frac{1}{h^2}\left(\gamma H_{z,y} + j\omega\epsilon E_{z,x}\right) \end{aligned}$$

Equations (11.3c) and (11.4c) then give the following equation:

$$-\gamma\left(\frac{E_{z,y}}{h^2}\right)_{,X} + j\omega\left(\frac{\mu H_{z,x}}{h^2}\right)_{,X} + \gamma\left(\frac{E_{z,x}}{h^2}\right)_{,Y} + j\omega\left(\frac{\mu H_{z,y}}{h^2}\right)_{,Y} + j\omega\mu H_z = 0 \quad (11.5a)$$

$$-\gamma\left(\frac{H_{z,y}}{h^2}\right)_{,X} + j\omega\left(\frac{\epsilon E_{z,x}}{h^2}\right)_{,X} + \gamma\left(\frac{H_{z,x}}{h^2}\right)_{,Y} - j\omega\left(\frac{\epsilon E_{z,y}}{h^2}\right)_{,Y} - j\omega\epsilon E_z = 0 \quad (11.5b)$$

These can be expended as:

$$\begin{aligned} &\frac{j\omega\mu}{h^2}\Delta H_z + j\omega\mu H_z + j\omega\left\{\left(\frac{\mu}{h^2}\right)_{,X} H_{z,x} + \left(\frac{\mu}{h^2}\right)_{,Y} H_{z,y}\right\} \\ &\quad + \gamma\left\{\left(\frac{1}{h^2}\right)_{,Y} E_{z,x} - \left(\frac{1}{h^2}\right)_{,X} E_{z,y}\right\} = 0 \end{aligned} \quad (11.6a)$$

and,

$$\begin{aligned}&\frac{j\omega\epsilon}{h^2}\Delta E_z + j\omega\epsilon E_z + j\omega\left\{\left(\frac{\epsilon}{h^2}\right)_{,X} E_{z,x} + \left(\frac{\epsilon}{h^2}\right)_{,Y} E_{z,y}\right\}\\ &\quad + \gamma\left\{\left(\frac{1}{h^2}\right)_{,X} H_{z,y} - \left(\frac{1}{h^2}\right)_{,Y} H_{z,x}\right\} = 0\end{aligned} \tag{11.6b}$$

or equivalently,

$$\begin{aligned}&(\Delta + h^2)E_z + \frac{h^2}{\epsilon}\left\{\left(\frac{\epsilon}{h^2}\right)_{,X} E_{z,x} + \left(\frac{\epsilon}{h^2}\right)_{,Y} E_{z,y}\right\}\\ &\quad + \frac{\gamma h^2}{j\omega\epsilon}\left\{\left(\frac{1}{h^2}\right)_{,X} H_{z,y} - \left(\frac{1}{h^2}\right)_{,Y} H_{z,x}\right\} = 0\end{aligned} \tag{11.7a}$$

$$\begin{aligned}&(\Delta + h^2)H_z + \frac{h^2}{\mu}\left\{\left(\frac{\mu}{h^2}\right)_{,X} H_{z,x} + \left(\frac{\mu}{h^2}\right)_{,Y} H_{z,y}\right\}\\ &\quad + \frac{\gamma h^2}{j\omega\mu}\left\{\left(\frac{1}{h^2}\right)_{,Y} E_{z,x} - \left(\frac{1}{h^2}\right)_{,X} E_{z,y}\right\} = 0\end{aligned} \tag{11.7b}$$

or in matrix notation with

$$\chi_e + \chi_m = \chi_1(x,y), \quad \chi_e\chi_m = \chi_2(x,y) \tag{11.7c}$$

this can be expended as:

$$\begin{aligned}&(\Delta + h^2)\begin{pmatrix}E_z\\ H_z\end{pmatrix} + k^2\delta\begin{pmatrix}\chi_1 & 0\\ 0 & \chi_1\end{pmatrix}\begin{pmatrix}E_z\\ H_z\end{pmatrix} + k^2\delta^2\begin{pmatrix}\chi_2 & 0\\ 0 & \chi_2\end{pmatrix}\begin{pmatrix}E_z\\ H_z\end{pmatrix}\\ &+ k^2\delta^2\begin{pmatrix}\chi_2 & 0\\ 0 & \chi_2\end{pmatrix}\begin{pmatrix}E_z\\ H_z\end{pmatrix}(\log\left(\frac{\epsilon}{h^2}\right), x\frac{\partial}{\partial x})\\ &+ (\log\left(\frac{\epsilon}{h^2}\right)_{,y}\frac{\partial}{\partial y}, \frac{-\gamma}{j\omega\epsilon}(\log h^2), x\frac{\partial}{\partial y}) + \frac{\gamma}{j\omega\epsilon}(\log h^2)_{,y}\frac{\partial}{\partial x}\frac{-\gamma}{j\omega\mu}(\log h^2)_{,y}\frac{\partial}{\partial x}\\ &+ (\log h^2)_{,x}\frac{\partial}{\partial y}, (\log\left(\frac{\mu}{h^2}\right)_{,x}\frac{\partial}{\partial x} + (\log\left(\frac{\mu}{h^2}\right)_{,y}\frac{\partial}{\partial y}\begin{pmatrix}E_z\\ H_z\end{pmatrix} = 0\end{aligned}$$

Note that $h^2(x,y) = h_0^2 + k^2\delta_{\chi_1}(x,y) + k^2\delta^2\chi_2(x,y)$.

We write $\gamma = \sqrt{h_0^2 - h^2}$ in (11.8) and then (11.8) can be expanded as:

$$(\Delta + h^2)\begin{pmatrix}E_z\\ H_z\end{pmatrix} + k^2\delta\begin{pmatrix}\chi_1 & 0\\ 0 & \chi_1\end{pmatrix}\begin{pmatrix}E_z\\ H_z\end{pmatrix} + k^2\delta^2\begin{pmatrix}\chi_2 & 0\\ 0 & \chi_2\end{pmatrix}\begin{pmatrix}E_z\\ H_z\end{pmatrix}$$

$$\begin{pmatrix} f_{11}(x,y,\lambda,\delta)\frac{\partial}{\partial x}+g_{11}(x,y,\lambda,\delta)\frac{\partial}{\partial y}, & f_{12}(x,y,\lambda,\delta)\frac{\partial}{\partial x}+g_{12}(x,y,\lambda,\delta)\frac{\partial}{\partial y} \\ f_{21}(x,y,\lambda,\delta)\frac{\partial}{\partial x}+g_{21}(x,y,\lambda,\delta)\frac{\partial}{\partial y}, & f_{22}(x,y,\lambda,\delta)\frac{\partial}{\partial x}+g_{22}(x,y,\lambda,\delta)\frac{\partial}{\partial y} \end{pmatrix}\begin{pmatrix} E_z \\ H_z \end{pmatrix}=0$$

where $\lambda = -h_0^2$,

$$f_{11}(x,y,\lambda,\delta) = (\log\left(\frac{\epsilon}{h^2}\right)_{,x}\, g_{11}(x,y,\lambda,\delta)(\log\left(\frac{\epsilon}{h^2}\right)_{,y}$$

$$\begin{aligned} f_{12}(x,y,\lambda,\delta) &= \frac{\gamma}{j\omega\epsilon}(\log h^2)_{,y} \\ g_{12}(x,y,\lambda,\delta) &= \frac{-\gamma}{j\omega\epsilon}(\log h^2)_{,x} \\ f_{21}(x,y,\lambda,\delta) &= \frac{-\gamma}{j\omega\mu}(\log h^2)_{,y} \\ g_{21}(x,y,\lambda,\delta) &= \frac{\gamma}{j\omega\mu}(\log h^2)_{,x} \\ f_{22}(x,y,\lambda,\delta) &= \left(\log\frac{\mu}{h^2}\right)_{,x} \\ g_{22}(x,y,\lambda,\delta) &= \left(\log\frac{\mu}{h^2}\right)_{,y} \end{aligned}$$

We therefore define the matrix differential part as follows:

$$\mu(\lambda,\delta) = \begin{pmatrix} f_{11}\frac{\partial}{\partial x}+g_{11}\frac{\partial}{\partial y} & f_{12}\frac{\partial}{\partial x}+g_{12}\frac{\partial}{\partial y} \\ f_{21}\frac{\partial}{\partial x}+g_{21}\frac{\partial}{\partial y} & f_{22}\frac{\partial}{\partial x}+g_{22}\frac{\partial}{\partial y} \end{pmatrix}$$

where $f_{\alpha\beta}, I_{\alpha\beta}$ are the functions of x, y, λ, δ. It is early to see that for small $\delta, f_{\lambda\beta}$ and $g_{\lambda\beta}$ can be expanded in power of δ with the series state from δ.

In other words,

$$\begin{aligned} f_{\alpha\beta}(x,y,\lambda,0) &= 0, \\ g_{\alpha\beta}(x,y,\lambda,0) &= 0. \end{aligned}$$

Writing therefore

$$f_{\alpha\beta}(x,y,\lambda,\delta) = \delta f_{\alpha\beta 1}(x,y,\lambda) + \delta^2 f_{\alpha\beta 2}(x,y,\lambda) + O(\delta^3)$$

and likewise,

$$g_{\alpha\beta}(x,y,\lambda,\delta) = \delta g_{\alpha\beta 1}(x,y,\lambda) + \delta^2 g_{\alpha\beta 2}(x,y,\lambda) + O(\delta^3)$$

We have

$$\begin{aligned}\mu(\lambda,\delta) &= F(x,y,\lambda,\delta)\frac{\partial}{\partial x} + G(x,y,\lambda,\delta)\frac{\partial}{\partial y} \\ &= \delta\left(F_1(x,y,\lambda)\frac{\partial}{\partial x} + G_1(x,y,\lambda)\frac{\partial}{\partial y}\right) \\ &\quad + \delta^2\left(F_2(x,y,\lambda)\frac{\partial}{\partial x} + G_2(x,y,\lambda)\frac{\partial}{\partial y}\right) + \mathrm{O}(\delta^3)\end{aligned}$$

where

$$\begin{aligned}F(x,y,\lambda,\delta) &= \left(\left(f_{\alpha\beta}(x,y,\lambda,\delta)\right)\right)_{|1\le\alpha,\beta\le 2|} \\ G(x,y,\lambda,\delta) &= \left(\left(g_{\alpha\beta}(x,y,\lambda,\delta)\right)\right)_{|1\le\alpha,\beta\le 2|}\end{aligned}$$

$$\begin{aligned}F_1 &= \left(\left(f_{\alpha\beta 1}\right)\right), \quad G_1 = \left(\left(f_{\alpha\beta 1}\right)\right), \\ F_2 &= \left(\left(f_{\alpha\beta 2}\right)\right), \quad G_2 = \left(\left(f_{\alpha\beta 2}\right)\right),\end{aligned}$$

Note that

$$\begin{aligned}F &= \delta F_1 + \delta^2 F_2 + \mathrm{O}(\delta^3) \\ G &= \delta G_1 + \delta^2 G_2 + \mathrm{O}(\delta^3)\end{aligned}$$

Thus,

$$\mu(\lambda,\delta) = \delta\mu_1(\lambda) + \delta^2\mu_2(\lambda) + \mathrm{O}(\delta^3)$$

where

$$\begin{aligned}\mu_1(\lambda) &= F_1(x,y,\lambda)\frac{\partial}{\partial x} + G_1(x,y,\lambda)\frac{\partial}{\partial y} \\ \mu_2(\lambda) &= F_2(x,y,\lambda)\frac{\partial}{\partial x} + G_2(x,y,\lambda)\frac{\partial}{\partial y}\end{aligned}$$

We define

$$\begin{aligned}v_1(\lambda) &= \mu_1(\lambda) + k^2\chi_1(x,y)I_2, \\ v_2(\lambda) &= \mu_2(\lambda) + k^2\chi_2(x,y)I_2,\end{aligned}$$

then

$$(\Delta - \lambda)\psi(x, y) + \left(\delta v_1(\lambda) + \delta^2 v_2(\lambda)\right)\psi(x, y) = 0$$

With neglecting higher order terms of $\mathrm{O}\left(\delta^3\right)$, where

$$\psi(x, y) = \begin{pmatrix} E_z(x, y) \\ H_z(x, y) \end{pmatrix}$$

The boundary conditions are $E_z = 0$; where $x = 0, a$ or $y = 0, b$, $H_x = 0$; where $x = 0, a$, $H_y = 0$; where $y = 0, b$. Using the expression for $H_x H_y$ in terms of E_z, H_z and the boundary conditions on E_z, it follows that the boundary conditions on H can be replaced by $H_{z,x} = 0$, where $x = 0, a$ and $H_{z,y} = 0$, where $y = 0, b$.

Where $\delta = 0$, (the homogeneous case), ψ satisfies $(\Delta - \lambda)\psi = 0$.

We denote the solution to this by $\psi_{mn}^{(0)}(x, y)$ since application of the boundary conditions leads to (after separation of variables)

$$\psi(x, y) = \begin{bmatrix} A(m, n)u_{mn}(x, y) \\ B(m, n)v_{mn}(x, y) \end{bmatrix}$$

where $A(m, n), B(m, n)$ are the complex constants and

$$\begin{aligned} u_{mn}(x, y) &= \frac{2}{\sqrt{ab}} \sin\left(\frac{m\pi x}{a}\right) \sin\left(\frac{n\pi y}{b}\right) \\ v_{mn}(x, y) &= \frac{2}{\sqrt{ab}} \cos\left(\frac{m\pi x}{a}\right) \cos\left(\frac{n\pi y}{b}\right) \end{aligned}$$

Note that,

$$\begin{aligned} \langle u_{mn}, u_{kl} \rangle &= \int_0^a \int_0^b u_{mn}(x, y) u_{kl}(x, y) \mathrm{d}x\mathrm{d}y \\ &= \delta_{mk}\delta_{nl} \\ \langle u_{mn}, u_{kl} \rangle &= \delta_{mk}\delta_{nl} \\ \langle u_{mn}, u_{kl} \rangle &= 0 \end{aligned}$$

The general solution to the $\delta = 0$ case (homogeneous medium) is

$$(\Delta - \lambda)\psi^{(0)} = 0, \quad \lambda = -h_0^2$$

With boundary condition, the equation can be written as:

$$\psi^{(0)}(x,y) = \psi^{(0)}_{(x,y)}(x,y) = \begin{bmatrix} A(m,n)u_{mn}(x,y) \\ B(m,n)v_{mn}(x,y) \end{bmatrix}$$

$$\lambda = \lambda^{(0)}_{mn} = -h^2[m,n] = -\left(\left(\frac{m\pi}{a}\right)^2 + \left(\frac{n\pi}{b}\right)^2\right)$$

where m, $n = 1, 2, \ldots$. The corresponding z-components of the electromagnetic field are given as follows:

$$\begin{pmatrix} E_z(t,x,y,z) \\ H_z(t,x,y,z) \end{pmatrix} = \sum_{m,n=1}^{\infty} \begin{bmatrix} A(m,n)u_{mn}(x,y) \\ B(m,n)v_{mn}(x,y) \end{bmatrix} \exp(-\gamma_{mn})z$$

$$= \sum_{m,n=1}^{\infty} \left[\psi^{(0)}_{m,n}(x,y)\right] \exp(-\gamma_{m,n}z)$$

The active z-component of electromagnetic fields in time at frequency ω are given as follows:

$$\begin{pmatrix} E_z(t,x,y,z) \\ H_z(t,x,y,z) \end{pmatrix} = \mathrm{Re}\left\{\sum_{mn} \psi^{(0)}_{mn}(x,y) \exp\{j\omega t - \gamma_{mn}z\}\right\}$$

the x and y components of the electromagnetic fields are easily obtained for $\delta = 0$.
Fields are easily obtained for $\delta = 0$:

$$E_x(x,y,z) = -\sum_{m,n}\left\{\frac{1}{h_0^2[m,n]}\left(\gamma_{mn}A(m,n)u_{m,n},x(x,y)\right)\right.$$
$$\left.+j\omega\mu_0 B(m,n)v_{m,n}X(x,y)\exp(-\gamma_{m,n}z)\right\},$$

$$E_y(x,y,z) = \sum_{m,n}\left\{\frac{1}{h_0^2[m,n]}\left(-\gamma_{mn}A(m,n)u_{m,n},x(x,y)\right)\right.$$
$$\left.+j\omega\mu_0 B(m,n)v_{m,n}Y(x,y)\exp(-\gamma_{m,n}z)\right\},$$

$$H_x(x,y,z) = \sum_{m,n}\left\{\frac{1}{h_0^2[m,n]}\left(j\omega\epsilon_0 A(m,n)u_{m,n},x(x,y)\right)\right.$$
$$\left.-\gamma_{mn}B(m,n)v_{m,n}X(x,y)\exp(-\gamma_{m,n}z)\right\},$$

$$H_y(x,y,z) = -\sum_{m,n}\left\{\frac{1}{h_0^2[m,n]}\left(j\omega\epsilon_0 A(m,n)u_{m,n},x(x,y)\right)\right.$$
$$\left.+\gamma_{mn}B(m,n)v_{m,n}Y(x,y)\exp(-\gamma_{m,n}z)\right\},$$

where

$$\gamma_{mn} = \sqrt{h_0^2[m,n] - k^2}$$
$$\gamma_{mn} = \sqrt{\lambda_{mn}^{(0)} - k^2}$$

Consider now $\delta > 0$ and δ is small. Then, we have to solve the following equation:

$$(\Delta - \lambda)\psi + \delta V_1(\lambda) + \delta^2 V_2(\lambda_1)\psi = 0$$

Let

$$\lambda = \lambda^{(0)} + \delta\lambda^{(1)} + \delta^2\lambda^{(2)} + \mathrm{O}(\delta^3),$$
$$\psi = \psi^{(0)} + \delta\lambda^{(1)} + \delta^2\lambda^{(2)} + \mathrm{O}(\delta^3)$$

Then, we get on equality coefficient of $\delta^0, \delta, \delta^2$ separately

$$\mathrm{O}(\delta^0) \cong \left(\Delta - \lambda^{(0)}\right)\psi^{(0)} = 0,$$
$$\mathrm{O}(\delta^1) \cong \left(\Delta - \lambda^{(0)}\right)\psi^{(1)} + V_1\left(\lambda^{(0)}\right)\psi^{(0)} - \lambda^{(1)}\psi^{(0)}$$
$$\mathrm{O}(\delta^2) \cong \left(\Delta - \lambda^{(0)}\right)\psi^{(2)} + V_1\left(\lambda^{(0)}\right)\psi^{(1)}$$
$$+ V_1\left(\lambda^{(0)}\right)\psi^{(0)} + \lambda^{(1)}V_1'\left(\lambda^{(0)}\right)\psi^{(0)} - \lambda^{(1)}\psi^{(1)} - \lambda^{(2)}\psi^{(0)} = 0$$

The solution to $\mathrm{O}\left(\delta^0\right)$ case has already been obtained:
For each m and n, we have two orthogonal solutions:

$$\psi_{mn}^{(0)1} = \begin{pmatrix} u_{mn}(x,y) \\ 0 \end{pmatrix} \quad \text{and} \quad \psi_{mn}^{(0)2} = \begin{pmatrix} 0 \\ v_{mn}(x,y) \end{pmatrix}.$$

These solutions are normalized as follows:

$$\int \left\|\psi_{mn}^{(0)K}\right\|^2 \mathrm{d}x\mathrm{d}y = 1, \quad k = 1,2 \quad \text{where} \quad \|\xi\|^2 = \xi^{\mathrm{T}}\xi,$$
$$\xi \in \mathbb{D}^2$$
$$0<x<a, \quad 0<y<b,$$

Both solutions have the same eigenvalues

$$\psi_{mn}^{(0)} = -\left(\left(\frac{m\pi}{a}\right)^2 + \left(\frac{n\pi}{b}\right)^2\right)$$

We can set $\lambda^{(0)} = \psi_{mn}^{(0)}$ in (α_z) and also

$$\begin{aligned}\psi_{mn}^{(0)} &= A\begin{pmatrix} u_{mn} \\ 0 \end{pmatrix} + B\begin{pmatrix} 0 \\ v_{mn} \end{pmatrix} \\ &= A\psi_{mn}^{(0)1} + B\psi_{mn}^{(0)2}\end{aligned}$$

To get

$$\left(\Delta - \lambda_{mn}^{(0)}\right)\psi^{(1)} - \lambda^{(1)}\psi_{mn}^{(0)} + V_1\left(\lambda_{mn}^{(0)}\right)\psi_{mn}^{(0)} = 0$$

Taking the inner product with both $\psi_{mn}^{(0)k}, k = 1, 2$ gives the following equation:

$$\begin{aligned}-\lambda^{(1)}A + \left\langle \psi_{mn}^{(0)1}, V_1\left(\lambda_{mn}^{(0)}\right)\psi_{mn}^{(0)} \right\rangle &= 0, \\ -\lambda^{(1)}B + \left\langle \psi_{mn}^{(0)2}, V_1\left(\lambda_{mn}^{(0)}\right)\psi_{mn}^{(0)} \right\rangle &= 0.\end{aligned}$$

This expands as:

$$\begin{pmatrix} \left\langle \psi_{mn}^{(0)1}, V_1\left(\lambda_{mn}^{(0)}\right)\psi_{mn}^{(0)1} \right\rangle & \left\langle \psi_{mn}^{(0)1}, V_1\left(\lambda_{mn}^{(0)}\right)\psi_{mn}^{(0)2} \right\rangle \\ \left\langle \psi_{mn}^{(0)2}, V_1\left(\lambda_{mn}^{(0)}\right)\psi_{mn}^{(0)1} \right\rangle & \left\langle \psi_{mn}^{(0)2}, V_1\left(\lambda_{mn}^{(0)}\right)\psi_{mn}^{(0)2} \right\rangle \end{pmatrix}$$

$$x\begin{pmatrix} A \\ B \end{pmatrix} = \lambda^{(1)}\begin{pmatrix} A \\ B \end{pmatrix}$$

Thus, the given unperturbed of eigenvalues $\lambda_{mn}^{(0)}$ split into two eigenvalues $\lambda_{mn}^{(0)} + \delta\lambda^{(1)}$.

Where $\lambda^{(1)}$ can be any one of the two eigenvalues of the secular matrix (2×2)

$$\left(\left(\left\langle \underline{\psi}_{mn}^{(0)\alpha}, V_1\left(\lambda_{mn}^{(0)}\right)\underline{\psi}_{mn}^{(0)\beta} \right\rangle\right)\right)_{1 \le \alpha,\beta \le 2}$$

Let $\begin{pmatrix} A_\alpha(m,n) \\ B_\alpha(m,n) \end{pmatrix}, \alpha = 1, 2$ be the corresponding normalized eigenvalues: $|A_\alpha(m,n)|^2 + |B_\alpha(m,n)|^2 = 1$.

We denote the corresponding eigenvalues by $\lambda_{mn}^{(1)}(\alpha), \alpha = 1, 2$.

The principal normalized eigenfunction of Δ corresponding to the eigenvalue $\lambda_{mn}^{(0)}$ is split into $\lambda_{mn}^{(0)} + \delta\lambda_{mn}^{(1)}(\alpha)$, $\alpha = 1, 2$, and

$$\lambda_{mn}^{(0)\alpha} = \begin{bmatrix} A_\alpha(m,n)u_{mn}(x,y) \\ B_\alpha(m,n)v_{mn}(x,y) \end{bmatrix}, \quad \alpha = 1, 2,$$

respectively.

The radiation of an antenna that feeds a waveguide is controlled through a robot (dynamic or moving probe).

The antenna orientation at time t is specified by the vector

$$\hat{l}(t) = R(\phi(t), \theta(t), \psi(t))\hat{z} \ \Delta R(t)\hat{z}$$

where

$$R(\phi, \theta, \psi) = R_z(\phi), R_x(\theta), R_z(\psi)$$

i.e., (ϕ, θ, ψ) are the Euler angles. The antenna is a rigid body (like a top) that carries a current density. $J(t, r)$ in its initial configuration. So, after some time the volume current density within the antenna body is given by

$$J_b(t, r) = J\left(t, R(t)^{-1}r\right), \quad r \epsilon R(t)(B)$$

where B is the antenna body space at the time $t = 0$, we wish to control the orientation angles $(\phi(t), \theta(t), \psi(t))$, $0 \le t \le T$ so that the radiation pattern of the antenna is as close as possible to a given pattern. Let us say that the pattern is specified by the vector potential $A_d(t, r)$ in space.

Then, the vector potential produced by the rotating antenna is as follows:

$$\begin{aligned} A_d(t, r) &= \frac{\mu}{4\pi} \int\limits_{R^3} \frac{J_b\left(t - \frac{|r-r'|}{c}, r'\right)}{|r - r'|} d^3 r' \\ &= \frac{\mu}{4\pi} \int\limits_{R(t)(B)} \frac{J_b\left(t - \frac{|r-r'|}{c}, R^{-1}\left(t - \frac{|r-r'|}{c} r'\right)\right)}{|r - r'|} d^3 r' \\ &= \frac{\mu}{4\pi} \int\limits_{B} \frac{\underline{J}\left(t - \frac{|r-R(t)\xi|}{c}, R^{-1}\left(t - \frac{|r-R(t)\xi|}{c} r'\right)(R(t)\xi)\right)}{|r - R(t)\xi|} d^3 \xi \end{aligned}$$

The far-field vector potential pattern, i.e., $|r| \gg |\xi|$ is given by

$$A(t, r) = \frac{\mu}{4\pi^r} \int\limits_{B} J\left(t - \frac{r}{c} + \frac{(\hat{r}, R(t)\xi)}{c}, R^{-1}(t)\xi - \frac{r}{c}\right) d^3 \xi$$

We wish to track $A_d(t, r)$ over the space-time region $[0, T] \times \xi$ where $\xi \in R^3$ is in a region of space in the far-field zone.

We assume that r is relatively constant in ξ (it is the distance of the center of ξ from the origin).

Then,

$$\int_{[0,T]\times\Omega_0} |A(t,r) - A_d(t,r)|^2 \mathrm{d}t \sin\theta' \mathrm{d}\theta' \mathrm{d}\phi'$$

is to be minimized where Ω_0 is the solid angle subtended at the origin by ξ shifting our time or region by $\frac{r}{c}$; the quantity to be minimized is $\left(\text{taking } \frac{\mu}{4\pi} = 1\right)$

$$\int_{[0,T]\times\Omega_0} \left| rA_d(t,r) - \int_B J(t + {(R^{-1}(t)\hat{r},\xi)}/{c}, R^{-1}(t)\xi d^3\xi \right|^2 \sin\theta' \mathrm{d}\theta' \mathrm{d}\phi'$$

where

$$(r,\theta',\phi') = \underline{r}, (\theta',\phi') \equiv \hat{r}$$

$$A_d(t,r) = A_d(t,r,\theta,\phi) \quad r \approx \text{constant}$$

Note that $R^{-1}(t) = R_z(-\psi(t))R_x(-\theta(t))R_z(-\theta(t))$.

Once the optimal trajectory $\{\phi_d(t), \theta_d(t)\psi_d(t)\}$ has been determined by optimizing this highly nonlinear functional, we decide how to apply machine torques $\{(\tau_d(t), \tau_d(t)\tau_d(t))\}, 0 \le t \le T$.

The rigid body carrying current so that the time $\{(\tau_d(t), \tau_d(t)\tau_d(t)), 0 \le t \le T\}$ is as close the desired trajectory as possible.

The kinetic energy of the top is given in terms of its principal moments of inertia by (gold's term classical mechanism)

$$T = \frac{1}{2}(I_1\omega_1^2 + I_2\omega_2^2 + I_3\omega_3^2)$$

where

$$\begin{aligned}
\omega_1 &= \dot{\theta}\cos\psi + \dot{\theta}\sin\psi\sin\theta, \\
\omega_2 &= -\dot{\theta}\sin\psi + \dot{\theta}\cos\psi\sin\theta, \\
\omega_3 &= \dot{\psi} + \dot{\theta}\cos\theta.
\end{aligned}$$

For a symmetric top (body carrying current), $I_1 = I_2$.
We get,

$$T = \frac{1}{2} I_1 \left(\dot{\theta}^2 + \dot{\phi}^2 \sin^2 \theta \right) + \frac{1}{2} I_3 \left(\dot{\psi} + \dot{\phi} \cos \theta \right)^2$$

Then, the Lagrangian after taking machine torque into account is given as:

$$\mathcal{L} = T - V + \tau_\phi(t)\dot{\theta} + \tau_\theta(t)\dot{\theta} + \tau_\psi(t)\dot{\psi}$$

where $V = mgl \cos\theta, l$ being the distance between the CM of the top and the origin.
The equation of motion is given as:

$$\frac{\mathrm{d}}{\mathrm{d}t}\frac{\partial \mathcal{L}}{\partial \dot{\phi}} = \frac{\partial \mathcal{L}}{\partial \phi}, \quad \frac{\mathrm{d}}{\mathrm{d}t}\frac{\partial \mathcal{L}}{\partial \dot{\theta}} = \frac{\partial \mathcal{L}}{\partial \theta}, \quad \frac{\mathrm{d}}{\mathrm{d}t}\frac{\partial \mathcal{L}}{\partial \dot{\psi}} = \frac{\partial \mathcal{L}}{\partial \psi} \text{ give}$$

$$\frac{\mathrm{d}}{\mathrm{d}t}\left(I_1 \dot{\phi} \sin^2 \theta + I_3 \cos\theta \left(\dot{\psi} + \dot{\theta} \cos\theta \right) \right) = \tau_\psi(t),$$

$$\frac{\mathrm{d}}{\mathrm{d}t}\left(I_1 \dot{\theta} \right) = -I_3 \dot{\phi} \sin\theta (\dot{\psi} + \dot{\phi} \cos\theta) + I_1 \dot{\phi}^2 \sin\theta \cos\theta + mgl \sin\theta + \tau_\theta(t),$$

$$\frac{\mathrm{d}}{\mathrm{d}t} I_3 \left(\dot{\psi} + \dot{\theta} \cos\theta \right) = \tau_\psi(t)$$

We define

$$F_\phi(t) = \int_0^t \tau_\phi(s)\mathrm{d}s,$$

$$F_\psi(t) = \int_0^t \tau_\psi(s)\mathrm{d}s,$$

and then, the equations of motion are as follows:

$$I_1 \dot{\phi}^2 \sin^2 \theta + I_3 \cos\theta \left(\dot{\psi} + \dot{\theta} \cos\theta \right) = F_\phi(t) \tag{11.8}$$

$$I_3 \left(\dot{\psi} + \dot{\theta} \cos\theta \right) = F_\psi(t) \tag{11.9}$$

$$I_1 \ddot{\theta} + I_3 \phi \sin\theta \left(\dot{\psi} + \dot{\theta} \cos\theta \right) - I_1 \dot{\phi}^2 \sin\theta \cos\theta \, mgl \sin\theta = \tau_\phi(t) \tag{11.10}$$

The total work done by the machine torque is given as:

$$W = \int_0^T \left(\tau_\phi(t)\dot{\phi}(t) + \tau_\theta(t)\dot{\theta}(t) + \tau_\psi(t)\dot{\psi}(t)\right)\mathrm{d}t$$
$$= \int \left(\dot{F}_\phi(t)\dot{\phi}(t) + \tau_\theta(t)\dot{\theta}(t) + \dot{F}_\psi(t)\dot{\psi}(t)\right)\mathrm{d}t$$

This must be minimized subject to the constraints robotic of the equation of motion (11.8)–(11.10) and that the final orientation defined by $\phi(t), \theta(t), \psi(t)$ is given.

Here, we calculate $\{F_\phi(t), F_\psi(t), F_\theta(t)\}, 0 \le t \le T$ and hence, $\{(\tau_d(t), \tau_d(t), \tau_d(t)), 0 \le t \le T\}$ by putting

$$\phi(t) = \phi_d(t)$$
$$\theta(t) = \theta_d(t)$$
$$\psi(t) = \psi_d(t)$$

This gives us an algorithm for calculating the machine torque to be supplied over the range [0, *T*], so that the top antenna follows a desired trajectory that will lead to a radiation pattern that matches a given radiation pattern as closely as possible in a given solid angle element Ω_0 over a given time interval [0, *T*]. Another way to design the machine torque is to minimize a weighted combination of the error energy between the desired trajectories $\{\phi_d(\cdot), \theta_d(\cdot), \psi_d(\cdot)\}$ and the actual trajectory, and the total work done by the torques over the duration [0, *T*] is minimized as follows:

$$S = \alpha \int_0^T \left(\dot{F}_\phi(t)\dot{\phi}(t) + \tau_\theta(t)\dot{\theta}(t) + \dot{F}_\psi(t)\dot{\psi}(t)\right)\mathrm{d}t + \beta \int_0^T \left((\phi(t) - \phi_d(t))^2 + (\theta(t) - \theta_d(t))\right)^2$$
$$+ \left((\psi(t) - \psi_d(t))^2\right)\mathrm{d}t$$

where $F_\phi(t), \tau_\theta(t)F_\psi(t)$ are given in terms of $(\phi,\ \theta,\ \psi)$ in above equation we can write the equation as:

$$S = \int_0^T \xi\left(\phi, \dot{\phi}, \ddot{\phi}, \theta, \dot{\theta}, \ddot{\theta}, \psi, \dot{\psi}, \ddot{\psi}\right)\mathrm{d}t$$

α, β are the weights $\alpha, \beta > 0$. The optimal trajectories (ϕ, θ, ψ) then satisfy the Euler trajectories equations:

$$\frac{\mathrm{d}}{\mathrm{d}t}\frac{\partial\xi}{\partial\dot{\phi}} - \frac{\mathrm{d}^2}{\mathrm{d}t^2}\frac{\partial\xi}{\partial\ddot{\phi}} = \frac{\partial\xi}{\partial\phi},$$
$$\frac{\mathrm{d}}{\mathrm{d}t}\frac{\partial\xi}{\partial\dot{\theta}} - \frac{\mathrm{d}^2}{\mathrm{d}t^2}\frac{\partial\xi}{\partial\ddot{\theta}} = \frac{\partial\xi}{\partial\theta},$$
$$\frac{\mathrm{d}}{\mathrm{d}t}\frac{\partial\xi}{\partial\dot{\psi}} - \frac{\mathrm{d}^2}{\mathrm{d}t^2}\frac{\partial\xi}{\partial\ddot{\psi}} = \frac{\partial\xi}{\partial\psi}.$$

Now support that the electromagnetic field generated by an antenna falls on the aperture $z = 0$, $0 < x < a$, $0 < y < b$, of a rectangular waveguide.

We wish to complete the fields inside the guide. Assuming the guide to have constant permittivity and permeability, we get for the phase fields inside the guide at a given frequency ω,

$$(\Delta + h_0^2)\begin{pmatrix} E_z(x,y) \\ H_z(x,y) \end{pmatrix} = 0$$
$$h_0^2 = \gamma^2 + \omega^2\epsilon_0\mu_0 = \gamma^2 + k^2$$

We have

$$E_x = -\frac{1}{h_0^2}\left(\gamma E_{z,x} + j\omega\mu H_{z,y}\right)$$
$$E_y = \frac{1}{h_0^2}\left(-\gamma E_{z,y} + j\omega\mu H_{z,x}\right)$$
$$H_x = \frac{1}{h_0^2}\left(-\gamma H_{z,x} + j\omega\epsilon E_{z,y}\right)$$
$$H_y = -\frac{1}{h_0^2}\left(-\gamma H_{z,y} + j\omega\epsilon E_{z,x}\right)$$

The general solution for the fields within the given satisfying boundary conditions is (as we've seen earlier) given by

$$\begin{pmatrix} E_z(x,y,z) \\ H_z(x,y,z) \end{pmatrix} = \sum_{m,n=1}^{\infty}\begin{pmatrix} A(m,n)u_{mn}(x,y) \\ B(m,n)v_{mn}(x,y) \end{pmatrix} e^{-\gamma_{mn}z}$$
$$u_{mn}(x,y) = \frac{2}{\sqrt{ab}}\sin\left(\frac{m\pi x}{a}\right)\sin\left(\frac{n\pi y}{b}\right),$$
$$v_{mn}(x,y) = \frac{2}{\sqrt{ab}}\cos\left(\frac{m\pi x}{a}\right)\cos\left(\frac{n\pi y}{b}\right)$$
$$\gamma_{mn} = \gamma_{mn}(\omega) = \sqrt{h_0^2[m,n] - \omega^2\mu_0\epsilon_0}$$

where $h_0^2[m,n] = \pi^2\left(\frac{m^2}{a^2} + \frac{n^2}{b^2}\right)$.

When an electromagnetic field is incident on the surface $z = 0$ of the guide, let $(E_{0x}(x,y))$, $(E_{0y}(x,y))$, be the incident electric field (tangential components to the surface). Then by continuity of the tangential components of the electric field, we have

$$E_x(x,y,0+) = E_{0x}(x,y), \quad E_y(x,y,0+) = E_{0y}(x,y),$$
$$0 \leq x \leq a, \quad 0 \leq y \leq b,$$

where

$$E_x(x,y,0+) = \sum_{m,n} -\frac{1}{h_0^2[m,n]}(\gamma_{mn}A(m,n)u_{mn}(x,y) + j\omega\mu_0 B(m,n)v_{mn}(x,y))$$

and

$$E_y(x,y,0+) = \sum_{m,n} \frac{1}{h_0^2[m,n]}(-\gamma_{mn}A(m,n)u_{mn}(x,y) + j\omega\mu_0 B(m,n)v_{mn}(x,y))$$

Thus

$$E_{0x}(x,y) = -\sum_{m,n}\left\{\frac{\gamma_{mn}}{h_0^2[m,n]}A(m,n)\frac{2}{\sqrt{ab}}\left(\frac{m\pi}{a}\right)\cos\left(\frac{m\pi x}{a}\right)\sin\left(\frac{n\pi y}{b}\right)\right.$$
$$\left.-\frac{j\omega\mu_0}{h_0^2[m,n]}\frac{2}{\sqrt{ab}}\left(\frac{n\pi}{b}\right)B(m,n)\cos\left(\frac{m\pi x}{a}\right)\sin\left(\frac{n\pi y}{b}\right)\right\}$$

$$E_{0y}(x,y) = \frac{2}{\sqrt{ab}}\sum_{m,n}\left\{\frac{-\gamma_{mn}}{h_0^2[m,n]}A(m,n)\left(\frac{n\pi}{b}\right)\sin\left(\frac{m\pi x}{a}\right)\cos\left(\frac{n\pi y}{b}\right)\right.$$
$$\left.+\frac{j\omega\mu_0}{h_0^2[m,n]}B(m,n)\left(\frac{m\pi}{a}\right)\sin\left(\frac{m\pi x}{a}\right)\cos\left(\frac{n\pi y}{b}\right)\right\}$$

or equivalently

$$E_{0x}(x,y) = \sum_{m,n}\left(\frac{-2\gamma_{mn}m\pi}{h_0^2[m,n]\sqrt{ab}}A(m,n) + \frac{2j\omega\mu_0 n\pi}{h_0^2[m,n]b\sqrt{ab}}B(m,n)\right)\cos\left(\frac{m\pi x}{a}\right)\sin\left(\frac{n\pi y}{b}\right)$$

and

$$E_{0y}(x,y) = \sum_{m,n}\left(\frac{-2\gamma_{mn}n\pi}{h_0^2[m,n]b\sqrt{ab}}A(m,n) + \frac{2j\omega\mu_0 m\pi}{ah_0^2[m,n]\sqrt{ab}}B(m,n)\right)\sin\left(\frac{m\pi x}{a}\right)\cos\left(\frac{n\pi y}{b}\right)$$

Thus, we get

$$\frac{2}{\sqrt{ab}}\int E_{0x}(x,y)\cos\left(\frac{m\pi x}{a}\right)\sin\left(\frac{n\pi y}{b}\right)\mathrm{d}x\mathrm{d}y$$
$$0<x<a, \quad 0<y<b$$
$$= \left(\frac{-m\pi\gamma_{mn}}{ah_0^2[m,n]}A(m,n)+\frac{j\omega\mu_0 n\pi}{h_0^2[m,n]b}B(m,n)\right)$$

These are the two simultaneous linear equations for the two variables $A(m,n)$ and $B(m,n)$ which are easily solved. 2 × 2 matrix notation supports the incident electric field is $E_0(x,y,z,t)$. Then, we take

$$\hat{E}_{0x}(x,y,\omega)=\int\limits_R E_{0x}(x,y,0,t)e^{-j\omega t}\mathrm{d}t$$
$$\hat{E}_{0y}(x,y,\omega)=\int\limits_R E_{0y}(x,y,0,t)e^{-j\omega t}\mathrm{d}t$$

Define

$$\phi_x(m,n,t)=\frac{2}{\sqrt{ab}}\int\limits_D E_{0x}(x,y,0,t)\cos\left(\frac{m\pi x}{a}\right)\sin\left(\frac{n\pi y}{b}\right)\mathrm{d}x\mathrm{d}y$$
$$\phi_y(m,n,t)=\frac{2}{\sqrt{ab}}\int\limits_D E_{0y}(x,y,0,t)\sin\left(\frac{m\pi x}{a}\right)\cos\left(\frac{n\pi y}{b}\right)\mathrm{d}x\mathrm{d}y$$

Then,

$$\hat{\phi}_x(m,n,\omega)\triangleq\int\limits_R \phi_x(m,n,t)e^{-j\omega t}\mathrm{d}t$$
$$\triangleq\frac{2}{\sqrt{ab}}\int\limits_D \hat{E}_{0x}(x,y,\omega)\cos\left(\frac{m\pi x}{a}\right)\sin\left(\frac{n\pi y}{b}\right)\mathrm{d}x\mathrm{d}y$$
$$\hat{\phi}_y(m,n,\omega)\triangleq\int\limits_R \phi_y(m,n,t)e^{-j\omega t}\mathrm{d}t$$
$$\triangleq\frac{2}{\sqrt{ab}}\int\limits_D \hat{E}_{0y}(x,y,\omega)\sin\left(\frac{m\pi x}{a}\right)\cos\left(\frac{n\pi y}{b}\right)\mathrm{d}x\mathrm{d}y$$

11.3 Applications: Hybrid Modes Generation Inside RDRA Can Be Used for Polarization Diversity

11.3.1 RF Measurements for Antenna Parameters

- S11
- Gain
- Radiation pattern
- VSWR
- Dispersion
- Polarization
- Permittivity and permeability
- Bandwidth
- S21
- Isolation
- Efficiency
- Directivity
- Resonant frequency
- Propagation constants
- Axial ratio
- Resonant mode
- Dominant mode
- Higher-order modes

Chapter 12
Case Studies

Abstract This chapter deals with case studies, where implemented cases have been discussed. Various type of antennas have been fabricated. Their dominating parameters are shown. These cases have either been developed using HFSS simulations or hardcore experimentations. The case study is based on the rectangular DRA using ceramics such as eccostock-500. Nomenclature and parameters obtained have been mentioned below the each figure. Emphasis is also given on geometry of antennas and their experimental results. The experimental results have been obtained under specific environmental conditions i.e. anechoic chambers.

Keywords Rectangular DRA · Simulated and experimental results · Isolated RDRA and RDRA with ground plane · Single feed · Double feed · Anechoic chamber measurements · Radiation pattern · Gain · S11, VSWR, Z11, and *E* and *H* fields distribution · RF absorbers · Test set for measurement · Prototype · Azimuth and elevation pattern · Manganese–manganese material · Bandwidth enhancement · Higher order resonant modes · Variable DRA height · Smith chart · Group delay · Rectangular wells · LHCP · RHCP · Circular polarization · Phase distortion · S_{21} · Ferrite RDRA · Slot variation · Permittivity variation effects · Hardware implementation · VNA calibration · Aperture coupled RDRA and probe fed RDRA

12.1 Structure and Hardware Experimentations

The case study is based on the rectangular DRA where the various designs of DRA have been presented and the nomenclature is given below each figure. These antennas have been simulated and fabricated. The results obtained have been presented graphically here. These antennas also have been placed inside anechoic chamber to minimize external effects that come during measurements. The figures indicate various measurement steps involved in this process. For simulated results, Ansoft HFSS 13.0 has been used. These graphs of radiation pattern, gain, S11, VSWR, Z11, and field distribution have been presented and their domains are mentioned below each figure (Figs. 12.1, 12.2, 12.3, 12.4, 12.5, 12.6 and 12.7).

R.S. Yaduvanshi and H. Parthasarathy, *Rectangular Dielectric Resonator Antennas*, DOI 10.1007/978-81-322-2500-3_12

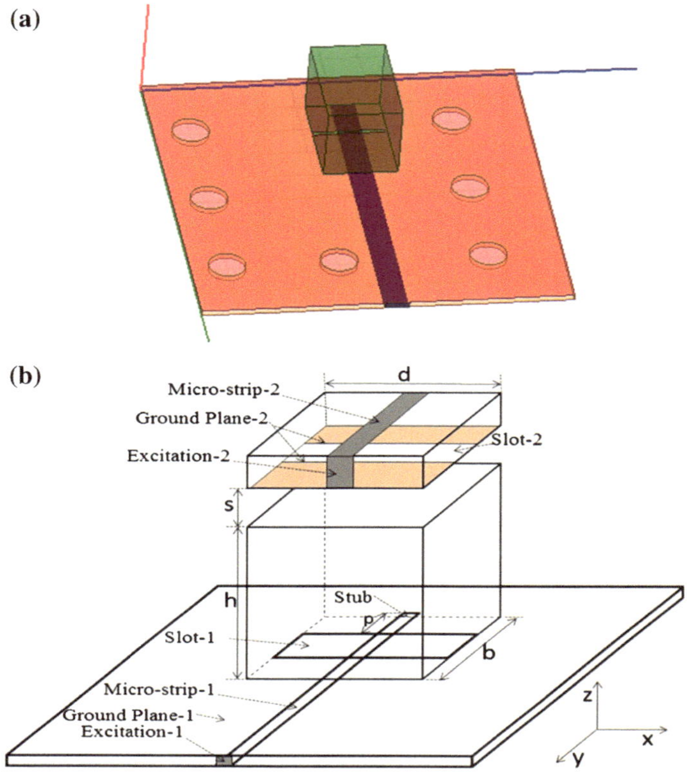

Fig. 12.1 **a** Model of electronic band gap (EBG) structure cavity rectangular dielectric resonator. **b** Diagrammatic representation of RDRA with top-loading DRA

Fig. 12.2 Positioning of RDRA antenna ready for test procedure setup

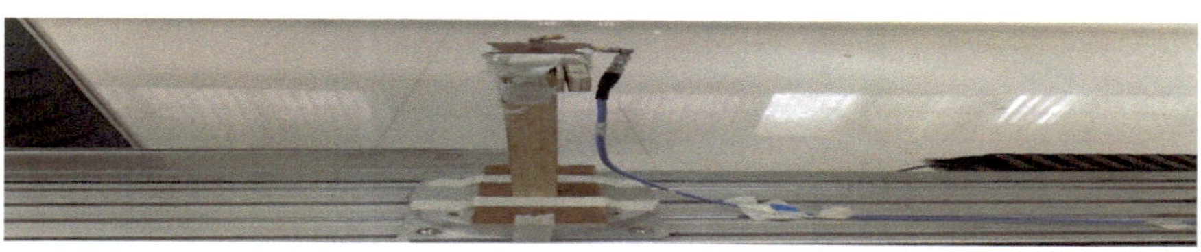

Fig. 12.3 Positioning of RDRA antenna under measurement setup

Fig. 12.4 RDRA antenna between RF absorbers inside chamber for gain testing

Fig. 12.5 View of RDRA antenna under test setup on sliding table

Fig. 12.6 View of RDRA antenna under test setup on sliding table

Fig. 12.7 RDRA antenna on sliding table with variation in position of RDRA

12.1.1 RDRA Antenna Results

See Figs. 12.8, 12.9, 12.10, 12.11, 12.12, 12.13, 12.14, 12.15, 12.16, 12.17, 12.18, 12.19, 12.20, 12.21, 12.22 and 12.23.

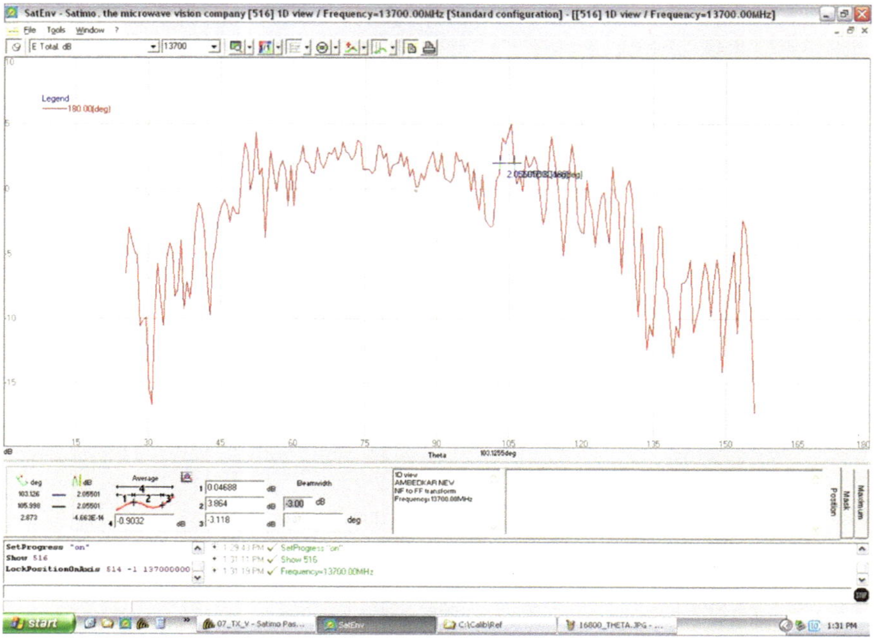

Fig. 12.8 *E*-plane radiation pattern at 13.7 GHz of simulated RDRA for electric field distribution

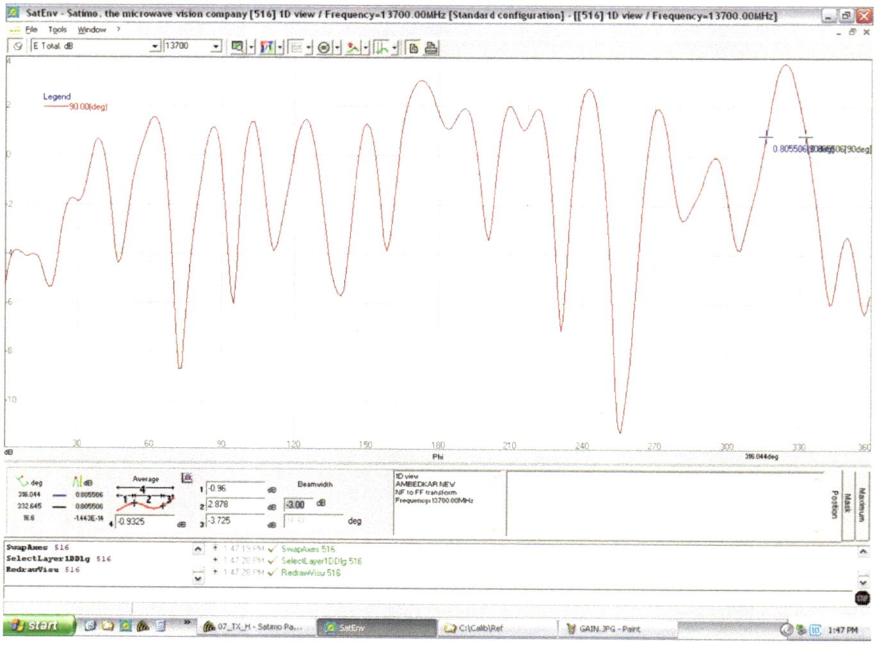

Fig. 12.9 *H*-plane radiation pattern of simulated RDRA for magnetic field distribution

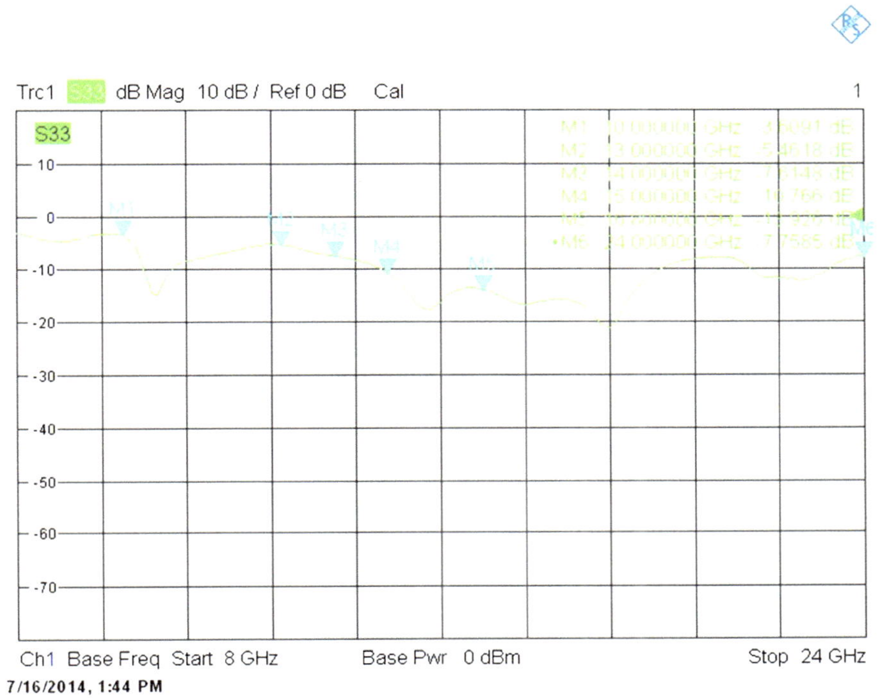

Fig. 12.10 VSWR measurement for measurement of reflected field strength of simulated RDRA

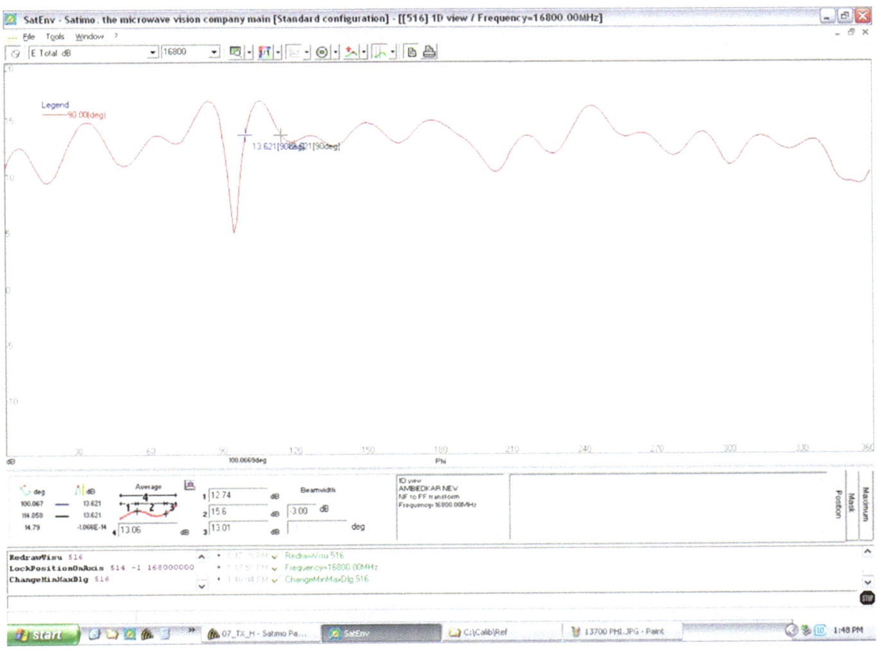

Fig. 12.11 *H*-plane radiation pattern at 16.8 GHz of simulated RDRA for magnetic field distribution

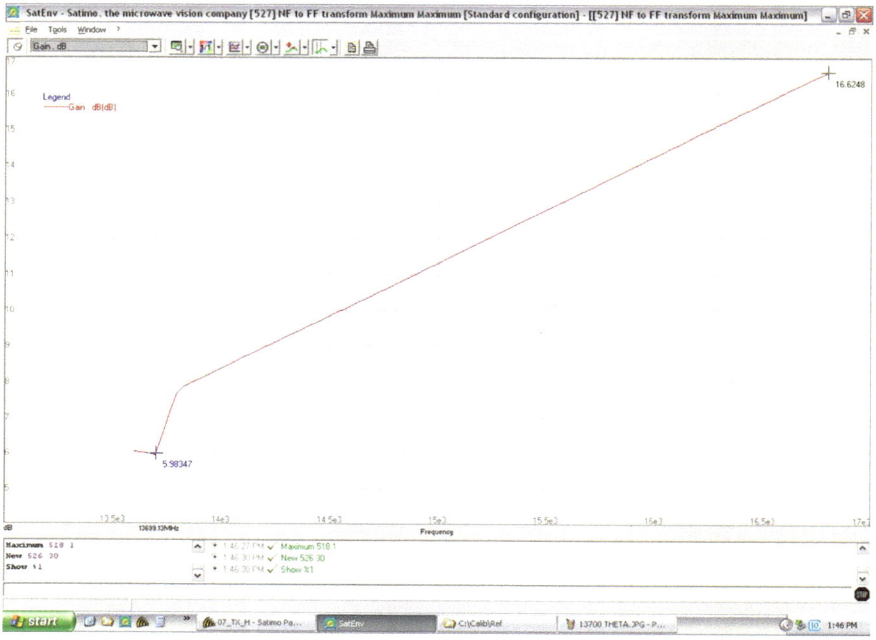

Fig. 12.12 Measurement of gain at 16.8 GHz of simulated RDRA

12.2 RDRA with Manganese–Manganese Material as Dielectric

In this case, the designing of RDRA using manganese–manganese material has been presented. This dielectric material shows the various effects on the parameters of the developed antenna. The bandwidth enhancement techniques have been implemented using two wells. The results obtained have been presented graphically here. These antennas also have been placed inside anechoic chamber to minimize external effects that come during measurements. The figures indicate various measurement steps involved in this process. These antennas have been simulated and fabricated. For simulated results, Ansoft HFSS 13.0 has been used. These graphs of radiation pattern, gain, S11, VSWR, Z11, and field distribution have been

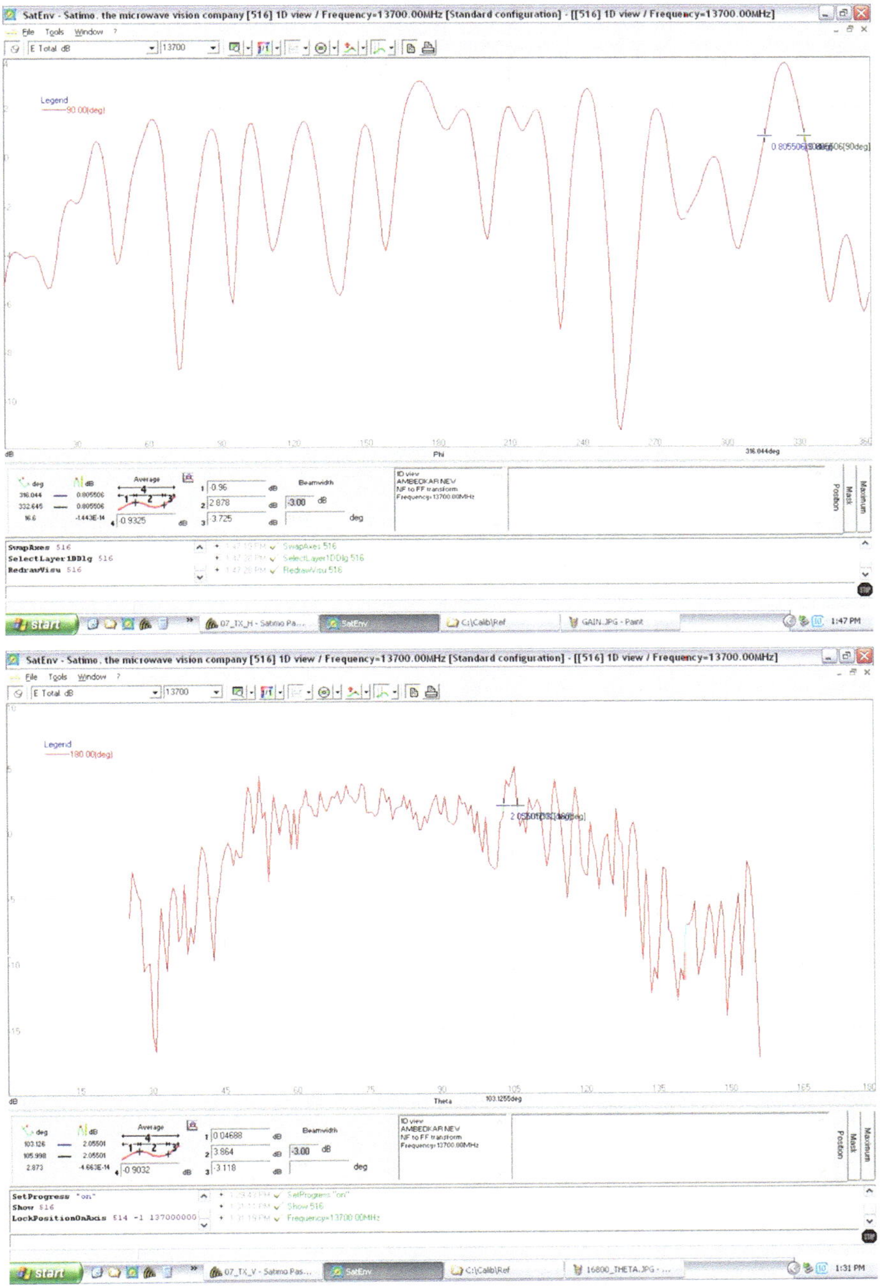

Fig. 12.13 Radiation pattern at 13.7 GHz for radiated field pattern of simulated RDRA

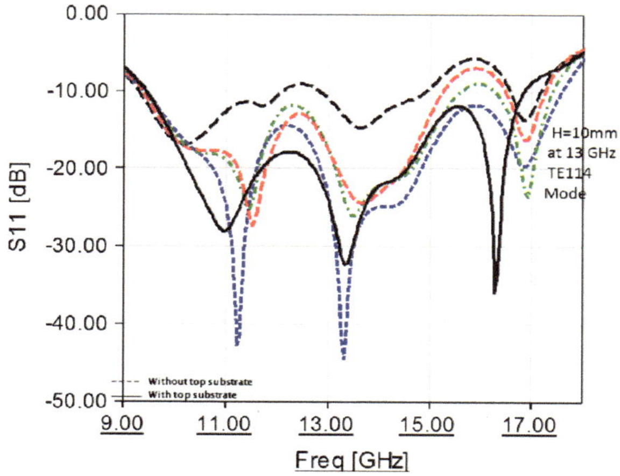

Fig. 12.14 Return loss S11 of higher-order modes excited inside simulated RDRA

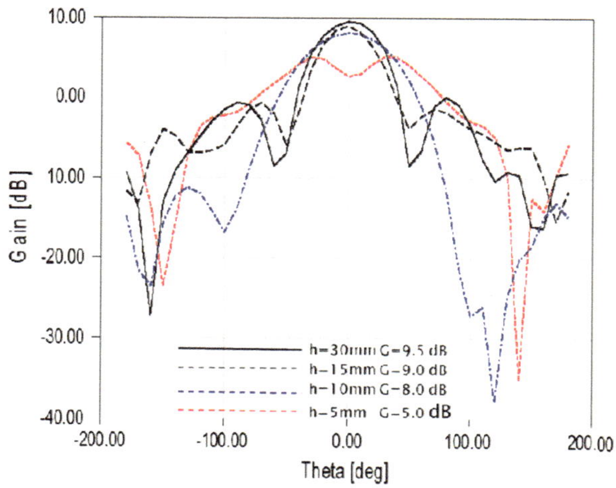

Fig. 12.15 Gain of simulated RDRA at various heights (with the excitation of higher-order modes)

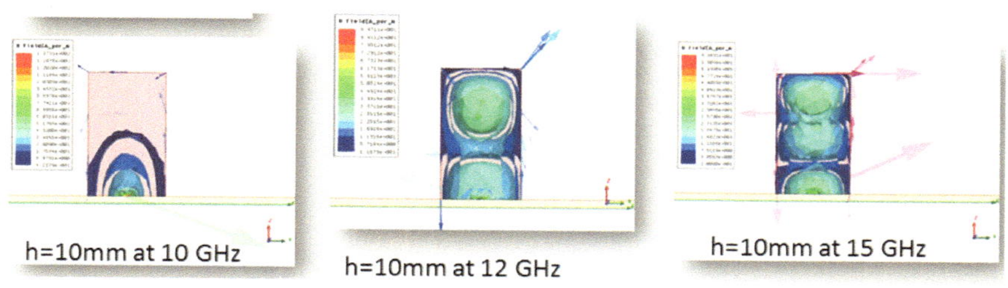

Fig. 12.16 Excitation of higher mode at RDRA 10 mm height with variable frequency of 10, 12, 15 GHz

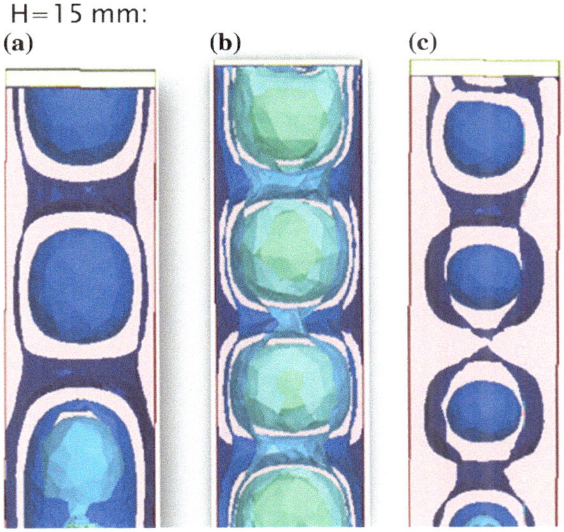

Fig. 12.17 RDRA height 15 mm, excitation frequency variable, generated higher modes. Even modes excitation with top excitation **a** TE114 at 11.7 GHz **b** TE116 at 13.7 GHz **c** TE118 at 16.7 GHz in RDRA

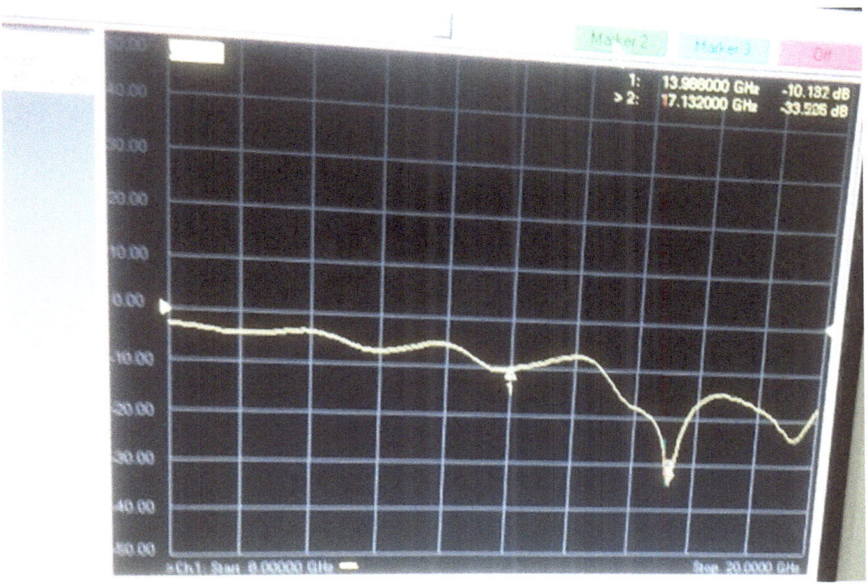

Fig. 12.18 Measured return loss S11 at −33.596 dB of RDRA

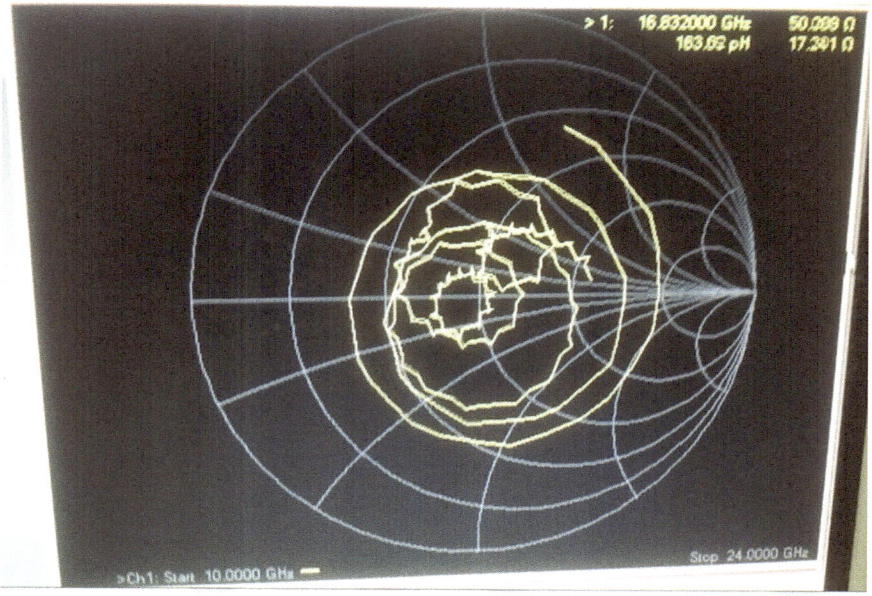

Fig. 12.19 Measured input impedance of 50.089 Ω by Smith chart of RDRA

Fig. 12.20 Prototype RDRA under test setup for measurements with VNA

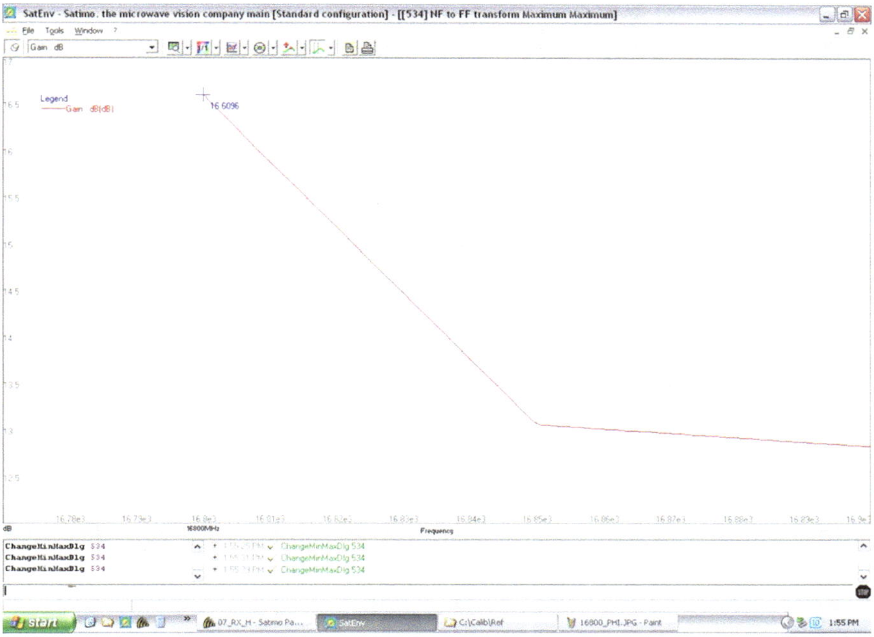

Fig. 12.21 Experimental gain of antenna of 16.624 dBi at 16.8 GHz RDRA inside microwave anechoic chamber

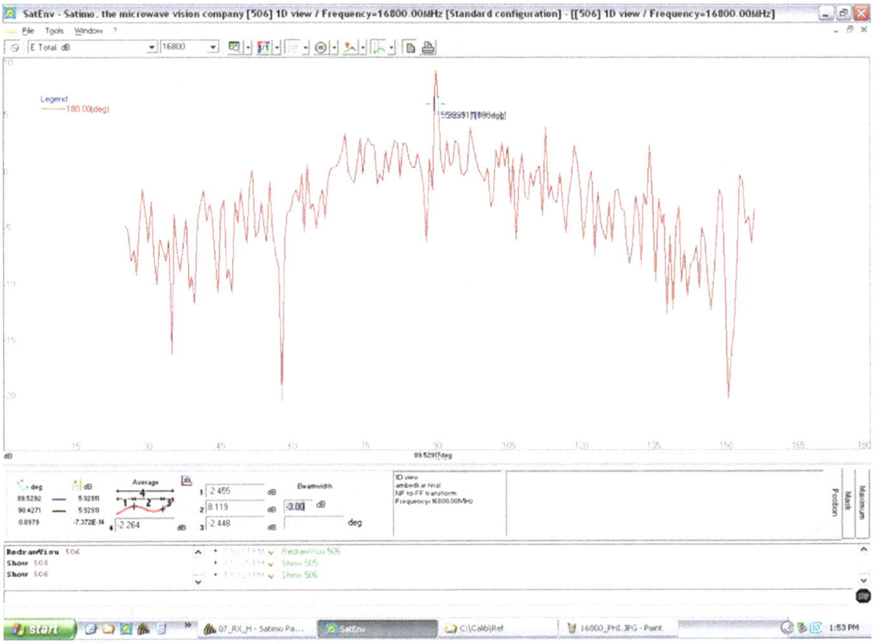

Fig. 12.22 *E*-plane radiation pattern at 16.8 GHz dBi inside microwave anechoic chamber amplitude versus theta

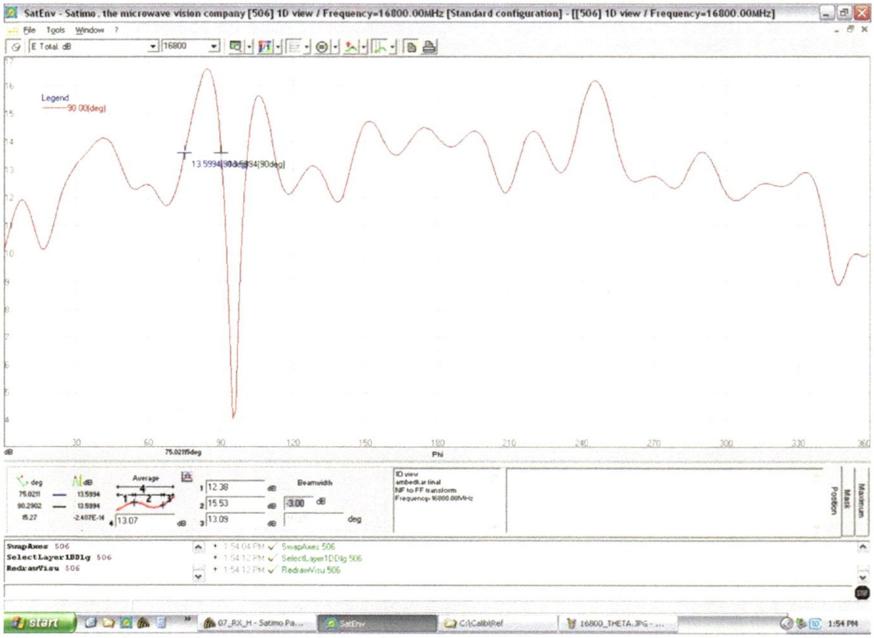

Fig. 12.23 *H*-plane radiation pattern at 16.8 GHz dBi inside microwave anechoic chamber amplitude versus theta

presented and their domains are mentioned below each figure. The phase versus frequency plots indicate distortions in the developed RDRA. The group delay, forward power has also been indicated in simulated results (Figs. 12.24, 12.25, 12.26, 12.27, 12.28, 12.29, 12.30, 12.31, 12.32, 12.33, 12.34, 12.35, 12.36, 12.37 and 12.38).

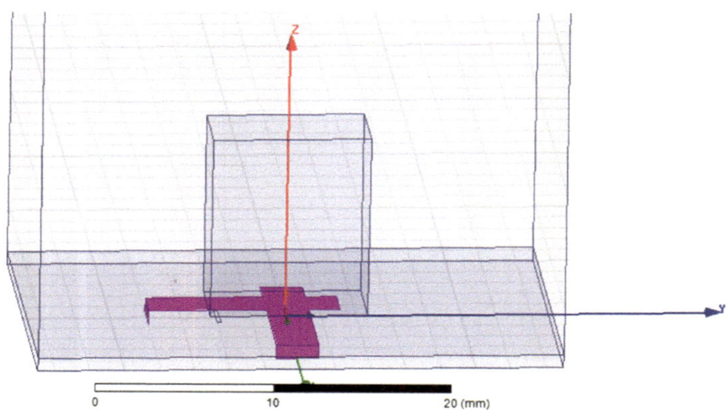

Fig. 12.24 HFSS model of dual-feed RDRA with circular polarization

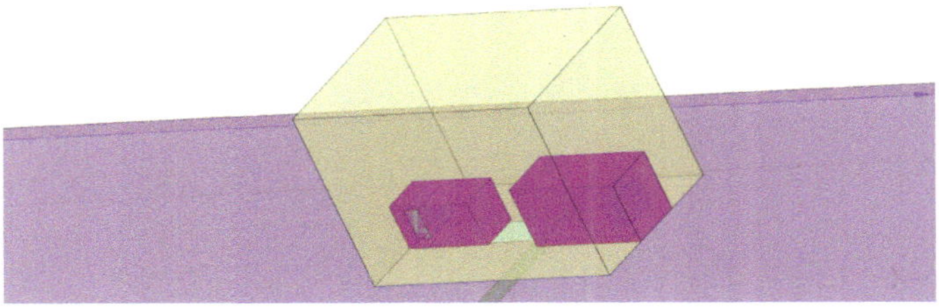

Fig. 12.25 HFSS model of DRA with two rectangular wells

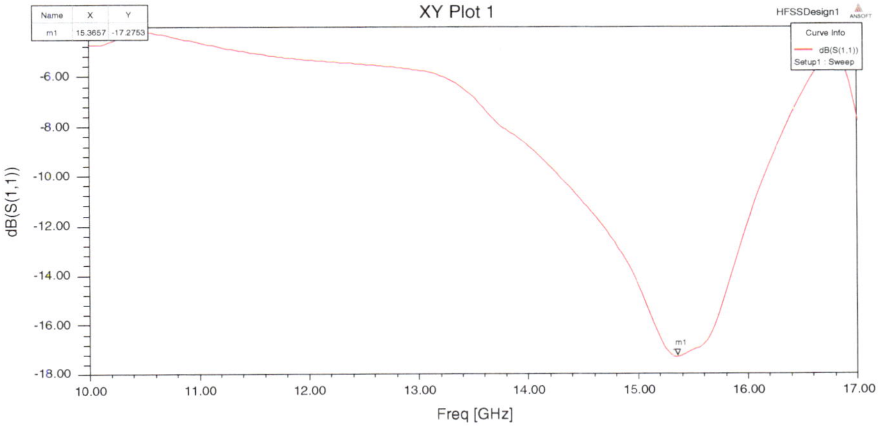

Fig. 12.26 Return loss curve S11 of simulated DRA at 15.5 GHz

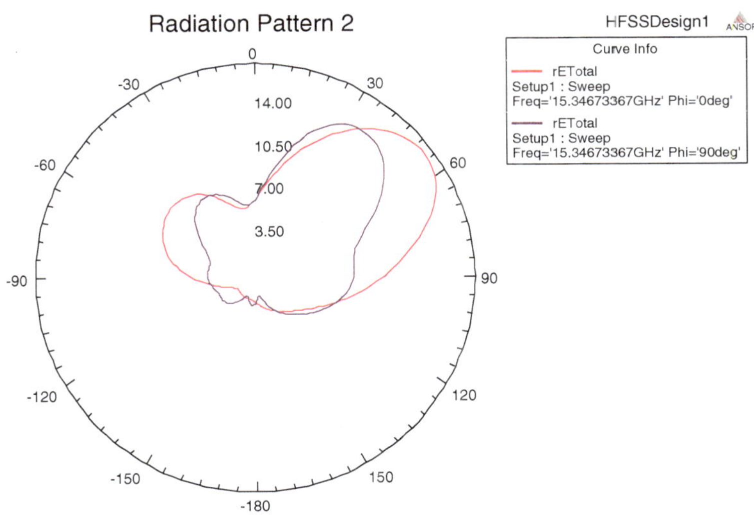

Fig. 12.27 Radiation pattern 2 of simulated DRA at 15.5 GHz

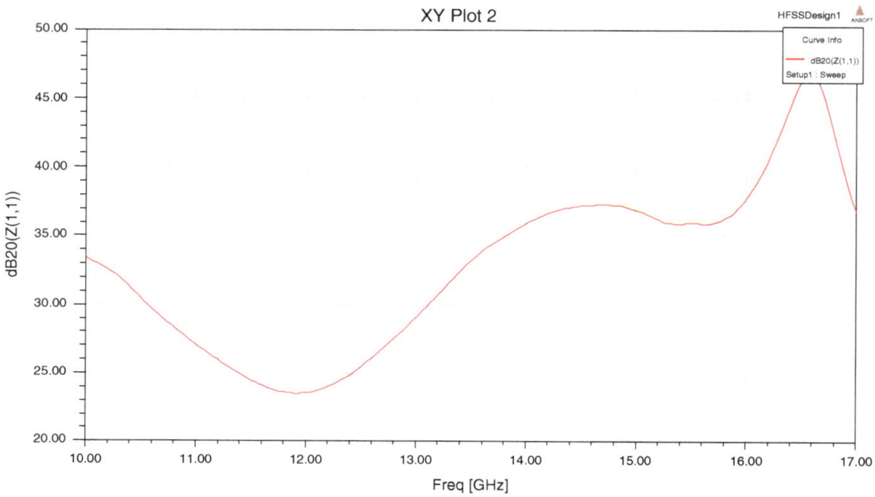

Fig. 12.28 Impedance plot Z11 of simulated DRA at 15.5 GHz

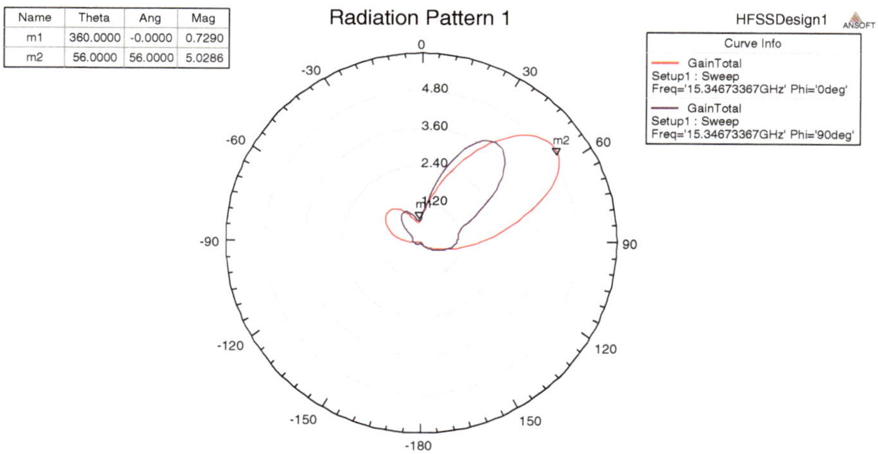

Fig. 12.29 Gain plot of 4.80 dBi of simulated DRA at 15.5 GHz

Ground plane	$20 \times 30\,\text{mm}^2$
Substrate ($\varepsilon = 2.2$)	$20 \times 30 \times 0.8\,\text{mm}^3$
DRA material ($\varepsilon = 12.2$)	12 (manganese – manganese)

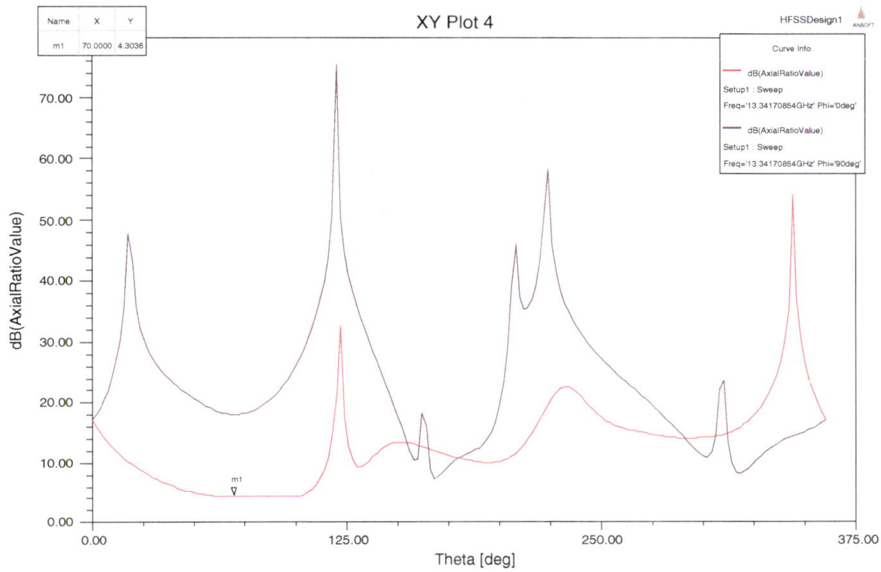

Fig. 12.30 Axial ratio magnitude for polarization inside DRA

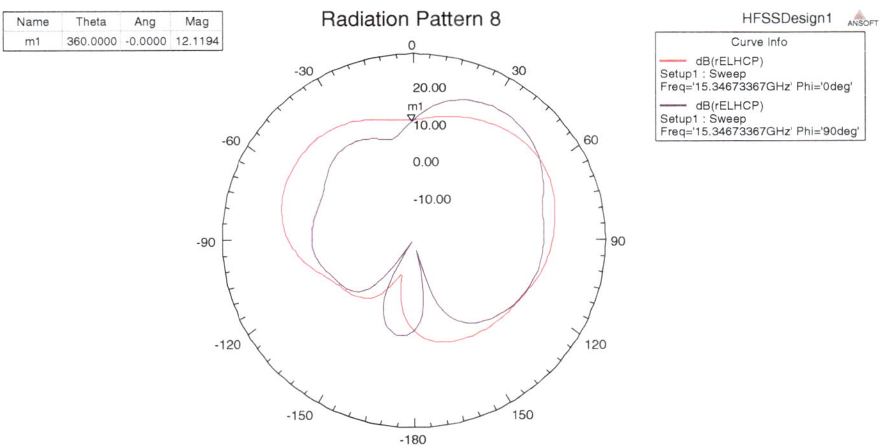

Fig. 12.31 *Left* Circular polarization radiation plot of simulated DRA

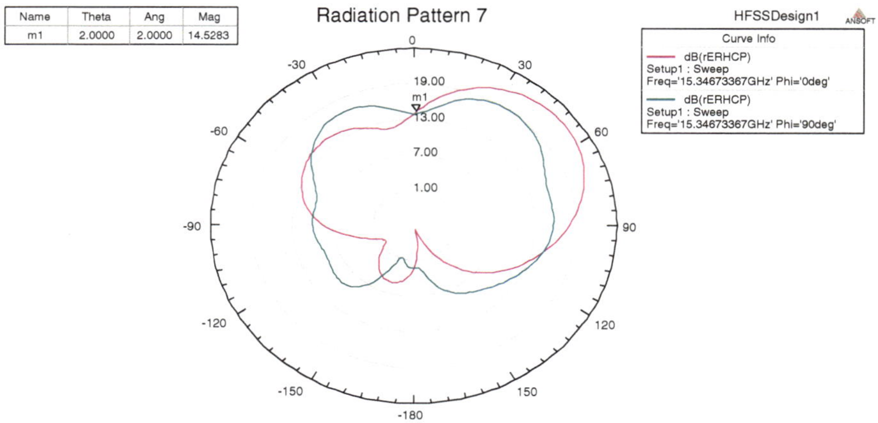

Fig. 12.32 *Right* Circular polarization of simulated DRA at 15.5 GHz

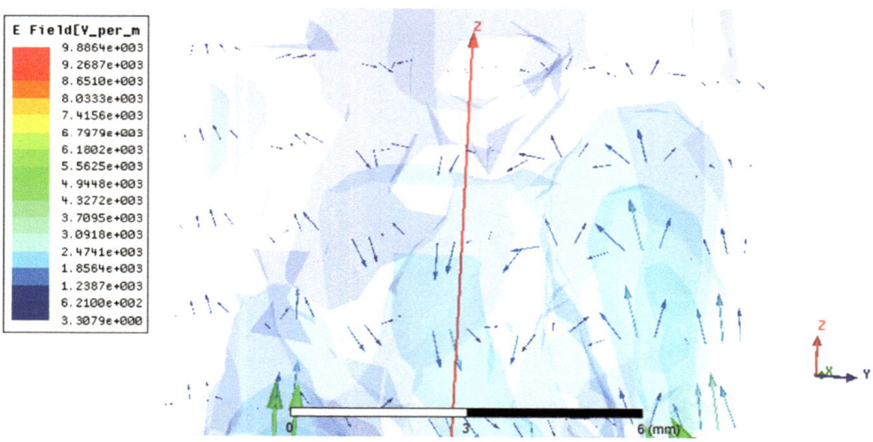

Fig. 12.33 *E* field pattern with dual feed of simulated DRA

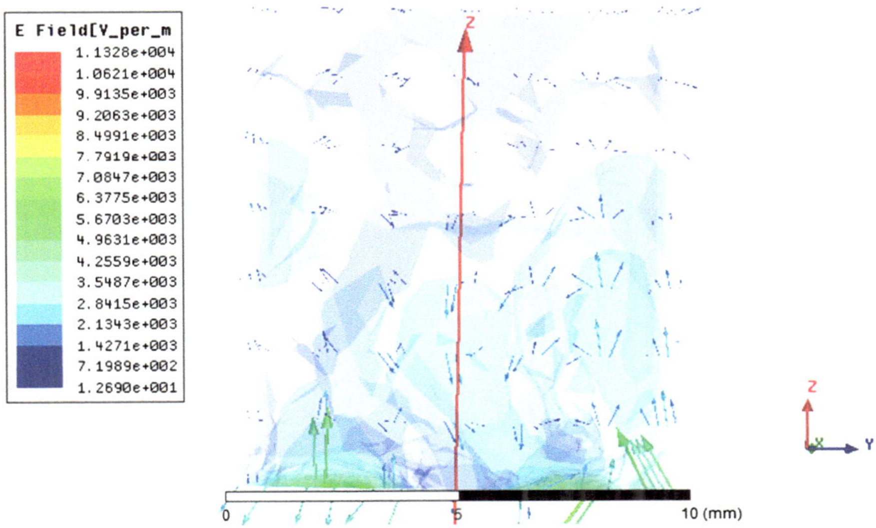

Fig. 12.34 *E* field pattern when single feed along *y*-axis of simulated DRA

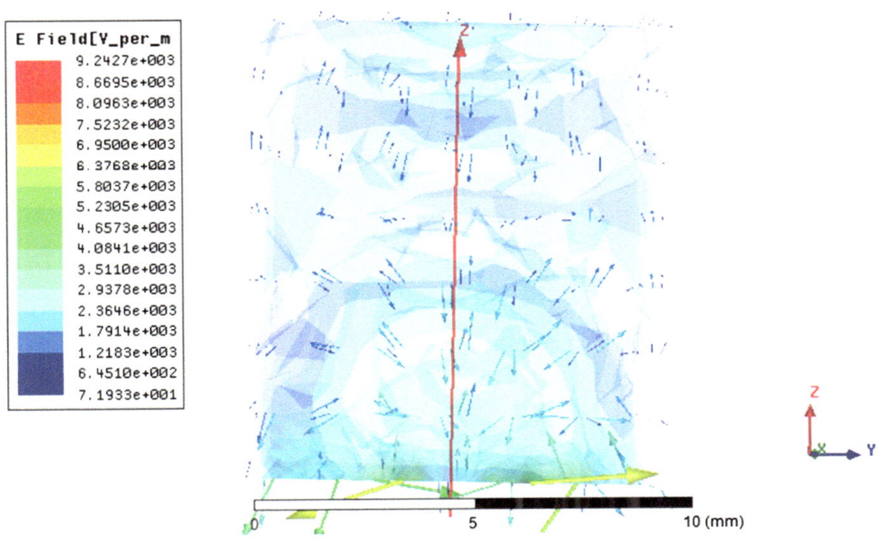

Fig. 12.35 *E* field pattern when single feed along *x*-axis applied of simulated DRA

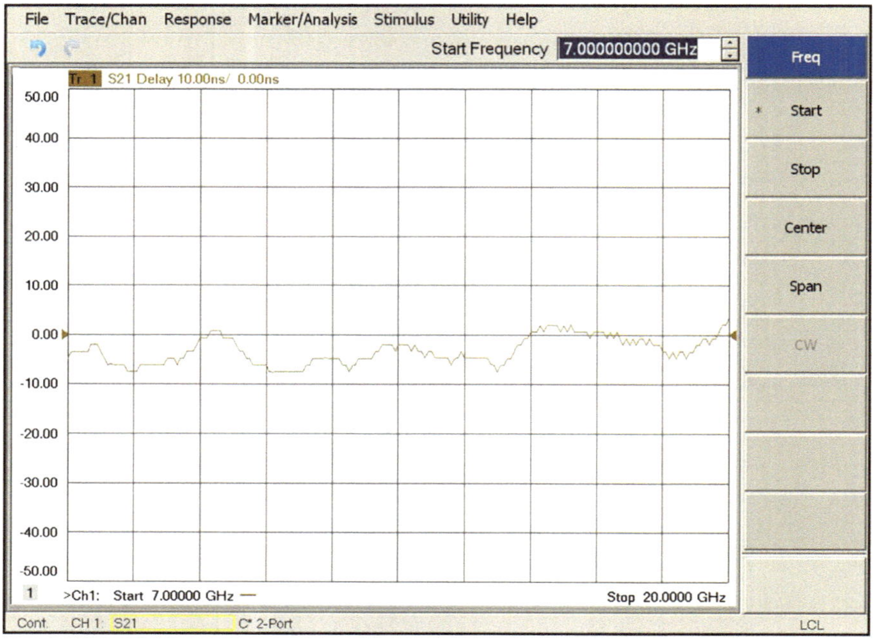

Fig. 12.36 Group delay measurements in RDRA dual feed of simulated DRA

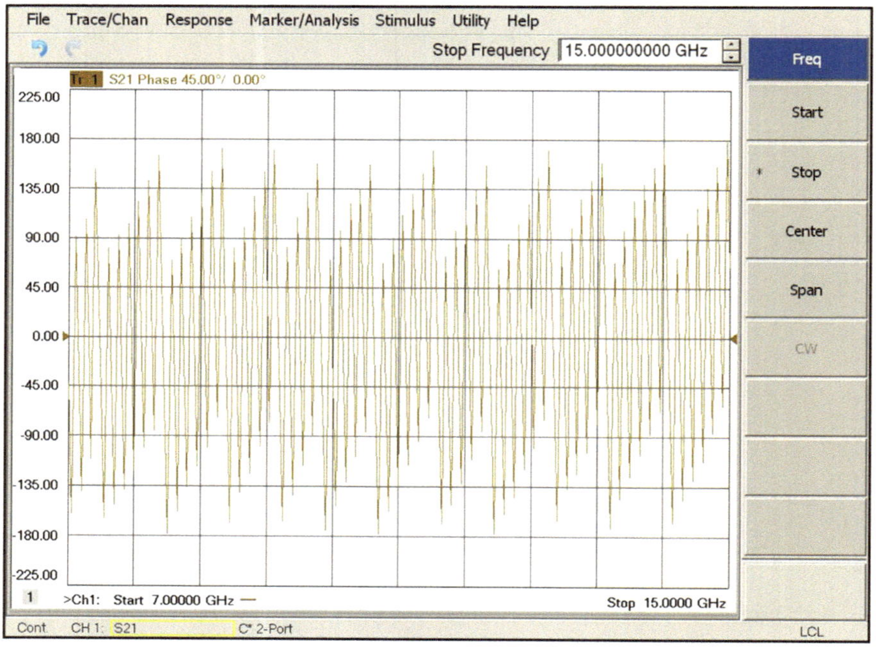

Fig. 12.37 Phase versus frequency plot (phase distortion) of simulated DRA

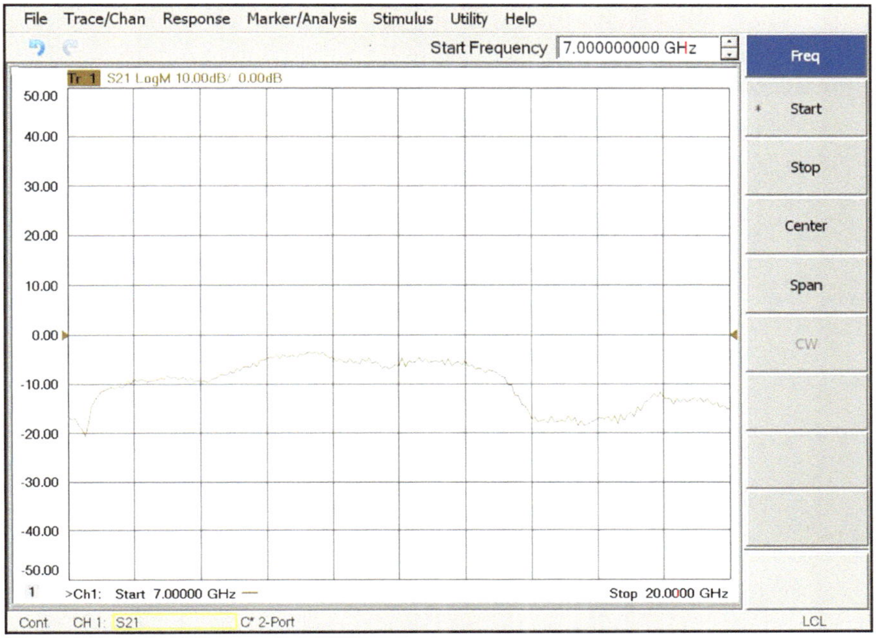

Fig. 12.38 S21 measurement plot in dual feed of simulated DRA

12.3 Dual-Feed RDRA with Measurements Results

In this case, the designing of RDRA dual-feed mechanism has been implemented for circular polarization. Ferrite DRA has been used for bandwidth enhancement using magnetization concept. The results obtained have been presented graphically. These antennas also have been placed inside anechoic chamber to minimize external effects that come during measurements. The figures indicate various measurement steps involved in this process. These antennas have been simulated and fabricated. For simulated results, Ansoft HFSS 13.0 has been used. These graphs of radiation pattern, gain, S11, VSWR, Z11, field distribution have been presented and their domains are mentioned below each figure. The impedance versus frequency has been presented. The hardware results using VNA for return loss S11 have been also included (Figs. 12.39, 12.40 and 12.41; Table 12.1).

Optimization of the feed position for impedance match to have maximum gain (Figs. 12.42, 12.43, 12.44, 12.45, 12.46, 12.47, 12.48, 12.49, 12.50, 12.51, 12.52, 12.53, 12.54, 12.55, 12.56, 12.57, 12.58, 12.59, 12.60, 12.61, 12.62 and 12.63).

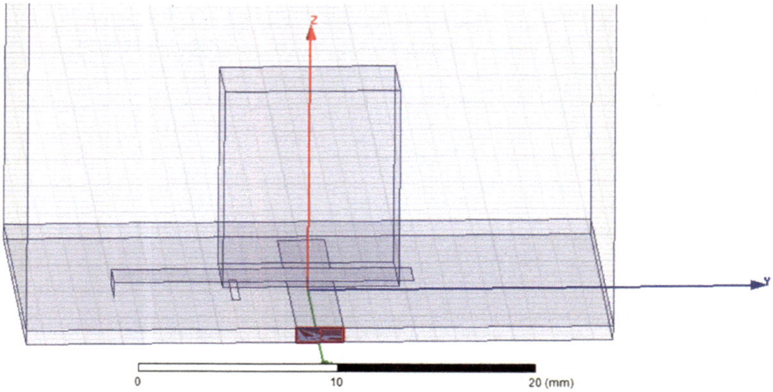

Fig. 12.39 Double-feed RDRA HFSS model

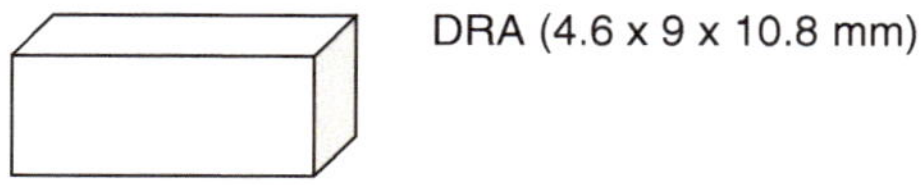

Fig. 12.40 Design dimensions of RDRA under design

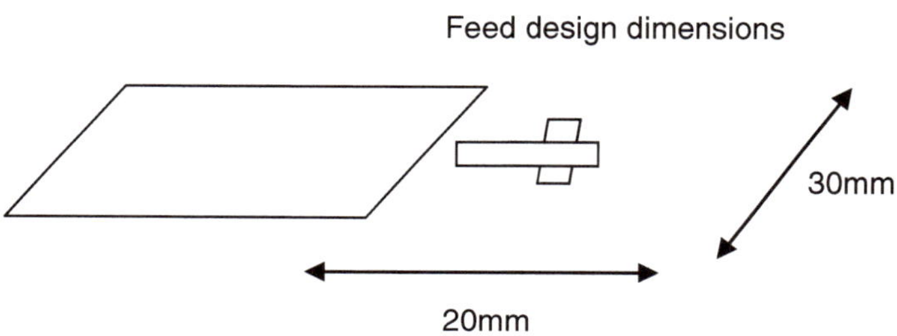

Fig. 12.41 Ground plane with slot/stub/micro-strip feed in RDRA

Table 12.1 Specification/dimensions

S. No.	Element	Dimension (mm)
1	Ground plane	20×30
2	Substrate ($\varepsilon_r 2.2$)	$20 \times 30 \times 0.8$
3	DRA ($\varepsilon_r 12.2$)	$4.6 \times 9 \times 10.8$
4	Width of micro-strip	2.4
5	Length of stub and micro-strip	18.693
6	Slot ($l \times w$)	3.743×0.404

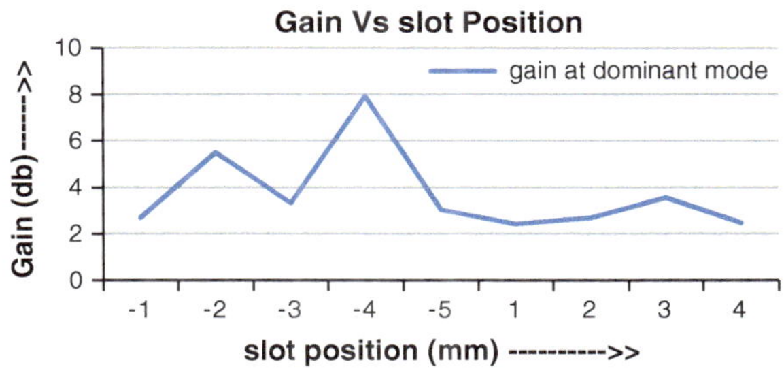

Fig. 12.42 Variation of slot position for maximum gain in RDRA

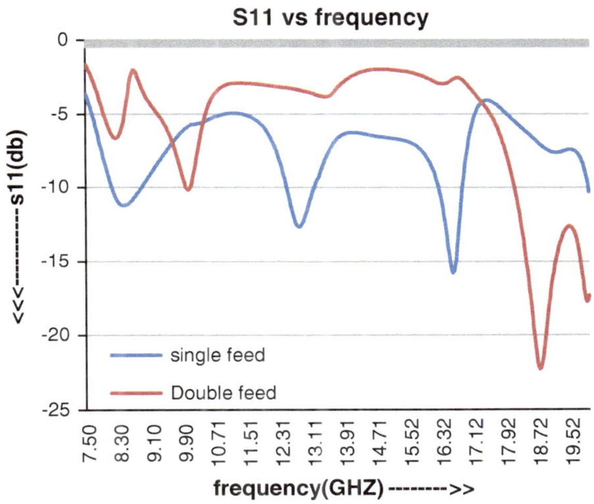

Fig. 12.43 Return loss plot S11 for single and double feed of RDRA

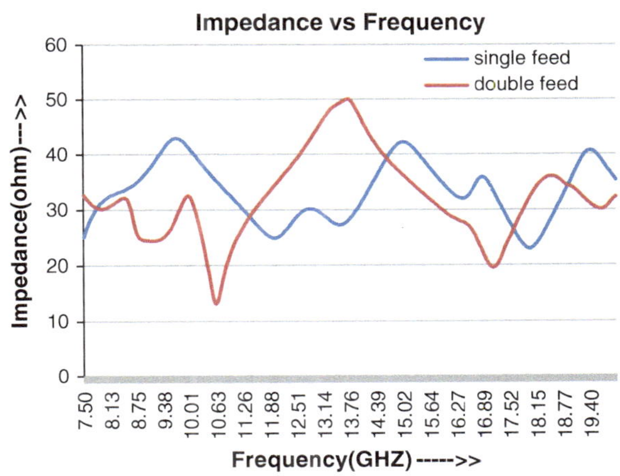

Fig. 12.44 Impedance match for single and double feed in RDRA

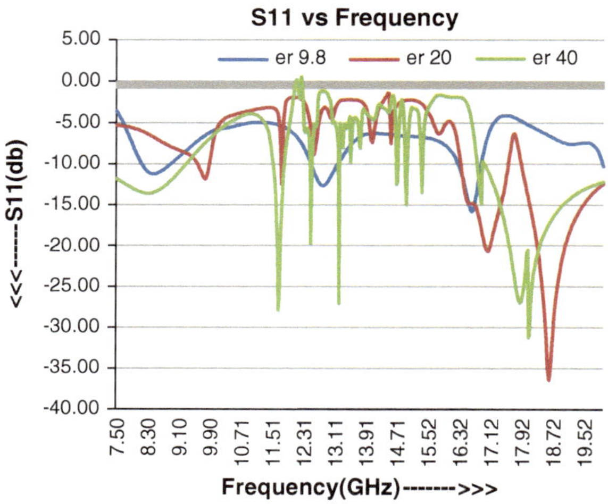

Fig. 12.45 Effect of variation in permittivity on return loss S11 of RDRA

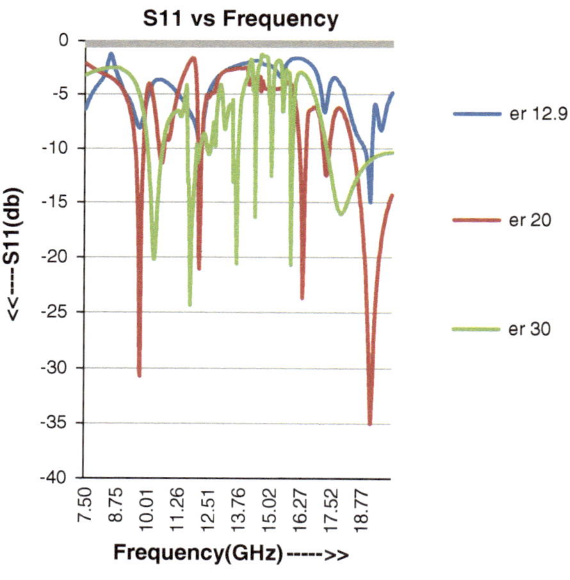

Fig. 12.46 Effect of variation in permittivity on return loss S11 of RDRA

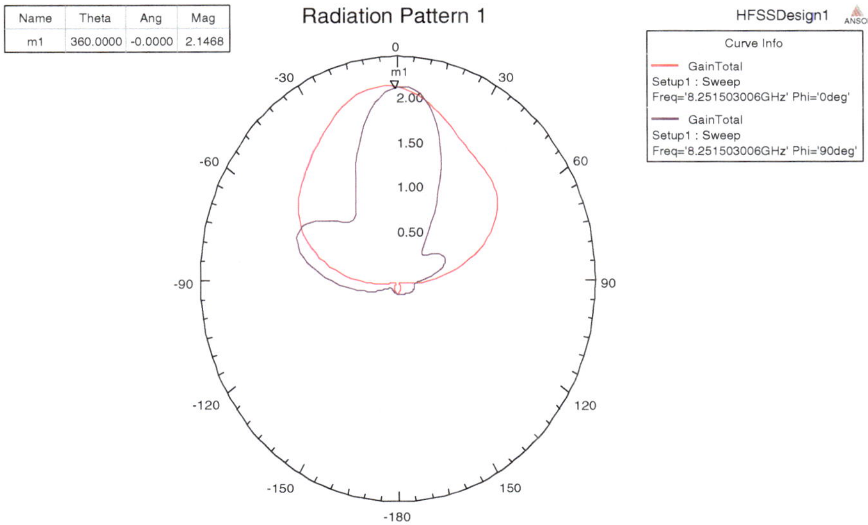

Fig. 12.47 Radiation pattern 1 of simulated RDRA for field strength

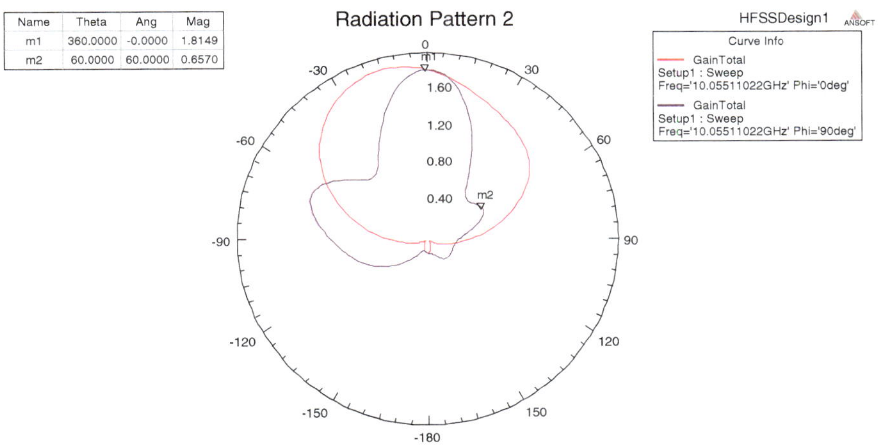

Fig. 12.48 Radiation pattern 2 of simulated RDRA for field strength

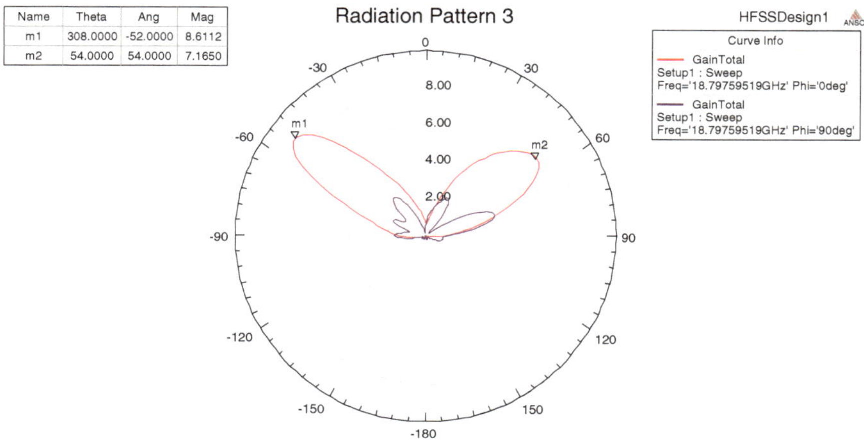

Fig. 12.49 Radiation pattern of simulated RDRA for field strength

Fig. 12.50 Front and rear view of hardware implemented of dual-feed RDRA

Fig. 12.51 Short, open, and load termination for calibration of VNA

Fig. 12.52 *Top* and *side* view of single- and double-feed aperture couple feed of RDRA

Fig. 12.53 Back side view of double-feed aperture couple feed of RDRA

Fig. 12.54 Fabricated ferrite RDRA

Fig. 12.55 Smith chart for impedance matching of RDRA

Fig. 12.56 Position of the slot in ground plane of single-feed RDRA

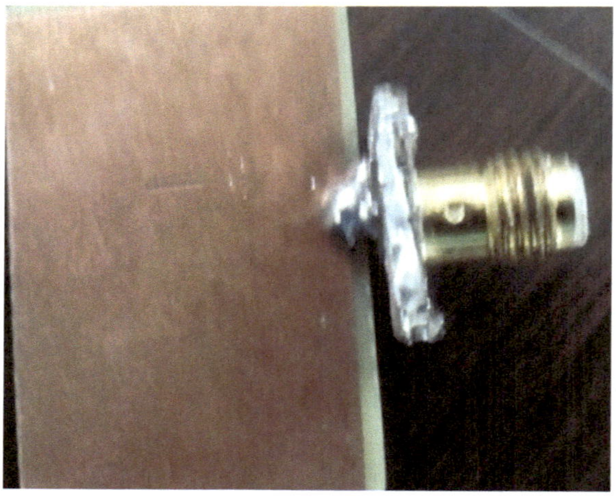

Fig. 12.57 RDRA investigation under testing setup

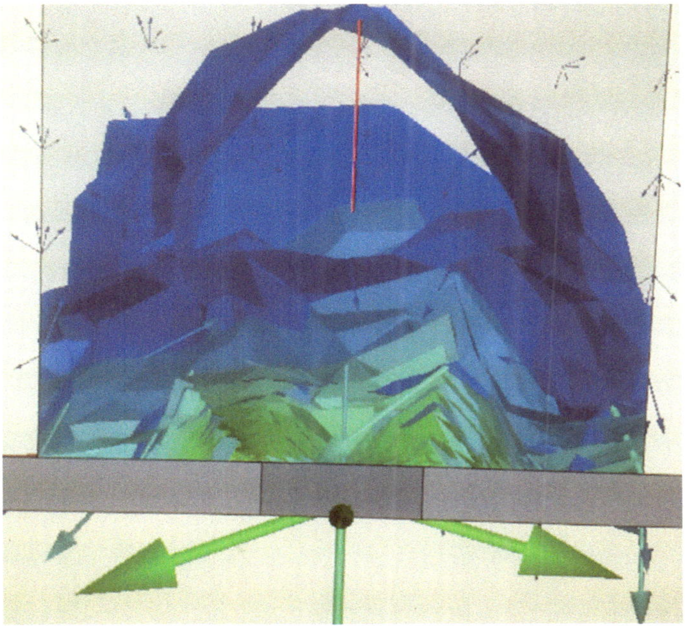

Fig. 12.58 RDRA *H* fields pattern showing magnetic field strength

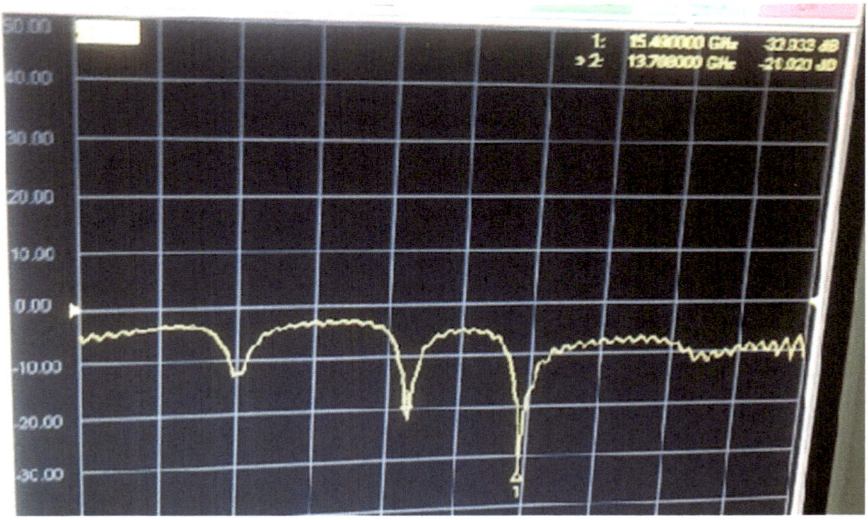

Fig. 12.59 VNA showing measured S11 of RDRA

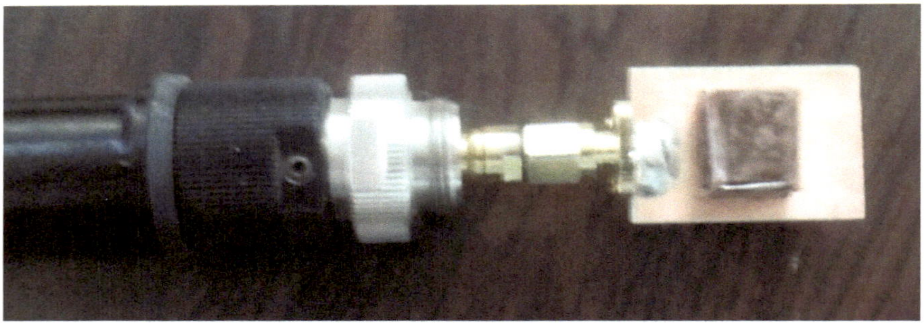

Fig. 12.60 Fabricated model of RDRA under test with single feed and slot

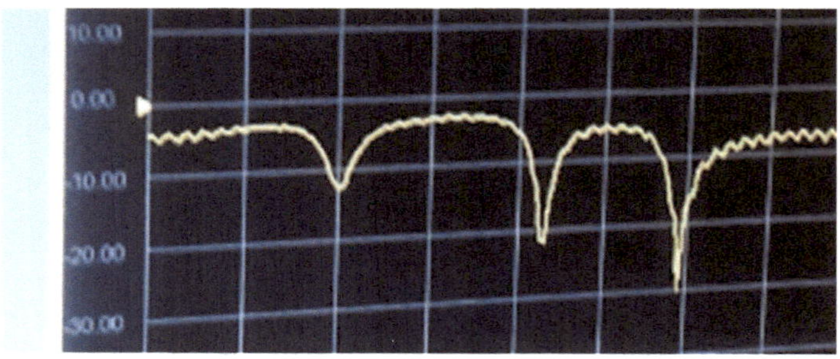

Fig. 12.61 Measurements of return loss S11 of fabricated RDRA

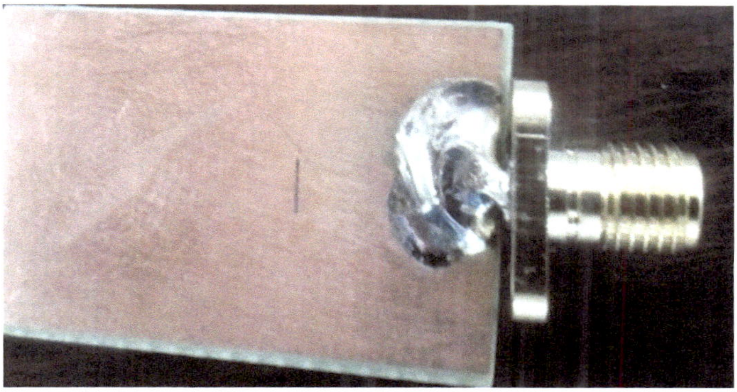

Fig. 12.62 Slot in ground plane of fabricated RDRA

Fig. 12.63 Top view with single feed and SMA connector of fabricated RDRA

12.4 Isolated and Grounded RDRA

Design of Isolated DRA: Isolated and grounded RDRA has different lengths due to image theory. Isolated RDRA is shown in Fig. 12.64. The RDRA is excited by a coax feed. Ground plane is absent in the first design. The rectangular DRA height can be reduced to half if we use ground plane of finite dimensions (Table 12.2).

12.4.1 S11 Plot

Return loss of isolated DRA is shown in Fig. 12.65. It has resonant frequency of 3.99 GHz with −41.74 return loss.

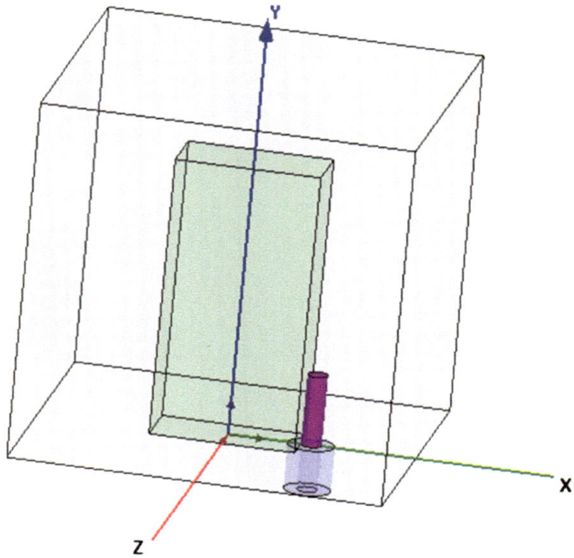

Fig. 12.64 Isolated DRA

Table 12.2 Dimensions of isolated DRA

Dimension of DRA in *X*-direction = 9.31 mm
Dimension of DRA in *Y*-direction = 18.62 mm
Dimension of DRA in *Z*-direction = 4.6 mm
Permittivity of DRA = 37.84

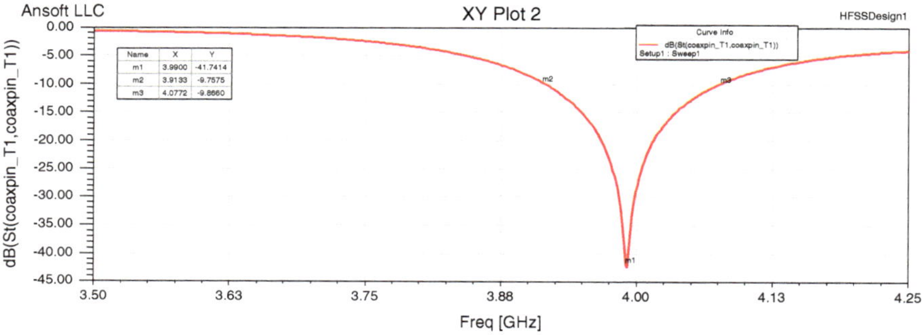

Fig. 12.65 Return loss for isolated DRA

12.4.2 Gain Plot

Gain plot is shown in Fig. 12.66. It shows that isolated DRA has 4.7 dB gain at resonant frequency.

12.4.3 Impedance (Z) Plot

Impedance plot of isolated DRA is shown in Fig. 12.67. This has the real impedance nearly 50 Ω at resonant frequency.

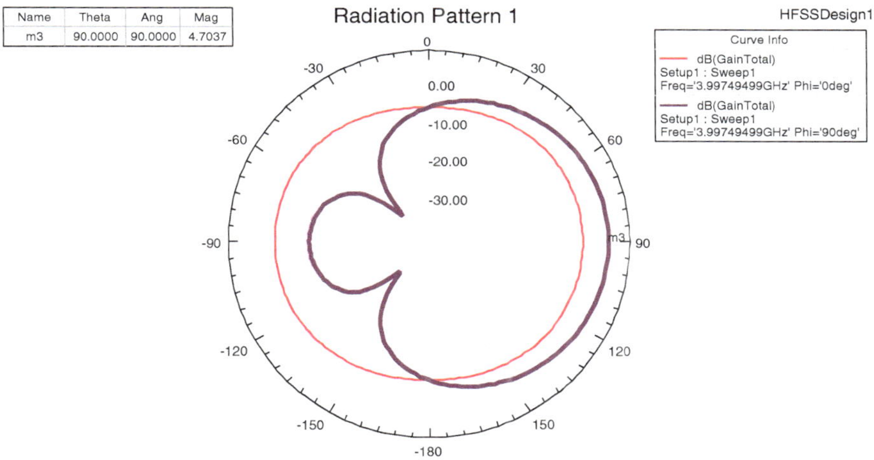

Fig. 12.66 Gain plot of isolated DRA

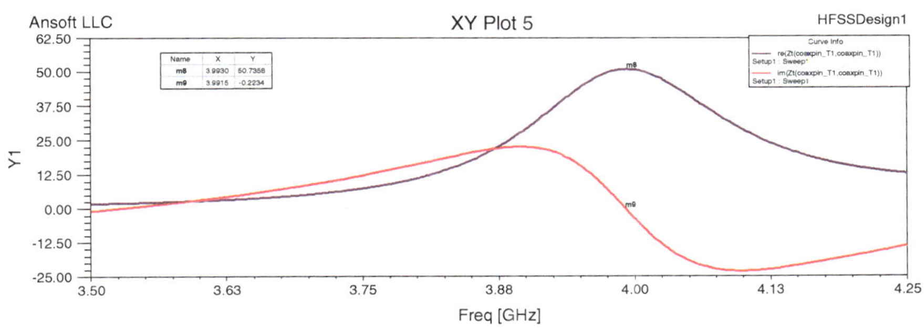

Fig. 12.67 Impedance plot of isolated DRA

12.4.4 Design of RDRA with Ground Plane

Image theory has been applied to reduce the height of the antenna. HFSS model of DRA with ground plane has been developed using excitation with coaxial feed. Here, the height of DRA has been reduced to half as compared to isolated DRA (Fig. 12.68; Table 12.3).

12.4.5 S11 Plot

Simulated return loss of DRA with ground is shown in Fig. 12.69. It has resonant frequency 4.18 GHz with −28 dB return loss.

12.4.6 Gain Plot

Gain plot is shown in Fig. 12.70. It shows that antenna radiates in the end fire direction and holds the value of gain 4.62 dB at resonant frequency.

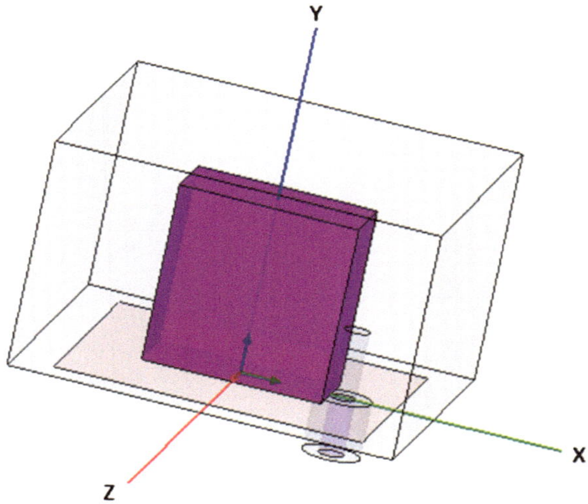

Fig. 12.68 HFSS model for DRA with ground

Table 12.3 Dimensions of DRA with ground plane

Dimension of DRA in *X*-direction = 9.31 mm
Dimension of DRA in *Y*-direction = 9.31 mm
Dimension of DRA in *Z*-direction = 4.6 mm
Permittivity of DRA = 37.84

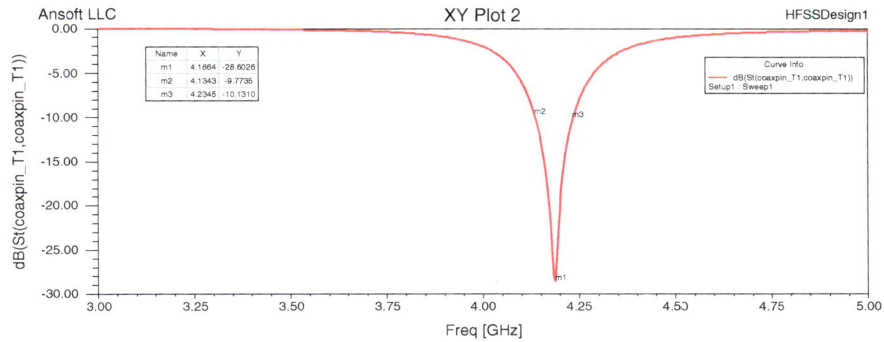

Fig. 12.69 Return losses for DRA with ground

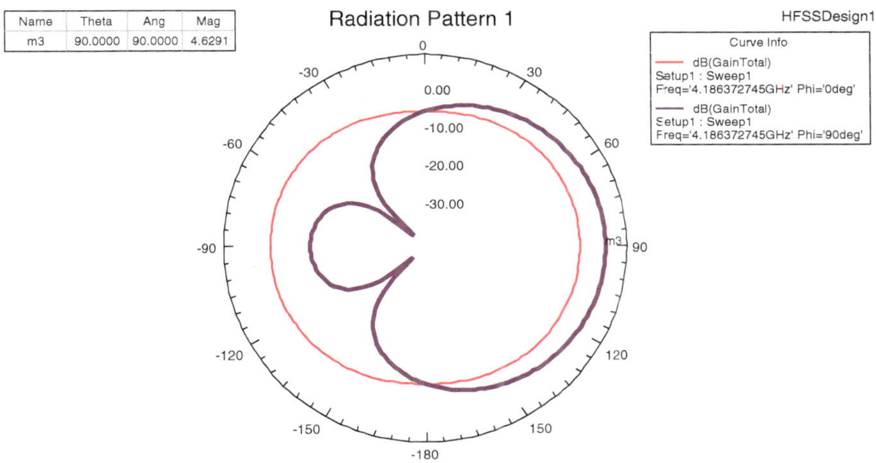

Fig. 12.70 Gain pattern for DRA with ground

12.4.7 Impedance Plot

Impedance plot of DRA with ground is shown in Fig. 12.71. This has the real impedance nearly 48.2 Ω and reactive part is 0.17 Ω at resonant frequency.

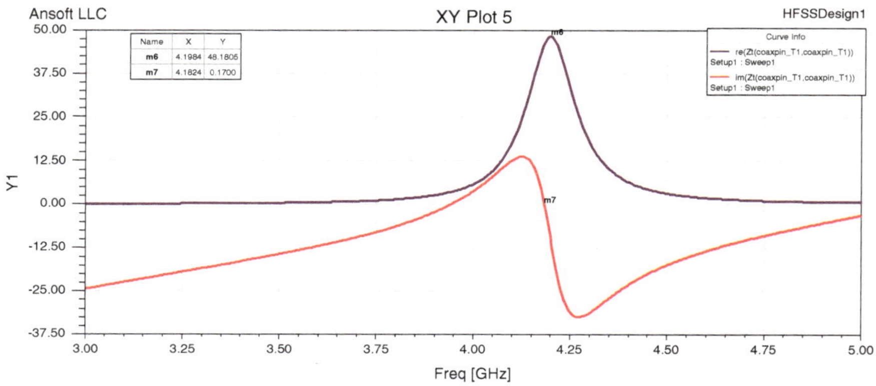

Fig. 12.71 Impedance plot for DRA with ground

12.4.8 Comparison of DRA With and Without Ground Plane

From the comparison in Table 12.4, it is seen that resonant frequency and gain for isolated DRA and DRA with ground are approximately same but return loss and impedance bandwidth are better in isolated DRA (Fig. 12.72).

Table 12.4 Comparison of DRA with and without ground plane

Parameter	Isolated	With ground
Resonant frequency (GHz)	3.99	4.16
Gain (dB)	4.7	4.629
B.W. (GHz)	0.17	0.1
Return loss (dB)	−41.04	−28

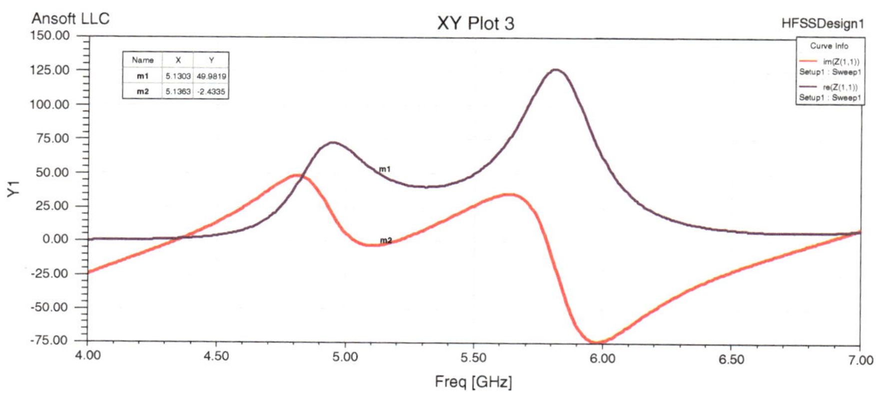

Fig. 12.72 Impedance plot of DRA

12.4.9 Detailed Design of Aperture-Coupled DRA

Figure 12.73 shows the view of a simple structure of rectangular DRA. The rectangular DRA of length L, width W, and height H is placed over a slot, cut at the center of ground plane of size $50 \times 50\ \text{mm}^2$. The micro-strip line of length L_m and width W_e is placed on the other side of the ground plane. The dielectric material used for substrate is having permittivity, $\varepsilon_r = 10.2$ and thickness 0.64 mm. The dielectric material used for DRA is Rogers RT/Duroid 6010/6010LM having permittivity $\varepsilon_r = 10.2$. All other the dimensions have been shown in Table 12.5 (Fig. 12.74).

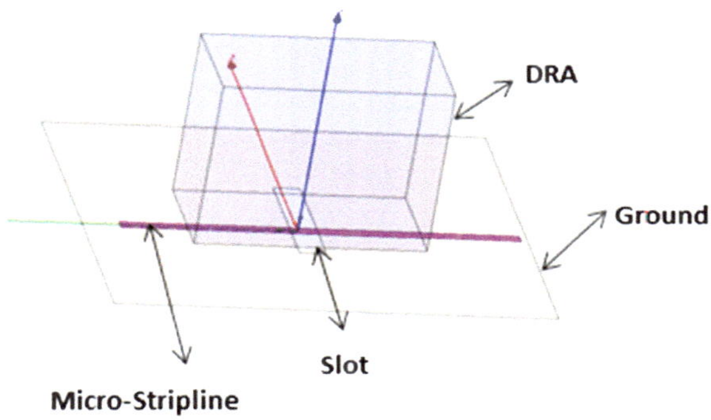

Fig. 12.73 HFSS model for reference antenna 1

Table 12.5 Dimensions of reference antenna

L (mm)	*W* (mm)	*H* (mm)	*Ls* (mm)	*Ws* (mm	*Lm* (mm)	*Wm* (mm)	*Er* (sub)	*er* (dra)	*H* (sub) (mm)
10.6	6	9.6	7.2	1.2	28	0.6	10.2	10.2	0.64

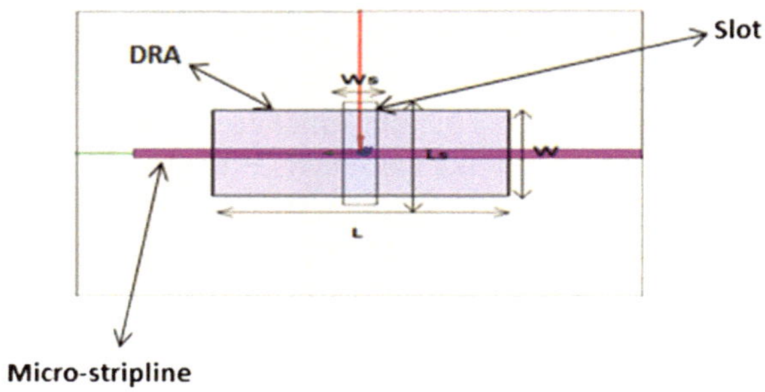

Fig. 12.74 Bottom view of aperture-coupled DRA

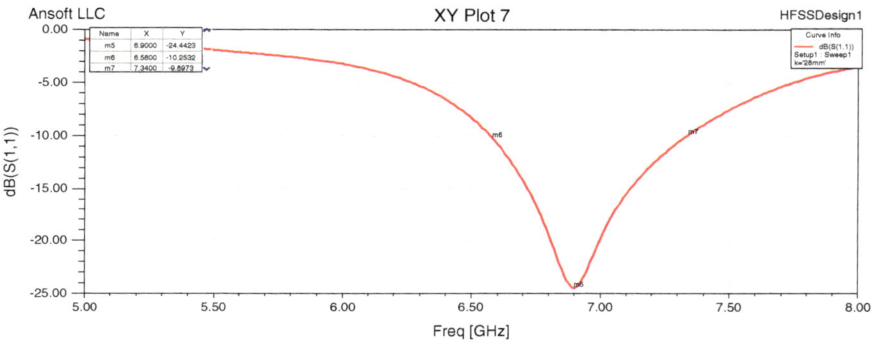

Fig. 12.75 Return loss of aperture-coupled DRA

12.4.10 Return Loss

Return loss of antenna 1 is shown in Fig. 12.75. Resonant frequency is 6.89 GHz with return loss −24 dB.

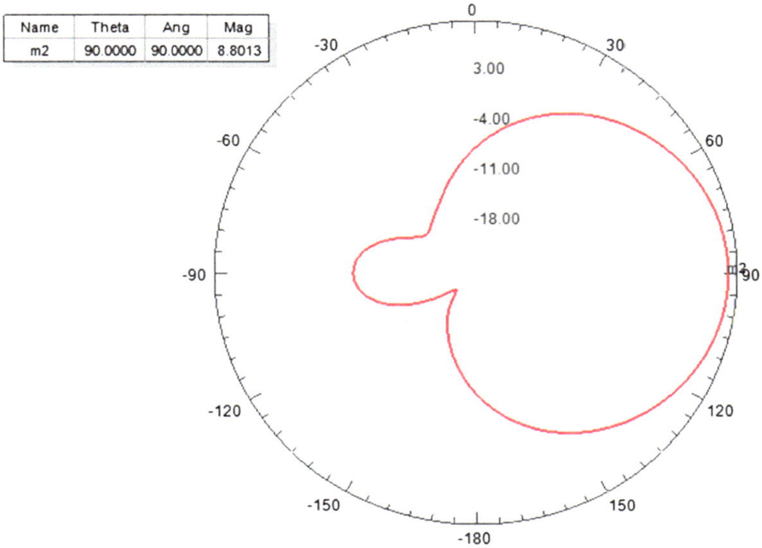

Fig. 12.76 Radiation pattern of gain for (antenna A) $\theta = 90°$ and $\Phi = 90°$

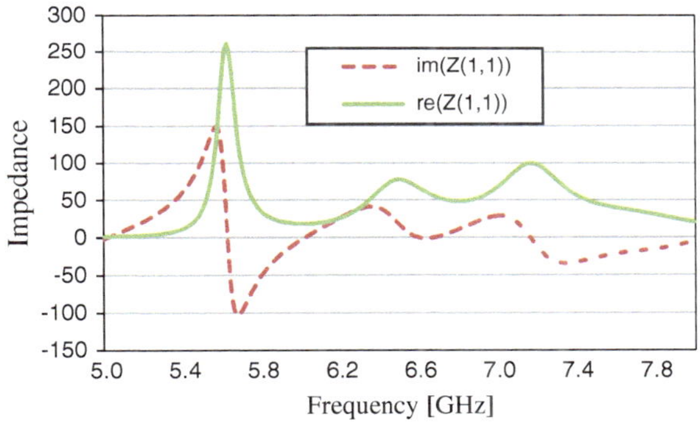

Fig. 12.77 Impedance matching plots

12.4.11 Radiation Pattern

Antenna radiates in end fire direction. Gain of antenna at $\theta = 90°$ and $\Phi = 90°$ is 8.796 dBi (Figs. 12.76 and 12.77).

Annexure-1

Details of the Dielectric Materials and Their Suppliers

S. No.	Material	Permittivity	Supplier or manufacturer
1.	$MgO–SiO_2$ (CD-6)	6.3	**Countis Laboratories** 12295 Charles Dr, Grass Valley, CA 95945, United States +1 530-272-8334 tcountis@countis.com
2.	$MgO–SiO_2–TiO_2$ (CD-9)	9.5	
3.	$MgO–TiO_2–SiO_2$ (CD-13)	13.0	
4.	$MgO–TiO_2$ (CD-15)	15.0	
5.	$MgO–TiO_2$ (CD-16)	16.0	
6.	$Mgo–CaO–TiO_2$ (CD-18)	18.0	
7.	$Mgo–CaO–TiO_2$ (CD-20)	20.0	
8.	$Mgo–CaO–TiO_2$ (CD-30)	30.0	
9.	$Mgo–CaO–TiO_2$ (CD-50)	50.0	
10.	$Mgo–CaO–TiO_2$ (CD-100)	100.0	
11.	$Mgo–CaO–TiO_2$ (CD-140)	140.0	

(continued)

R.S. Yaduvanshi and H. Parthasarathy, *Rectangular Dielectric Resonator Antennas*, DOI 10.1007/978-81-322-2500-3

S. No.	Material	Permittivity	Supplier or manufacturer
12.	Boron nitride (ECCOSTOCK@)	4.0	**Emerson & Cuming Microwave Products N.V.** A unit of Laird Technologies Hong Kong Holdings (4) Ltd. Unit 2507-8, 25/F, Office Tower, Langham Place, 8 Argyle Street, Mongkok Kowloon, Hong Kong Tel: +852-2923 0600 Call: +852-2923 0605 Email: sales@hk.eccosorb.com
13.	Beryllium oxide (ECCOSTOCK@)	6.0	
14.	Magnesium oxide (ECCOSTOCK@)	9.0	
15.	Magnesium titanate (ECCOSTOCK@)	10.0	
16.	Zirconia (ECCOSTOCK@)	20.0	
17.	Titanium dioxide (rutile) (ECCOSTOCK@)	50.0	
18.	Strontium titanate (ECCOSTOCK@)	>100.0	
19.	Magnesium manganese Aluminum iron ferrite	9.2	**Hiltek Microwave Limited** 15200 Shady Grove Road Suite 350 Rockville, Maryland 20850 United States (301) 670-2833 (301) 670-2831 Fax www.hiltek.com
20.	Magnesium titanate	16.0	
21.	Lithium ferrite	20.0	
22.	Zirconium tin titanate	37.0	
23.	Titania ceramic	80–100	
24.	MgSi (Steatile) (D6)	6.0	**Morgan advanced Materials** 150 Kampong ampat 05-06a Ka Centre Singapore 368324 t +65 6595 0000 F +65 6595 0005 asia.mc@morganplc.com
25.	CaMgTi (Mg, Ca titanate) (D20)	20.0	
26.	ZrTiSn (Zr, Sn titanate) (D36)	37.0	
27.	BaSmTi (Ba, Sm titanate) (D37)	76.5	

(continued)

S. No.	Material	Permittivity	Supplier or manufacturer
28.	Titanate with other ingredients (PD-6)	6.5	**Pacific Ceramics, Inc.** Advanced Microwave Ceramic Materials 824 San Aleso Ave Sunnyvale, CA 94085 USA (408) 747-4600 info@pceramics.com
29.	Titanate with other ingredients (PD-9)	9.5	
30.	Titanate with other ingredients (PD-12)	12.0	
31.	Titanate with other ingredients (PD-13)	13.0	
32.	Titanate with other ingredients (PD-15)	15.0	
33.	Titanate with other ingredients (PD-16)	16.0	
34.	Titanate with other ingredients (PD-18)	18.0	
35.	Titanate with other ingredients (PD-25)	25.0	
36.	Titanate with other ingredients (PD-38)	38.0	
37.	Titanate with other ingredients (PD-50)	50.0	

(continued)

S. No.	Material	Permittivity	Supplier or manufacturer
38.	Titanate with other ingredients (PD-100)	98.0	
39.	Titanate with other ingredients (PD-160)	160.0	
40.	Titanate with other ingredients (PD-270)	270.0	
41.	Zr Sn Ti oxide (E2000)	37.0	**Temex Components & Temex Telecom, USA** **Supplier 1** SM CREATIVE No 845, 2nd Cross, 7th Main HAL 2nd Stage Indiranagar, Bangalore, 560 038 India sundar@smcel.com +91 (80) 25210268 +91 (80) 41255492 Mobile Phone Number: +91 (98) 45410417 http://www.smcel.com S M Creative Electronics Ltd #10, Electronic City, Sector-18 Gurgaon 122 015, Haryana Tel: +91 124-4909850 Fax: +91 124-2455 212 smcel@smcel.com **Supplier 2** SIMAL # 60 & 60/1, 18th Cross, 4th Main Malleswaram, Bangalore, 560 055 India agencies@simal.com.sg +91 (80) 41532079 +91 (80) 23444410 Mobile Phone Number: +91 (99721) 24165 http://www.simal.com.sg
42.	E3000	34.0	
43.	Ba Zn Ta oxide (E4000)	30.0	
44.	Ba Sm Ti oxide (E5000)	78.0	
45.	Ti Zr Nb Zn oxide (E6000)	45.0	
46.	Cordierite (Mg, Al, silicate) (D-4)	4.5	**Trans-Tech** **Skyworks Solutions, Inc.** 5520 Adamstown Road Adamstown, MD 21710 **Supplier** SM Electronic Technologies Pvt. Ltd. #1790, 5th Main, 9th Cross, RPC Layout Vijayanagar 2nd Stage Bangalore 560 040 India
47.	Forsterite (Mg, Si, oxide) (D-6)	6.3	
48.	Mg–Ti (D-15)	15.0	
49.	Mg–Ti (D-16)	16.0	

(continued)

S. No.	Material	Permittivity	Supplier or manufacturer
50.	Ba, Zn, Ta-oxide (D-29)	29.0–30.7	Mr. Manjunath +91-80-23301030 smgroup@vsnl.com
51.	Ba, Zn, Ta-oxide (perovskite) (D-87)	29.5–31.0	
52.	BaZnCoNb (D-83)	35.0–36.5	
53.	Zirconium titanate based (D-43)	44.7–46.2	
54.	E-11	11.0	**T-CERAM, RF & Microwave** Okružní 1144 500 03 Hradec Králové Czech Republic, EU sales@t-ceram.com www.t-ceram.com +420 774 406 438 CZ 42196078
55.	E-20	20.0	
56.	E-37	37.0	
57.	TE-21	21.0	**Token Electronics Industry Co., Ltd.** No. 137, Sec. 1, Chung Shin Rd., Wu Ku Hsiang, Taipei Hsien, Taiwan, R.O.C TEL: 886-2-2981 0109; FAX: 886-2-2988 7487 http://www.token.com.tw rfq@token.com.tw
58.	TE-30	30.0	
59.	TE-36	36.0	
60.	TE-45	45.0	
61.	TE-80	80.0	
62.	TE-90	90.0	
63.	Mg–Ca–Ti (MDR20)	20.0	**MCV Microwave** 6640 Lusk Blvd, Suite A102 San Diego, CA 92121 Tel: 858-450-0468 Fax: 858-869-8404 www.mcv-microwave.com
64.	Ta with other ingredients (MDR24)	24.0	
65.	Ta with other ingredients (MDR30)	30.0	
66.	Zn–Sn–Ti (MDR38)	38.0	
67.	La–Ba–Ti (MDR45)	45.0	
68.	DR-30	30.0	**TCI Ceramics, Inc.** 18450 Showalter Rd., Hagerstown, MD 21742 Ph: 301-766-0560 Fax: 301-766-0566 E-mail: sales@tciceramics.com www.tciceramics.ccm
69.	DR-36	36.0	
70.	DR-45	45.0	
71.	DR-80	80.0	
72	RT6010, RT-6002	10.2	
73	MCT-25 SMAT $BaTiO_3$	25 27 14	TRANS TECH
74	ECCOSTOCK'SHIK	10,20,30,40	

Annexure-2

Two-Dimensional Mathematical Model of Resonant Modes in Cavity Resonator

See Figs. A2.1, A2.2 and A2.3.

Probe inserted dl

Characteristic equation of RDRA is given below:

$$k_x^2 + k_y^2 + k_z^2 = \varepsilon_r k_{mn}^2 \tag{A2.1}$$

The field E_z can be expressed as follows:

$$\left(\nabla_\perp^2 + h^2\right)(H_z \,\text{or}\, E_z) = 0;\ \text{Helmholtz equation}$$

$$E_z = \sum C_{mn} \cdot \sin(m\pi x/a) \cdot \sin(n\pi y/b) \cdot e^{-\gamma_{mn} z} e^{j\omega t}$$

In the above equation, C_{mn} are the amplitude coefficients and wave is propagating in z-direction

$$\therefore \quad \gamma_{mn}(\text{Propagation constant}) = \sqrt{h_{mn}^2 - k^2} = \sqrt{h_{mn}^2 - \omega^2 \mu\epsilon}$$

where $h_{mn} = k_c = n\pi/b$; are possible eigenvalues.

Hence, computation of field E_z *when* all the four sides of resonator are transparent and magnetic walls (PMC walls) and top and bottom walls are PEC (Electrical walls). We are well versed that $H_z = 0$ *at* magnetic walls and $E_z = 0$ at electric walls.

The feed probe is inserted into rectangular DRA at point (a/2, b/2) in z-direction. $I(t)$ Current can be expressed in terms of *magnetic* vector potential A_z.

R.S. Yaduvanshi and H. Parthasarathy, *Rectangular Dielectric Resonator Antennas*, DOI 10.1007/978-81-322-2500-3

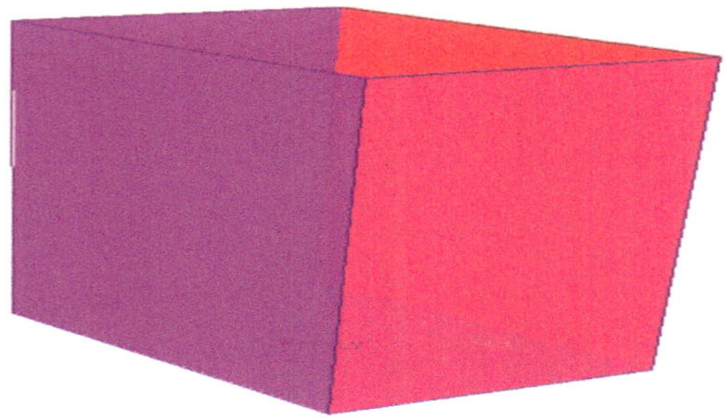

Fig. A2.1 RDRA without ground plane

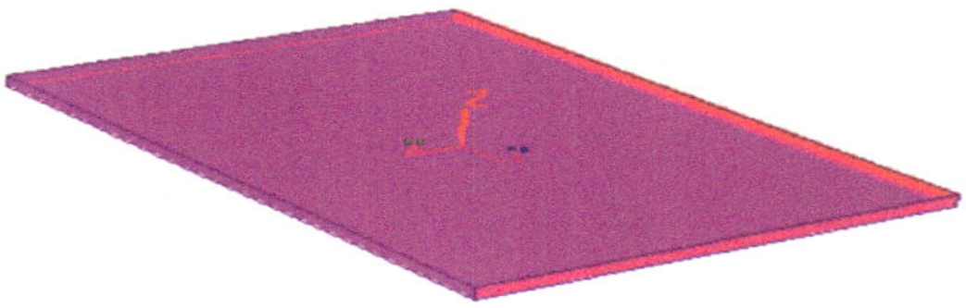

Fig. A2.2 Ground plane of RDRA

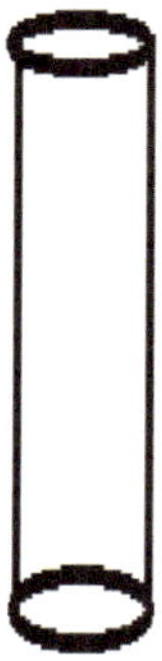

Fig. A2.3 RF feed

$$
\begin{aligned}
\underline{A}_z &= \frac{\hat{z}\ \mu I \mathrm{d}l}{4\pi r} e^{-jkr}, \quad r \text{ is far field point.} \\
\text{div } \underline{A} &= -\frac{jk(\hat{r}\ \cdot \hat{z})\ \mu I \mathrm{d}l - e^{-jkr}}{4\pi r} \\
\text{div } \underline{A} &= -\frac{jk \cos\theta\ \mu I \mathrm{d}l - e^{-jkr}}{4\pi r}; \\
&= -\frac{j\omega \nexists}{c^2}
\end{aligned}
$$

Let,

$$\nexists = \sqrt{a^2 + b^2}\frac{kc}{\omega}\ \frac{\mu I \mathrm{d}l}{4\pi r}\ e^{-jkr}$$
$$= \frac{\cos\theta \mu I \mathrm{d}l}{4\pi r} e^{-jkr}$$

Now,

$$E = -\nabla ø - j\omega A;\ \text{Lorentz's gauge condition}$$

$$E_r = 0 + O\left(\frac{1}{r^2}\right)$$
$$E_\theta = j\omega A_\theta = j\omega Az, \quad \text{at } z = 0$$
$$E_ø = -j\omega Aø = 0 \quad \text{at } z = 0$$

Hence,

$$E_\theta = E_z = j\omega Az$$
$$= \frac{j\omega\mu I \mathrm{d}l}{4\pi r} e^{-jkr} \cdot \frac{2}{\sqrt{ab}} \sin\left(\frac{m\pi x}{a}\right) \sin\left(\frac{n\pi y}{b}\right)$$

where $r = \sqrt{x^2 + y^2}$

$\delta m, m', \delta n, n' = \langle u_{nm}, u_{m'n'} \rangle$; where $z = 0$; (property of orthogonality as product of basis function becomes zero)

$$E_z = \sum_{mn} C_{mn} u_{mn}(x, y) e^{-\gamma_{mnz}}$$

$$h_{mn} \leq \omega$$

At $z = 0$

$$E_z \Rightarrow \sum_{mn} C_{mn} u_{mn}(x, y) \frac{j\omega\mu I \mathrm{d}l}{4\pi\sqrt{x^2 + y^2}} e^{-jk\sqrt{x^2+y^2}};$$

Hence, amplitude coefficient

$$C_{mn} = \frac{j\omega\mu I \mathrm{d}l}{4\pi} \int_0^a \int_0^b \frac{u_{mn(x,y)}}{\sqrt{x^2 + y^2}} e^{-jk} \sqrt{\left(x - {a}/{2}\right)^2 + \left(x - {b}/{2}\right)^2} \mathrm{d}x\mathrm{d}y;$$

$$C_{mn} = \frac{j\omega\mu I \mathrm{d}l}{4\pi} \int_0^a \int_0^b \frac{u_{mn(x,y)}}{\sqrt{\left(x - a/2\right)^2 + \left(y - b/2\right)^2}} e^{-jk\sqrt{\left(x - a/2\right)^2 + \left(x - b/2\right)^2}} \mathrm{d}x\mathrm{d}y \tag{A2.2}$$

$$\text{Hence } \pi^2\left[\left(\frac{m}{a}\right)^2 + \left(\frac{n}{b}\right)^2\right] \leq \omega^2$$

$$\text{if } a > b \quad \text{and} \quad m = 1, 2, 3, \ldots \quad n = 1, 2, 3, \ldots$$

$$\pi^2\left(\frac{1}{a^2} + \frac{1}{b^2}\right) \leq \omega^2 < \pi^2\left(\frac{2}{a^2} + \frac{1}{b^2}\right)$$

$$\sqrt{\frac{1}{a^2} + \frac{1}{b^2}} \leq \frac{\omega}{\pi} < \sqrt{\frac{2}{a^2} + \frac{1}{b^2}}$$

$$\gamma_{mn} = \frac{j\pi p}{d}$$

$$k^2 + \gamma_{mn}^2 = h_{mn}^2$$

$$\text{hence, } k^2 = h_{mn}^2 + \frac{\pi^2 p^2}{d^2}$$

C_{mn} Fourier coefficients of modes;
u_{mn} depends on input excitation;
h_{mn} resonant mode (cut off frequency); and
k-propagation constant.

Generation of modes or characteristics frequencies $\omega(mnp)$ e.m. of electromagnetic fields oscillations inside the cavity resonator has been discussed. The basic Maxwell's theory can be applied with boundary conditions to express resonator fields as superposition of these characteristics frequencies.

The fields

$$E_z(x, y, z, t) = \sum_{mnp} \mathrm{Re} \int C_{mnp} e^{j\omega(mnp)t} u_{mnp}(x, y, z)$$

or

$$\sum_{mnp} \left|C_{mnp}\right| u_{mnp}(x, y, z) \cos(\omega(mnp) + \Phi(mnp));$$

where $u_{mn}(x,y) = \frac{2}{\sqrt{ab}} \sin\left(\frac{m\pi x}{a}\right) \sin\left(\frac{n\pi y}{b}\right)$; $\omega(mnp)$ is the characteristic frequency and $\omega(mnp)$ is the phase of current applied. The rectangular cavity resonator is excited at the centre with an antenna probe carrying current $I(t)$ of some known frequency $\omega(mnp)$. This generates the field E_z inside the cavity of the form given below:

$$E_z(x,y,\delta,t) = \int G(x,y) \frac{j\omega\mu I \mathrm{d}l(x^2+y^2)}{4\pi\left(x^2+y^2+\delta^2\right)^{3/2}} e^{\left(j\omega t - \frac{\omega}{c}\sqrt{x^2+y^2+\delta^2}\right)} I(\omega) e^{j\omega t} \mathrm{d}\omega$$

where $G(x, y)$ are the constant terms associated with current.

Equating resonator field with the antenna current fields at $z = \delta$ plane;

Antenna or resonator radiation current or fields

$$= \sum_p |C_{mnp}| \sqrt{\frac{2}{d}} \sin\left(\frac{p\pi\delta}{d}\right) \cos(\omega(mnp)t + \phi(mnp))$$

Antenna probe current

$$= \int G(x,y) \frac{j\omega\mu I \mathrm{d}l(x^2+y^2)}{4\pi\left(x^2+y^2+\delta^2\right)^{3/2}} I(\omega)$$

$$e^{j\omega t} \mathrm{d}\omega \left(e^{\left(j\omega t - \frac{\omega}{c}\sqrt{x^2+y^2+\delta^2} + \psi_{mnp}\right)} u_{mn}(x,y) \mathrm{d}x\mathrm{d}y; \right.$$

Multiply both sides by $e^{-j\omega(mnp)t}$ and then taking time averaging (KAM) gives us the following

$$|C_{mnp}| \sqrt{\frac{2}{d}} \sin\left(\frac{p\pi\delta}{d}\right) e^{j\varnothing(mnp)} = \lim_{T\to\infty} \frac{1}{2\pi} \cdot \int_{-T}^{T} G(x,y) \frac{j\omega\mu I \mathrm{d}l(x^2+y^2)}{4\pi\left(x^2+y^2+\delta^2\right)^{3/2}} e^{-j\omega(mnp)t}$$

$$I(\omega) e^{j\omega t} \mathrm{d}\omega \left(e^{\left(j\omega t - \frac{\omega}{c}\sqrt{x^2+y^2+\delta^2} + \psi_{mnp}\right)} u_{mn}(x,y) \mathrm{d}t \right.$$

It is clear that for these two expressions to be equal, the probe current can be defined as

$$I(\omega) = \frac{1}{2} \sum_{mnp} |I(mnp)| \left[\delta(\omega - \omega(mnp) e^{j\varnothing(mnp)} + e^{j\varnothing(mnp)} \delta\{\omega - \omega(mnp)\} \right]$$

The antenna probe current must contain only the resonator characteristics frequencies $\omega(mnp)$, then

$$\sum_p |C_{mnp}|\sqrt{\frac{2}{d}}\sin\left(\frac{p\pi\delta}{d}\right)\cos(\omega(mnp)t+\phi(mnp)$$
$$=\int G(x,y)\frac{j\omega\mu I\mathrm{d}l(x^2+y^2)}{4\pi\left(x^2+y^2+\delta^2\right)^{3/2}}I(\omega) \tag{A2.3}$$
$$e^{j\omega t}\mathrm{d}\omega(e^{\left(j\omega t-\frac{\omega}{c}\sqrt{x^2+y^2+\delta^2}+\psi_{mnp}\right)}u_{mn}(x,y)\mathrm{d}x\mathrm{d}y$$

Antenna probe current = Resonator radiated current or magnetic fields, as per the law of conservation of energy. The modes' diagrams are given below (Figs. A2.4, A2.5, A2.6, A2.7, A2.8, A2.9, A2.10, A2.11, A2.12, A2.13, A2.14 and A2.15):

Fig. A2.4 Mode diagram

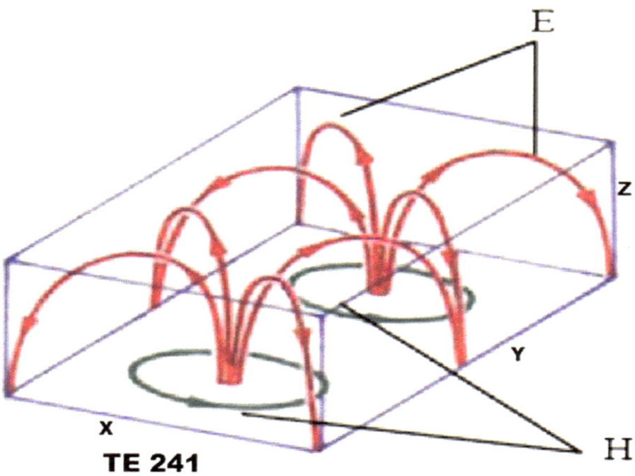

Fig. A2.5 TE_{112}

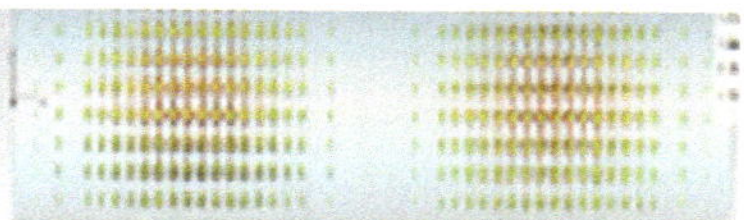

Fig. A2.6 TE_{113}

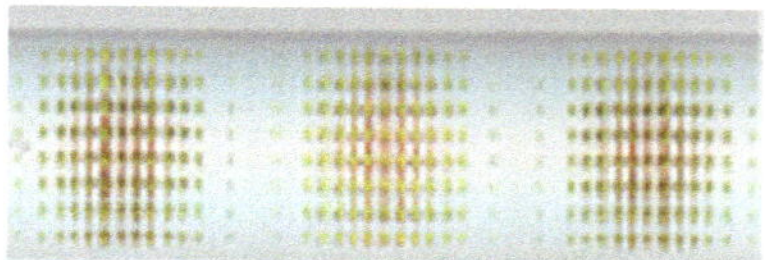

Fig. A2.7 TE_{114}

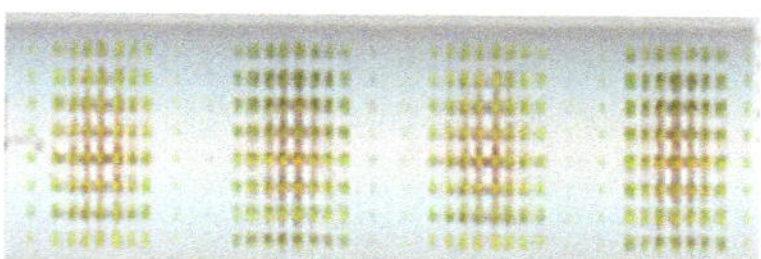

Fig. A2.8 TE_{111}

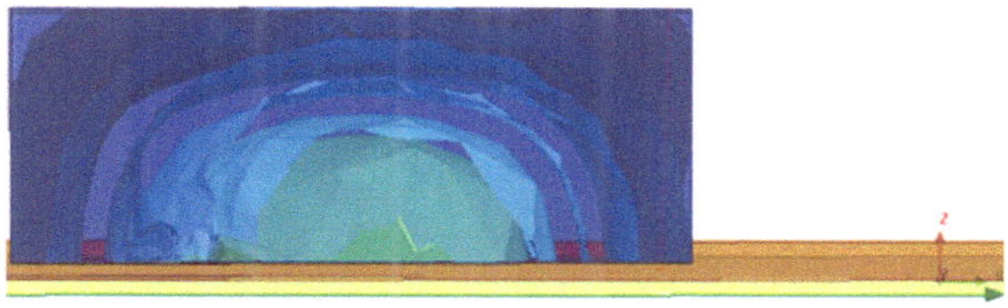

Fig. A2.9 TE_{112}

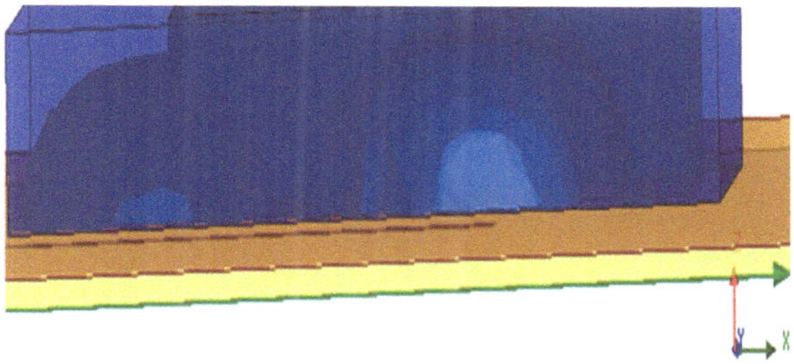

Fig. A2.10 TE_{113}

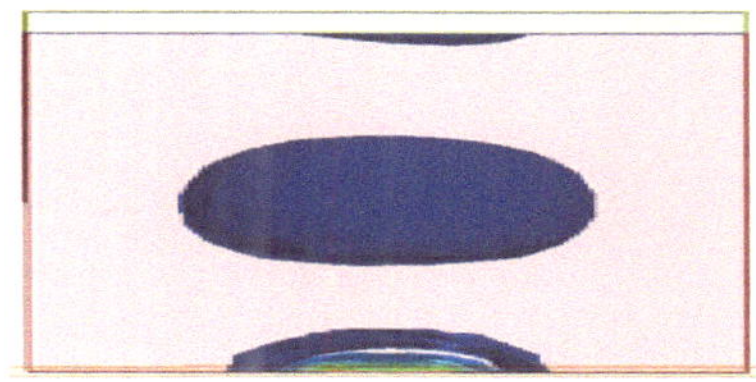

Fig. A2.11 TE_{115}

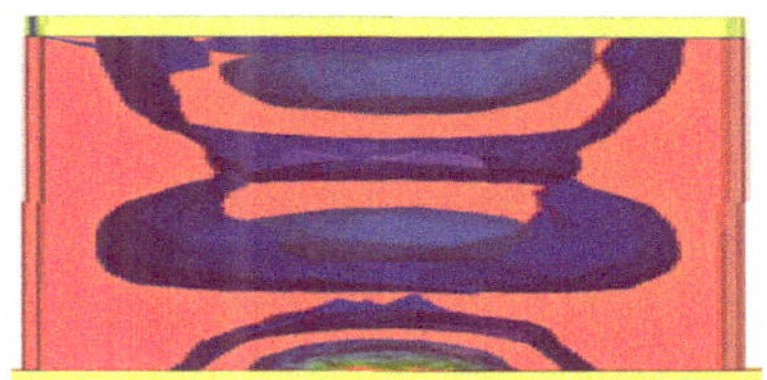

Fig. A2.12 TE_{116}

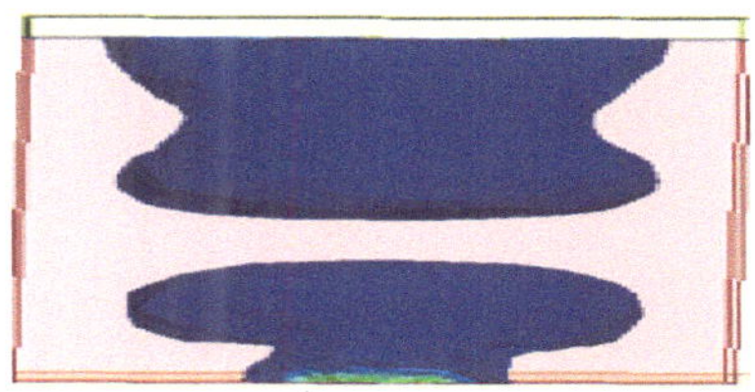

Mode sketch

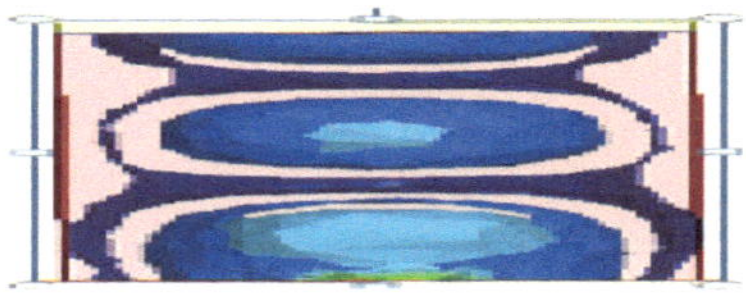

Fig. A2.13 TE_{114}

Fig. A2.14 TE_{118}

Fig. A2.15 TE_{114}

Rectangular design

$\omega_{mn} = \frac{\pi}{\sqrt{\mu\varepsilon}}\left(\frac{m^2}{a^2} + \frac{n^2}{b^2}\right)^{\frac{1}{2}}$, in two-dimensional case

$$\frac{\partial^2 X}{\partial^2 x} + k_x^2 \cdot x = 0; \quad \frac{\partial^2 Y}{\partial^2 y} + k_y^2 \cdot y = 0; \quad \frac{\partial^2 Z}{\partial^2 z} + k_z^2 \cdot z = 0$$

$$\varepsilon_r k_0^2 = k_x^2 + k_y^2 + k_z^2; \quad \text{where } k \text{ is wave number}$$

$h(k_x x)$ is the harmonic function and can be written as follows: $\sin(k_x x)$ or $\cos(k_x x)$.

These are solution of wave function and if boundary conditions are applied, then eigenvalues can be defined as follows:

$$k_0 = \frac{2\pi f_0}{c}, \quad k_y \tan(k_y d/2) = \sqrt{(\varepsilon_r - 1)k_0^2 - k_y^2}$$

$$k_x^2 + k_y^2 + k_z^2 = \varepsilon_r k_0^2, \text{ Resonant frequency} \quad f_0 = \frac{c}{2\pi\sqrt{\varepsilon_r}}\sqrt{k_x^2 + k_y^2 + k_z^2}$$

where $k_x = m\frac{\pi}{a}, k_y = n\frac{\pi}{b}$, and $k_z \operatorname{tag}\left(\frac{k_z d}{2}\right) = \sqrt{(\varepsilon_r - 1)k_0^2 - k_z^2}$.

The resonance frequency of this antenna can be estimated using the approximate analytical expressions for the resonance frequency of TE_{111} mode in the a rectangular resonator (three dimensional) given by

$$f_{111} = \frac{c}{2\pi\sqrt{\varepsilon_r}}\sqrt{\left(\frac{\pi}{a}\right)^2 + \left(\frac{\pi}{2b}\right)^2 + \left(\frac{\pi}{d}\right)^2},$$

Three-dimensional case

$$\text{Propagation constant, } \gamma^2 = k^2 - k_c^2$$

A2.1 Fourier Series

$$f(x) = \frac{a_0}{2} + \sum_{n=1}^{\infty}\left[a_n \cos\left(\frac{2n\pi}{a}x\right) + b_n \sin\left(\frac{2n\pi}{a}x\right)\right]$$

$$a_n = \frac{2}{a}\int_0^a f(x)\cos\left(\frac{2n\pi}{a}x\right)\mathrm{d}x$$

$$b_n = \frac{2}{a}\int_0^a f(x)\sin\left(\frac{2n\pi}{a}x\right)\mathrm{d}x$$

Half-wave Fourier analysis will have odd or even terms, i.e., sine–sine or cosine–cosine.

If $f(x) = f(-x)$, then even harmonics will take place and only cosine terms will occur, i.e.,

$$f(x) = \sum_{n=1}^{\infty} C_n \cos\left(\frac{\pi n x}{a}\right)$$

where $C_n = \frac{2}{a}\int_0^a f(x)\cos\left(\frac{n\pi}{a}x\right)\mathrm{d}x$

Similarly for odd terms, $f(x) \neq f(-x)$,

$$f(x) = \sum_{n=1}^{\infty} B_n \sin\left(\frac{\pi n x}{a}\right)$$

where $B_n = \frac{2}{a}\int_0^a f(x)\sin\left(\frac{n\pi}{a}x\right)\mathrm{d}x.$

A2.2 Spectral Resolution of EM Waves

Every wave can be subjected to the process of spectral resolution, i.e., can be represented as a superposition of monochromatic waves of various frequencies. The character of this expansion varies according to the character of the time dependence of the fields.

One category consists of those cases where the expansion contains frequencies forming a discrete sequence of values. The simplest case of this type arises in the resolution of a purely periodic field. This is the usual expansion in Fourier series. It contains the frequencies which are integral multiples of the "fundamental" frequency $\omega_0 = {}^{2\pi}/_{T}$, where T is the period of the field. We therefore write it in the form as follows:

$$f = \sum_{n=-\infty}^{\infty} f_{ne^{-j\omega_0 nt}}$$

where f is any of the quantities describing the field. The quantities f_n are defined in terms of the function f by the integrals

$$f_n = \frac{1}{T} \int_{-T/2}^{T/2} f(t) e^{jn\omega_0 t\mathrm{d}t}.$$

Because $f(t)$ must be real

$$f_n = f_n^*.$$

in more complicated cases, the expansion may contain integral multiples of several different incommensurable fundamental frequencies. When the sum is squared and averaged over the time, the product of terms with different frequencies is given zero because they contain oscillating factors.

Only terms of the form $f_n f_{-n} = |f_n|^2$ remain. Thus, the average of the square of the field, i.e., the average intensity of the wave, is the sum of the intensities of its monochromatic components.

$\overline{f^2} = \sum_n = -\infty^\infty |f_n|^2 = 2\sum_{n=1}^{\infty} |f_n|^2$, where it is assumed that the average of the function f over a period is zero. Another category consists of fields which are expandable in a Fourier integral containing a continuous distribution of different frequencies. For this to be possible, the function $f(t)$ must satisfy certain definite conditions; usually we consider functions which vanish for $t \to \pm\infty$.

Similarly, $f_{-\omega} = f_{\omega}^*$; let us express the total intensity of the wave, i.e., the integrals of f^2 over all time, in terms of the intensity of the Fourier components. Now, we have

$$\int_{-\infty}^{\infty} f^2 \mathrm{d}t = \int_{-\infty}^{\infty} \left\{ f \int_{-\infty}^{\infty} f_\omega e^{-j\omega t} \frac{\mathrm{d}\omega}{2\pi} \right\} \mathrm{d}t = \int_{-\infty}^{\infty} \left\{ f_\omega \int_{-\infty}^{\infty} f e^{-j\omega t} \mathrm{d}t \right\} \frac{\mathrm{d}\omega}{2\pi}$$

$$= \int_{-\infty}^{\infty} f_\omega f_{-\omega} \frac{\mathrm{d}\omega}{2\pi},$$

or

$$\int_{-\infty}^{\infty} f^2 \mathrm{d}t = \int_{-\infty}^{\infty} |f_\omega|^2 \frac{d\omega}{2\pi} = 2 \int_{0}^{\infty} |f_\omega|^2 \frac{d\omega}{2\pi}.$$

$f(t) = \frac{1}{2\pi} \int_{-\infty}^{\infty} f_\omega e^{-j\omega t} \mathrm{d}\omega$, where the Fourier components are given in terms of the function $f(t)$ by the integrals, $f_\omega = \int_{-\infty}^{\infty} f(t) e^{j\omega t} \mathrm{d}t$.

A2.3 Coordinate System and Their Transformations

Rectangular (x, y, z), cylindrical (ρ, ϕ, z), and spherical (r, θ, ϕ) coordinates can be expressed as follows:

$$x \rho \cos\phi = r \sin\theta \cos\phi.$$

$$y = \rho \sin\phi = r \sin\theta \sin\phi \cdot z = r \cos\theta.$$

$$\rho = \sqrt{x^2 + y^2} = \mathrm{r}\ \sin\theta.$$

$$\theta = \tan^{-1} \frac{\sqrt{x^2 + y^2}}{z} = \tan^{-1} \frac{\rho}{z}$$

Transformations of the coordinate components of a vector among the three coordinate systems are given by

$$\begin{aligned} A_z &= A_\rho \cos\phi - A_\phi \sin\phi \\ &= A_r \sin\theta \cos\phi + A_\theta \cos\theta \cos\phi - A_\phi \sin\phi \end{aligned}$$

$$\begin{aligned} A_y &= A_\rho \sin\phi - A_\phi \cos\phi \\ &= A_r \sin\theta \sin\phi + A_\theta \cos\theta \sin\phi - A_\phi \cos\phi \end{aligned}$$

$$\begin{aligned} A_z &= A_r \cos\theta - A_\theta\ \sin\theta \\ A_\rho &= A_z \cos\phi + A_y \sin\phi = A_r \sin\theta + A_\theta \cos\theta \\ A_\phi &= -A_x \sin\phi + A_y \cos\phi \end{aligned}$$

$$\begin{aligned}A_r &= A_x \sin\theta\cos\phi + A_y \sin\theta\sin\phi + A_z\cos\theta\\ &= A_\rho \sin\theta + A_z\cos\theta\end{aligned}$$

$$\begin{aligned}A_\theta &= A_x \cos\theta\cos\phi + A_y \cos\theta\sin\phi - A_z\sin\theta\\ &= A_\rho \cos\theta - A_z\sin\theta\end{aligned}$$

unit vector in the three systems are denoted by $(u_x, u_y, u_z), (u_\rho, u_\phi, u_z)$, and (u_r, u_θ, u_ϕ)

$$\mathrm{d}r = \mathrm{d}x\mathrm{d}y\mathrm{d}z = \rho\mathrm{d}\rho\mathrm{d}\phi\mathrm{d}z = r^2\sin\theta\mathrm{d}r\mathrm{d}\theta\mathrm{d}\phi$$

$$\begin{aligned}\text{Area}\,(D_s) &= u_x\mathrm{d}y\mathrm{d}z + u_y\mathrm{d}x\mathrm{d}z + u_z\mathrm{d}x\mathrm{d}y\\ &= u_\rho\rho\mathrm{d}\phi\mathrm{d}z + u_\phi\mathrm{d}\rho\mathrm{d}z + u_z\rho\mathrm{d}\rho\mathrm{d}\phi\\ &= u_r r^2\sin\theta\mathrm{d}\theta\mathrm{d}\phi + u_\theta r\sin\theta\mathrm{d}r\mathrm{d}\phi + u_\phi r\mathrm{d}r\mathrm{d}\theta\\ \text{Length}\,(L) &= u_x\mathrm{d}x + u_y\mathrm{d}y + u_z\mathrm{d}z\\ &= u_\rho\mathrm{d}\rho + u_\phi\rho\mathrm{d}\phi + u_z\mathrm{d}z\end{aligned}$$

$$u_r\mathrm{d}r + u_\theta r\mathrm{d}\theta + u_\phi r\sin\theta\mathrm{d}\phi$$

Scalar multiplication is defined by

$$A\cdot B = A_1B_1 + A_2B_2 + A_3B_3$$

$$\nabla\cdot\nabla v = \nabla^2 v$$

$$\nabla\times\nabla\times A = \nabla(\nabla\cdot A) - \nabla^2 A$$

$$\mathrm{Re}(re^{j\o}) = r\cos(\omega t + \theta)$$

$$\mathrm{Im}(re^{j\o}) = r\sin(\omega t + \theta)$$

Kronecker Tensor $\otimes$

$f = \frac{1}{2\pi\sigma^2}e^{\frac{-(n-a)^2}{2\sigma^2}}$ where a is the mean and σ is the variance and vector multiplication can be defined as:

$$A\times B = \begin{vmatrix} u_1 & u_2 & u_3\\ A_1 & A_2 & A_3\\ B_1 & B_2 & B_3\end{vmatrix}$$

The differential operators are the gradient $(\nabla\omega)$,

$$\text{Divergence}\,(\nabla \cdot A),$$

$$\text{curl}\,(\nabla \times A)$$

$$\text{Laplacian operator}\,(\nabla_{\omega}^{2})$$

In rectangular coordinates, we can think of del (∇) as the vector operator

$$\nabla = u_x \frac{\partial}{\partial x} + u_y \frac{\partial}{\partial y} + u_z \frac{\partial}{\partial z}$$
$$\nabla_{\omega} = u_x \frac{\partial \omega}{\partial x} + u_y \frac{\partial \omega}{\partial y} + u_z \frac{\partial \omega}{\partial z}$$
$$\nabla \cdot A = \frac{\partial A_x}{\partial x} + \frac{\partial A_y}{\partial x} + \frac{\partial A_z}{\partial x}$$

$$\nabla \times A = \begin{vmatrix} u_x & u_y & u_z \\ \frac{\partial}{\partial x} & \frac{\partial}{\partial y} & \frac{\partial}{\partial z} \\ A_z & A_y & A_z \end{vmatrix}$$

$$\nabla_{\omega}^{2} = \frac{\partial_{\omega}^{2}}{\partial_{x}^{2}} + \frac{\partial_{\omega}^{2}}{\partial_{y}^{2}} + \frac{\partial_{\omega}^{2}}{\partial_{z}^{2}}$$

In cylindrical coordinates, we have

$$\nabla \omega = u_\rho \frac{\partial \omega}{\partial \rho} + u_\phi \frac{1}{\rho}\frac{\partial \omega}{\partial \phi} + u_z \frac{\partial \omega}{\partial z}$$
$$\nabla \cdot A = \frac{1}{\rho}\frac{\partial}{\partial \rho}(\rho A_\rho) + \frac{1}{\rho}\frac{\partial A_\phi}{\partial \phi} + \frac{\partial A_z}{\partial z}$$
$$\nabla \times A = u_r \left(\frac{1}{\rho}\frac{\partial A_z}{\partial \phi} - \frac{\partial A_\phi}{\partial z}\right) + u_\phi \left(\frac{\partial A_\rho}{\partial z} - \frac{\partial A_z}{\partial \rho}\right) + u_z \left[\frac{1}{\rho}\frac{\partial}{\partial_\rho}(\rho A_\phi) - \frac{1}{\rho}\frac{\partial A_\rho}{\partial \phi}\right]$$
$$\nabla_{\omega}^{2} = \frac{1}{\rho}\frac{\partial}{\partial \rho}\left(\rho \frac{\partial \omega}{\partial \rho}\right) + \frac{1}{\rho^2}\frac{\partial_{\omega}^{2}}{\partial_{\phi}^{2}} + \frac{\partial_{\omega}^{2}}{\partial_{z}^{2}}$$

In spherical coordinates, we have

$$\nabla_{\omega} = u_r \frac{\partial \omega}{\partial r} + u_{\phi} \frac{1}{r} \frac{\partial \omega}{\partial \phi} + u_{\phi} \frac{1}{r \sin \theta} \frac{\partial \omega}{\partial \phi}$$

$$\nabla \cdot A = \frac{1}{r^2} \frac{\partial}{\partial r} \left(r^2 A_r\right) + \frac{1}{r \sin \theta} \frac{\partial}{\partial \theta} (A_{\theta} \sin \theta) + \frac{1}{r \sin \theta} \frac{\partial A_{\phi}}{\partial \phi}$$

$$\nabla \times A = \mathrm{u}_r \frac{1}{r \sin \theta} [\frac{\partial}{\partial \theta} (A_{\theta} \sin \theta) - \frac{\partial A_{\theta}}{\partial \phi}$$

$$+ u_{\theta} \frac{1}{r} [\frac{1}{\sin \theta} \frac{\partial A_r}{\partial \phi} - \frac{\delta}{\delta r} \left(r A_{\phi}\right) + u_{\varnothing} \frac{1}{r} \left(\frac{\partial}{\partial r} (r A_{\theta}) - \frac{\partial A_r}{\partial \theta} \right)$$

$$\nabla_{\omega}^2 = \frac{1}{r^2} \frac{\partial}{\partial r} \left(r^2 \frac{\partial \omega}{\partial r} \right) + \frac{1}{r^2 \sin \theta} \frac{\partial}{\partial \theta} \left(\sin \theta \frac{\partial \omega}{\partial \theta} \right) + \frac{1}{r^2 \sin^2 \theta} \frac{\partial_{\omega}^2}{\partial_{\phi}^2}$$

$$R = u_x x + u_y y + u_z z$$

And the "source coordinates" by

$$r' = u_x x' + u_y y' + u_z z'$$

$$|r - r'| = \sqrt{(x - x')^2 + (y - y')^2 (z - z')^2}$$

$$A = \frac{Il e^{-jk|r - r'|}}{4\pi |r - r'|}$$

To emphasize that A is evaluated at the field point (x, y, z) and Il is situated at the source point (x', y', z') (Table A2.1),

$$A(r) = A = \frac{Il(r') e^{-jk|r - r'|}}{4\pi |r - r'|}$$

Table A2.1 Frequency in H_z

Frequency	Symbol	Frequency in H_z
Tera	T	10^{12}
Giga	G	10^{9}
Mega	M	10^{6}
Kilo	K	10^{3}
Hecto	H	10^{2}
Deca	Da	10^{1}
Deci	d	10^{-1}
Centi	cm	10^{-2}
Milli	mm	10^{-3}
Micro	μ	10^{-6}
Nano	n	10^{-9}
Pico	p	10^{-12}
Femto	f	10^{-15}

Annexure-3

Design Steps of RDRA Using ADS Software

Steps →

1. Export the model from HFSS and save in G drive or any file (without Path).
2. Now right click the ADS icon and click run as administration.
3. Then, click the yes button.
4. Then, click the cancel and go create new project.
5. Now on schematic will open, then go to layout button, then go to create update layout.
6. Then go to file button and then go to import button, the layout model is complete.
7. Now click on line which is connected to the patch on layout model then delete it.
8. Then go to the view button then go to layer view then go to by name. Then go to conductor 2 button, now then drag the feed or patch and date it.
9. Now go the each capacitor then click double and give it value according to the formula.

   $C_{(v)} = 26f$ at $v = 0$ $\quad C_i = 1.298\mu f$
   $C_f = 0f$ and add each capacitor by line by clicking on line icon.

10. Now go to S-parameter then click on termination which also given in Fig. A3.1.
11. Now go to the S.P (S-parameter) button and put on schematic window then click the S-parameter which is on the schematic window and put frequency 1 to 3 by stepping 1 MHz frequency then ok.
12. Now go to simulate button and simulate it then after completing the button.
13. Now then go to EDS model, then go to substrate and create update then go to open button put substrate (RT Duroid-5880) then put the thickness of the substrate (1.524 mm) loss tangent (0.001) then go to apply and then go to ok.
14. Now again go to EDS model then go to component. Now go to create update then put start frequency and stop frequency 3 GHz.
15. Now put the port on the patch by single clicking on the patch from port Ze on.
16. Now minimize it.

R.S. Yaduvanshi and H. Parthasarathy, *Rectangular Dielectric Resonator Antennas*, DOI 10.1007/978-81-322-2500-3

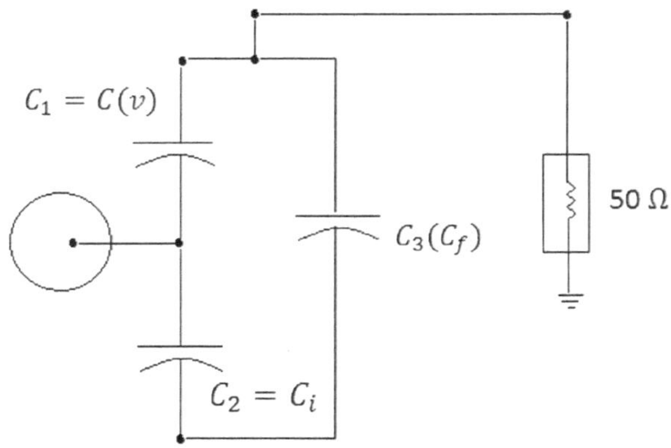

Fig. A3.1 Then go to ⏚ button and connected to the termination

17. Now go to schematic window and then go to library file and click on anywhere on schematic window.
18. Now go to lumped element and select on capacitor and put three by pressing control button.
    PCB manufacturing from HFSS model
    1. Save HFSS model bottom as view .dxf file after going to modeler and exporting it
    2. open .dfx in AutoCAD to generate .pdf or image as .jpg format.
    3. use butter paper to place this design on to PCB
    4. now connect SMA connectors and it is ready for testing antenna parameters.

II. HFSS design steps:

APPLY MAGNETIC AND ELECTRIC BIAS TO MHD ANTENNA
MAGNETIC BIASING STEPS WITH HFSS:

1. **MAKE THREE SLOTS**
2. **SLOTS SHOULD BE ENCLOSING MICRO-STRIP FEED LINE**
3. **THE UPPER EDGE OF ALL THE SLOTS SHOULD TOUCH EACH OTHER**
4. **THE SUBS AND SLOTS SHOULD NOT INTERSECT**
5. **UNITE ALL THE SLOTS**
6. **SELECT MATERIAL**
7. **(A) FERRITE**
8. **(B) MAGNETIC SATURATION EG 500 TESLA**
9. **GO TO BOX → ASSIGN EXCITATION → MAGNETIC BIAS—**
10. **NEXT**
11. **PERMEABILITY**
12. ***X*, *Y*, *Z* VALUE-DESIRED**
13. **FINISH**
14. **CHECK FOR VALIDATION**

15. **RELOCATE SLOT IF REQUIRED**
16. **SIMULATE**

ELECTRIC BIASING STEPS WITH HFSS:

1. **INSERT TWO BOXES OF COPPER INSIDE THE DRA OVER THE SLOT**

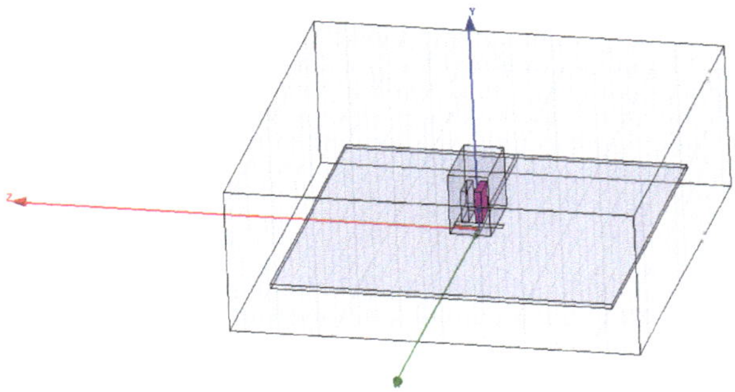

2. **NOW APPLY VOLTAGE BIAS BY RIGHT CLICK AND APPLY +15 V**

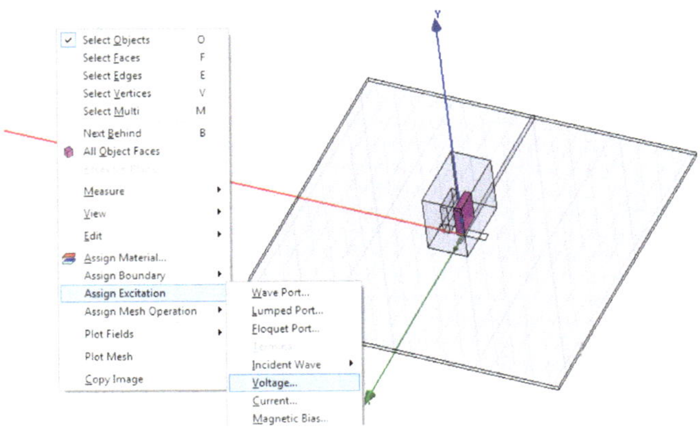

3. **WHEN CLICK ON VOLTAGE, THIS WINDOW COME WHERE WE ENTER VOLTAGE AND E FIELD DIRECTION**

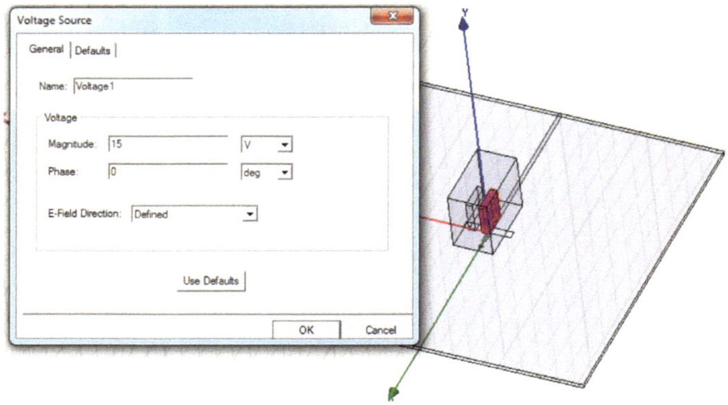

4. **IN THE SAME WAY, WE APPLY ELECTRIC BIAS TO SECOND ELECTRODE**

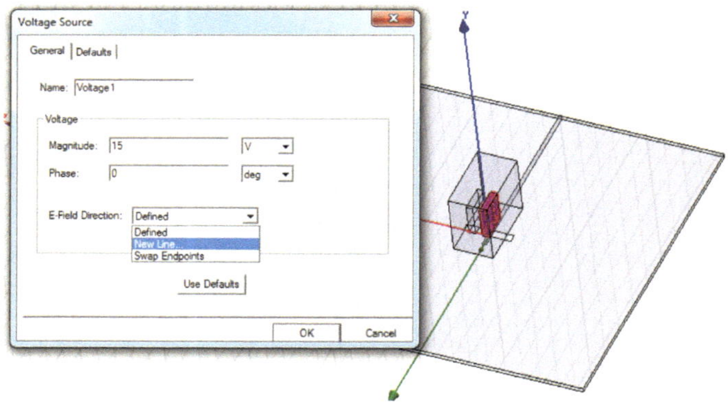

Annexure-4

Resonating Modes in Rectangular Resonators

See Fig. A4.1.

Rectangular waveguide solution:
Helmholtz equation

$$\nabla^2\psi + k^2\psi = 0 \text{ (source less medium)}$$
$$\nabla^2\psi + k^2\psi = -\mu j \text{ (medium with source)}$$

Maxwell's equations

$$\nabla \times E = -\mu \frac{\partial H}{\partial t}$$

$$\nabla \times H = j + \frac{\partial E}{\partial t}$$

Solving LHS of both sides first

$$\nabla \times E = \begin{vmatrix} i & j & k \\ \frac{\partial}{\partial x} & \frac{\partial}{\partial y} & \frac{\partial}{\partial z} \\ E_x & E_y & E_z \end{vmatrix} = i\left(\frac{\partial E_z}{\partial y} - \frac{\partial E_y}{\partial z}\right) - j\left(\frac{\partial E_z}{\partial x} - \frac{\partial E_x}{\partial z}\right) + k\left(\frac{\partial E_y}{\partial x} - \frac{\partial E_x}{\partial y}\right)$$

$$\nabla \times H = \begin{vmatrix} i & j & k \\ \frac{\partial}{\partial x} & \frac{\partial}{\partial y} & \frac{\partial}{\partial z} \\ H_x & H_y & H_z \end{vmatrix} = i\left(\frac{\partial H_z}{\partial y} - \frac{\partial H_y}{\partial z}\right) - j\left(\frac{\partial H_z}{\partial x} - \frac{\partial H_x}{\partial z}\right) + k\left(\frac{\partial H_y}{\partial x} - \frac{\partial H_x}{\partial y}\right)$$

Comparing with RHS in both equations and getting value of H_x, H_y, H_z and E_x, E_y, E_z, we get

R.S. Yaduvanshi and H. Parthasarathy, *Rectangular Dielectric Resonator Antennas*, DOI 10.1007/978-81-322-2500-3

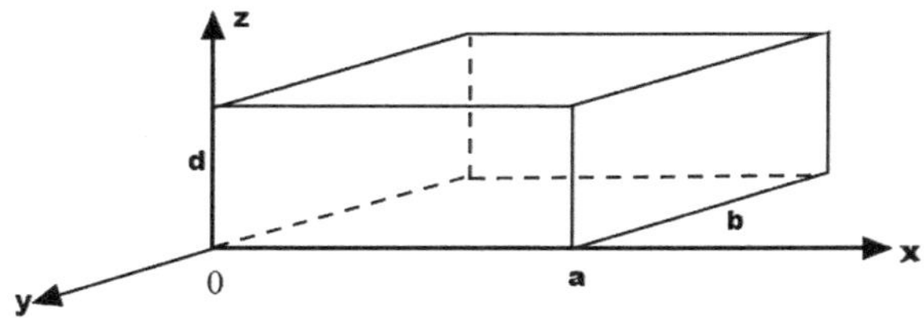

Fig. A4.1 Resonating modes in rectangular resonators

$$H_x = \frac{1}{-j\omega\mu}\left(\frac{\partial E_z}{\partial y} - \frac{\partial E_y}{\partial z}\right) \tag{A4.1}$$

$$H_y = \frac{1}{j\omega\mu}\left(\frac{\partial E_z}{\partial x} - \frac{\partial E_x}{\partial z}\right) \tag{A4.2}$$

$$H_z = \frac{1}{-j\omega\mu}\left(\frac{\partial E_y}{\partial x} - \frac{\partial E_x}{\partial y}\right) \tag{A4.3}$$

$$E_x = \frac{1}{j\omega\varepsilon}\left(\frac{\partial H_z}{\partial y} - \frac{\partial H_y}{\partial z}\right) \tag{A4.4}$$

$$E_y = \frac{1}{-j\omega\varepsilon}\left(\frac{\partial H_z}{\partial x} - \frac{\partial H_x}{\partial z}\right) \tag{A4.6}$$

$$E_z = \frac{1}{j\omega\varepsilon}\left(\frac{\partial H_y}{\partial x} - \frac{\partial H_x}{\partial y}\right) \tag{A4.7}$$

Substituting: $-\frac{\partial}{\partial z} = \gamma$;

$$H_x = \frac{j\omega\varepsilon\frac{\partial E_z}{\partial y} + \gamma\frac{\partial H_z}{\partial x}}{\gamma^2 + \omega^2\mu\varepsilon} \qquad E_x = \frac{-j\omega\mu\frac{\partial H_z}{\partial y} + \gamma\frac{\partial E_z}{\partial x}}{\gamma^2 + \omega^2\mu\varepsilon}$$

$$H_y = \frac{-j\omega\frac{\partial E_z}{\partial x} + \gamma\frac{\partial H_z}{\partial x}}{\gamma^2 + \omega^2\mu\varepsilon} \qquad E_y = \frac{j\omega\mu\frac{\partial H_z}{\partial x} + \gamma\frac{\partial E_z}{\partial y}}{\gamma^2 + \omega^2\mu\varepsilon}$$

$$-(\gamma^2 + \omega^2\mu\varepsilon) \qquad H_z = \frac{\delta^2 H_z}{\delta x^2} + \frac{\delta^2 H_z}{\delta y^2}$$

$$-(\gamma^2 + \omega^2\mu\varepsilon) \qquad E_z = \frac{\delta^2 E_z}{\delta x^2} + \frac{\delta^2 E_z}{\delta y^2}$$

On looking above equations, we get that H_z, E_z in 2-D Helmholtz equation Now, rewriting Helmholtz equation for source-free medium

$$\nabla^2\psi + k^2\psi = 0$$

Here, k is the wave number

$$\Psi = X(x)Y(y)Z(z)$$

$$\left(\frac{1}{X}\left(\frac{\mathrm{d}^2X}{\mathrm{d}x^2}\right) + \frac{1}{Y}\left(\frac{\mathrm{d}^2Y}{\mathrm{d}y^2}\right) + \frac{1}{Z}\left(\frac{\mathrm{d}^2Z}{\mathrm{d}z^2}\right)\right) + k^2 = 0$$

Separating the independent terms, we get

$$\frac{1}{X}\left(\frac{\mathrm{d}^2X}{\mathrm{d}x^2}\right) = -k_x^2$$

$$\frac{1}{Y}\left(\frac{\mathrm{d}^2Y}{\mathrm{d}y^2}\right) = -k_y^2$$

$$\frac{1}{Z}\left(\frac{\mathrm{d}^2Z}{\mathrm{d}z^2}\right) = -k_z^2$$

$$k^2 = k_x^2 + k_y^2 + k_z^2$$

$$\Psi = \left\{(A\sin k_x \cdot x + B\cos k_x \cdot x)\left(C\sin k_y \cdot y + D\cos k_y \cdot y\right)\right\}e^{-jk_z z}$$

Solving above function and keeping propagation in +z-direction only, we get TE mode

$$H_z = \sum_{mn}\left\{C_{mn}\left(\cos\frac{m\pi x}{a}\right)\left(\cos\frac{n\pi y}{b}\right)\right\}e^{-jk_z z}; \quad C_{mn} \text{ Fourier Coefficients} \quad \text{(A4.8)}$$

TM mode

$$E_z = \sum_{mn}\left\{D_{mn}\left(\sin\frac{m\pi x}{a}\right)\left(\sin\frac{n\pi y}{b}\right)\right\}e^{-jk_z z}; \quad D_{mn} \text{ Fourier Coefficients} \quad \text{(A4.9)}$$

These Fourier coefficients are resultant of mode amplitude and propagation constant at any instant.

Let $\gamma = -jk_z$ and m, n are integers and a, b are dimensions;

$$\left(\frac{m\pi}{a}\right)^2 + \left(\frac{n\pi}{b}\right)^2 = (k_c)_{mn}; \quad \text{cut off frequency}$$

$$k_z^2 = \omega^2\mu\varepsilon - \left(\left(\frac{m\pi}{a}\right)^2 + \left(\frac{n\pi}{b}\right)^2\right)$$

Hence, EM wave will propagate in z-direction if:

$$\omega^2\mu\varepsilon - \left(\left(\frac{m\pi}{a}\right)^2 + \left(\frac{n\pi}{b}\right)^2\right) > 0$$

This gives cutoff frequency as follows:

$$\omega_c = \frac{1}{\sqrt{\mu\varepsilon}}\sqrt{\left(\left(\frac{m\pi}{a}\right)^2 + \left(\frac{n\pi}{b}\right)^2\right)}$$

It means, waveguide will support all waves having ω greater than ω_c to propagate.

Now, rewriting H_z and E_z

$$H_z = \sum_{mn}\left\{C_{mn}\left(\cos\frac{m\pi x}{a}\right)\left(\cos\frac{n\pi y}{b}\right)\right\}e^{-jk_z z}$$

$$E_z = \sum_{mn}\left\{D_{mn}\left(\sin\frac{m\pi x}{a}\right)\left(\sin\frac{n\pi y}{b}\right)\right\}e^{-jk_z z}$$

Here C_{mn} and D_{mn} are coefficients of H_z and E_z fields

$E_{ix(x,y)}$ Incident EM wave in x-direction;

$E_{iy(x,y)}$ Incident EM wave in y-direction;

$$E_{ix(x,y)} = \sum\left[\frac{j\omega\mu D_{mn}\left(\frac{n\pi}{b}\right) + \gamma_{mn}C_{mn}\left(\frac{m\pi}{a}\right)}{h^2_{m,n}}\right]\cos\left(\frac{m\pi x}{a}\right)\sin\left(\frac{n\pi y}{b}\right)\exp(-\gamma_{mn}z);$$

Similarly,

$$E_{iy(x,y)} = \sum\left[\frac{j\omega\mu D_{mn}\left(\frac{m\pi}{a}\right) + \gamma_{mn}C_{mn}\left(\frac{n\pi}{b}\right)}{h^2_{m,n}}\right]\cos\left(\frac{m\pi x}{a}\right)\sin\left(\frac{n\pi y}{b}\right)\exp(-\gamma_{mn}z);$$

On simplification

$$E_{ix(m,n)} = \frac{j\omega\mu D_{mn}\left(\frac{n\pi}{b}\right) + \gamma_{mn}C_{mn}\left(\frac{m\pi}{a}\right)}{h^2_{m,n}}$$

Similarly

$$E_{iy(m,n)} = \frac{j\omega\mu D_{mn}\left(\frac{m\pi}{a}\right) + \gamma_{mn}C_{mn}\left(\frac{n\pi}{b}\right)}{h_{m,n}^2}$$

$$\begin{bmatrix} E_{ix(m,n)} \\ E_{iy(m,n)} \end{bmatrix} = \begin{bmatrix} \frac{m\pi}{a}\frac{\gamma_{(m,n)}}{h_{m,n}^2} & \frac{n\pi}{b}\frac{j\omega\mu}{h_{m,n}^2} \\ -\frac{\gamma_{(m,n)}}{b}\frac{n\pi}{h_{m,n}^2} & \frac{-j\omega\mu}{a}\frac{m\pi}{h_{m,n}^2} \end{bmatrix} \begin{bmatrix} C_{mn} \\ D_{mn} \end{bmatrix};$$

we can now get the value of C_{mn}, D_{mn} after substitution of $E_{ix(m,n)}, E_{iy(m,n)}$ values.

Where $h_{m,n}^2 = \left(\frac{m}{a}\right)^2 + \left(\frac{n}{b}\right)^2$ and

$$\gamma_{mn} = \sqrt{h_{m,n}^2 - \omega^2\mu\varepsilon}$$

Hence, C_{mn} and D_{mn} gives us relative amplitudes of E_z and H_z fields in TM or TE modes.

Hence, we get solution of possible amplitudes and phase of wave propagating through rectangular waveguide called as modes of propagation.

Half-wave Fourier expansion in waveguide is given as follows:

$$f_{mn} = \int_0^a \cos\left(\frac{m\pi x}{a}\right)\cos\left(\frac{m\pi x}{a}\right)dx = \int_0^b \sin\left(\frac{n\pi y}{b}\right)\sin\left(\frac{n\pi y}{b}\right)dy;$$

even or odd terms, i.e., $f(x) = f(-x)$ for even term (all cosine terms) or even modes.

Where m, m' and n, $n' \geq 1$

$$E_{ix(m,n)} = \frac{2}{ab}\int_0^a\int_0^b E_{ix(x,y)}\left(\cos\frac{m\pi x}{a}\right)\left(\cos\frac{n\pi y}{b}\right)dxdy$$

$$E_{iy(m,n)} = \frac{2}{ab}\int_0^a\int_0^b E_{iy(x,y)}\left(\sin\frac{m\pi x}{a}\right)\left(\sin\frac{n\pi y}{b}\right)dxdy$$

Half-wave Fourier analysis will have odd or even terms, i.e., sine–sine or cosine–cosine.

If $f(x) = f(-x)$, even harmonics will take place and only cosine terms will occur, i.e.,

$$f(x) = \sum_{n=1}^{\infty} C_n \cos\left(\frac{\pi n x}{a}\right)$$

where

$$C_n = \frac{2}{a}\int_0^a f(x)\cos\left(\frac{n\pi}{a}x\right)\mathrm{d}x$$

Similarly for odd terms, $f(x) \neq f(-x)$;

$$f(x) = \sum_{n=1}^{\infty} B_n \sin\left(\frac{\pi n x}{a}\right)$$

where

$$B_n = \frac{2}{a}\int_0^a f(x)\sin\left(\frac{n\pi}{a}x\right)\mathrm{d}x$$

Solving wave equation with boundary conditions $E_{\tan} = 0$, we find E fields and then H fields. Now shape and size of resonator is given, wave equation shall give solution of characteristic frequencies $\omega(mnp)$ called eigenvalues or eigenfrequencies of e-m oscillations of cavity resonator.

Lowest eigenfrequency ω_1 is $\frac{c}{l}$; where l is the dimension of resonator.

Higher frequency $\left(\omega \gg \frac{c}{l}\right)$; then ω is $\frac{v\omega^2}{2\pi^2 c^3}$.

Hence, it depends on volume and net on shape of resonator.

$$\text{For resonator:} \sum_{mnr} f_{mnr} \sin\left(\frac{\pi m x}{a}\right)\sin\left(\frac{\pi n y}{b}\right)\sin\left(\frac{\pi r z}{\mathrm{d}}\right) = f(x,y,z)$$

$$\left(\frac{\partial^2}{\partial x^2}+\frac{\partial^2}{\partial y^2}+\frac{\partial^2}{\partial z^2}\right)\psi(x,y,z) + k^2\psi(x,y,z) = f(x,y,z);$$

Helmholtz equation

$$\psi(x,y,z) = \sum_{mnr} C_{mnr}\sin\left(\frac{\pi m x}{a}\right)\sin\left(\frac{\pi n y}{b}\right)\sin\left(\frac{\pi r z}{d}\right)$$

$$\left[k^2 - \pi^2\left(\frac{m^2}{a^2}+\frac{h^2}{b^2}+\frac{r^2}{d^2}\right)\right]C_{mnr} = f_{mnr}$$

$$\text{Amplitude coefficient, } C_{mnr} = \frac{f_{mnr}}{k^2 - \left(\frac{m^2}{a^2}+\frac{h^2}{b^2}+\frac{r^2}{d^2}\right)\Big/\omega_{mnr}}$$

$$k = \omega\sqrt{\mu\epsilon}$$

$\omega = \omega(mnp) + \delta$; where δ small deviation and r is different from p.
Hence,

$$C_{mnr} = \frac{f_{mnr}}{(\omega(mnp) + \delta)^2 - \omega(mnr)^2}$$

$$= \frac{f_{mnr}}{\delta(\omega(mnp) - \omega(mnr)}$$

A4-3 Solution of Single-String Resonator

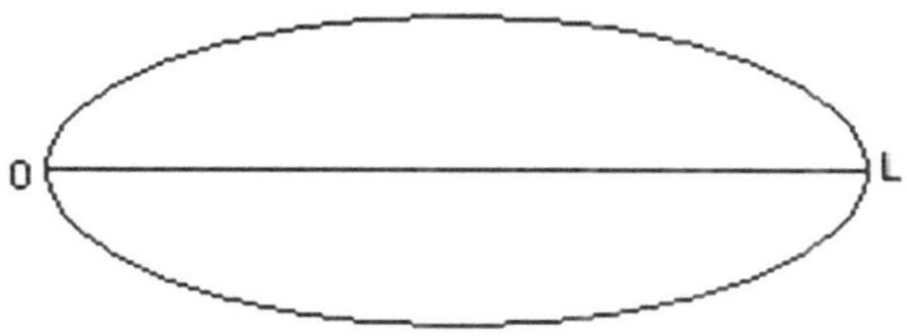

$$x''(t) + \omega_0^2 x(t) = Be^{j\omega t}$$

$$x(t) = Ae^{j\omega t}$$

$$(\omega_0^2 - \omega^2)A = B$$

Hence,

$$A = \frac{B}{\omega_0^2 - \omega^2}$$

$x(t) = \frac{Be^{j\omega t}}{\omega_0^2 - \omega^2}$, if $\omega_0 = \omega$; then $x(t)$ will be ∞

Now, $\omega = \omega_0 + \delta$ when δ is small deviation

$$= \frac{Be^{j\omega t}}{(\omega_0 + \omega)(\omega_0 - \omega)}$$

Hence, the solution of spring resonator is in one dimension

$$= \frac{Be^{j\omega t}}{\delta(2\omega_0)}$$

$$\left(\frac{\partial^2}{\partial x^2} - \frac{1}{c^2}\frac{\partial^2}{\mathrm{d}t^2}\right)u(x,t) = 0; \text{at boundaries}$$

$$u(0,t) = 0 \quad \text{and} \quad u(L,t) = 0$$

Taking Fourier transform of above equation

$$\left(\frac{\partial^2}{\partial x^2} + \frac{\omega^2}{c^2}\right)\hat{u}(x,\omega) = 0$$

Writing above terms in sine and cosine form, we have

$$A\sin\left(\frac{\omega x}{c}\right) + B\cos\left(\frac{\omega x}{c}\right) = 0$$

$$\hat{u}(0,\omega) = 0$$

$$\hat{u}(L,\omega) = 0$$

$\sin\left(\frac{\omega L}{c}\right) = 0$; Hence $k\,L = n\pi$; sine values to be zero.
$\omega = kc = \frac{n\pi c}{L}$, when n = 1, 2, 3 where $k = \omega/c$;
when $2L$, it is fundamental frequency ω_1
when L, the frequency is $2\omega_1$
when $2L/3$, the frequency $3\omega_1$;
which can be generalized as:

$$\sum_n C(n)\sin\left(\frac{n\pi x}{L}\right)$$

A-4 Solution of Two-Dimensional Resonator

General Helmholtz equation is given below (Fig. A4.2):

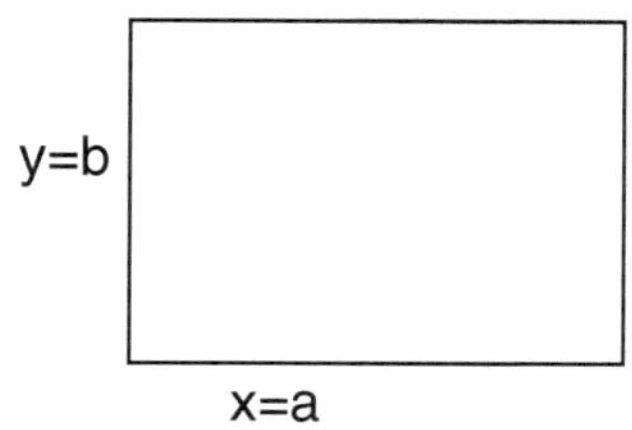

Fig. A4.2 Rectangular resonator

$$\frac{\partial^2\psi(x,y,t)}{\partial x^2}+\frac{\partial^2\psi(x,y,t)}{\partial y^2}-\frac{1}{c^2}\frac{\partial^2\psi(x,y,t)}{\partial t^2}=0$$

Applying boundary conditions

$$\psi(0,y,t)=\psi(a,y,t)=0$$

$$\psi(x,0,t)=\psi(x,b,t)=0$$

Let input excitation be some tension T

$$\sigma \mathrm{d}x\mathrm{d}y\frac{\partial^2\psi}{\partial t^2}=\frac{\partial}{\partial x}\left(T\mathrm{d}y\frac{\partial\psi}{\partial x}\right)\mathrm{d}x+\frac{\partial}{\partial y}\left(T\mathrm{d}x\frac{\partial\psi}{\partial y}\right)\mathrm{d}y$$

$$\frac{Y''}{Y}=-k_Y^2;\quad \frac{X''}{X}=-k_X^2;$$

Now from Helmholtz equation:

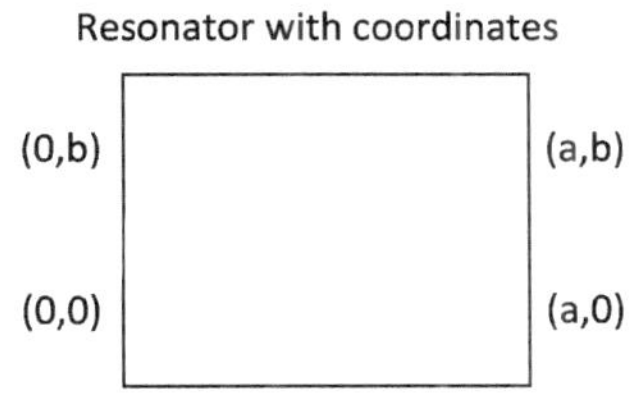

$$\frac{\partial^2\psi}{\partial t^2}-c^2\nabla^2\psi=0$$

Using separation of variables:

$$\psi(x,y,t)=X(x)Y(y)T(t)$$

$$-\omega^2=\frac{T''(t)}{T(t)}=c^2\left(\frac{X''(x)}{X(x)}+\frac{Y''(y)}{Y(y)}\right) \quad \text{(A4.10)}$$

Let

$$X(x)=\sin(k_x x)$$

$$Y(y)=\sin(k_y y)$$

$$k_x^2 + k_y^2 = \frac{\omega^2}{c^2}$$

where k_x and k_y can be

$$k_x = \frac{m\pi}{a}; \quad k_y = \frac{n\pi}{b}$$

Equation (A4.1) can be written as $\omega(mn) = c\pi\sqrt{\left(\frac{m}{a}\right)^2 + \left(\frac{n}{b}\right)^2}$.
From Fourier series analysis

$$\omega(m,n) = \sum_{mn=1}^{\infty} \sin\left(\frac{m\pi x}{a}\right) \sin\left(\frac{n\pi y}{b}\right) \left[C(m,n)e^{j\omega(m,n)t} + D(m,n)e^{-j\omega(m,n)t}\right] \tag{A4.12}$$

At $t = 0$; $\psi(x,y,0) = \psi_0(x,y)$
On differentiating equation $\psi_0(x,y)$, we get $\psi_1(x,y,0) = \psi_1(x,y)$.
When $t \neq 0$;

$$\psi_0(x,y) = \sum_{mn=1}^{\infty} (C(m,n) + D(m,n)) \sin\left(\frac{m\pi x}{a}\right) \sin\left(\frac{n\pi y}{b}\right) \tag{A4.13}$$

$$\psi_1(x,y) = \sum_{mn=1}^{\infty} j\omega(m,n)(C(m,n) - D(m,n)) \sin\left(\frac{m\pi x}{a}\right) \sin\left(\frac{n\pi y}{b}\right) \tag{A4.14}$$

$$\frac{2}{\sqrt{ab}} \int_0^a \int_0^b \psi_0(x,y) \sin\left(\frac{m\pi x}{a}\right) \sin\left(\frac{n\pi y}{b}\right) \mathrm{d}x\mathrm{d}y = [C(m,n) + D(m,n)] \tag{A4.15}$$

Similarly,

$$j\omega(m,n)\frac{2}{\sqrt{ab}} \int \psi_1(x,y) \sin\left(\frac{m\pi x}{a}\right) \sin\left(\frac{n\pi y}{b}\right) \mathrm{d}x\mathrm{d}y = [C(m,n) - D(m,n)] \tag{A4.16}$$

Hence, obtain the value of $C(m,n), D(m,n)$ from Eqs. (A4.3) and (A4.4)

$$[C(m,n), D(m,n)] = \frac{1}{\sqrt{ab}}\left[\iint \psi_0(x,y)\sin\left(\frac{m\pi x}{a}\right)\sin\left(\frac{n\pi y}{b}\right)\mathrm{d}x\mathrm{d}y \pm \frac{1}{j\omega(m,n)}\int \psi_,(x,y)\sin\left(\frac{m\pi x}{a}\right)\sin\left(\frac{n\pi y}{b}\right)\mathrm{d}x\mathrm{d}y\right] \tag{A4.17}$$

Hence, from Eq. (A4.17),

$$\psi_0(x,y) = A\sin\left(\frac{m_0\pi x}{a}\right)\sin\left(\frac{n_0\pi y}{b}\right)$$

$$\psi_1(x,y) = B\sin\left(\frac{m_0\pi x}{a}\right)\sin\left(\frac{n_0\pi y}{b}\right)$$

due to force, perturbation occurs (Fig. A4.3)
Solving equation (A4.17)

$$\begin{aligned}(C(m,n), D(m,n)) &= \delta[m-m_0]\delta[n-n_0]\\ &= \left(\frac{A}{\sqrt{ab}}\left(\frac{a}{2}\times\frac{b}{2}\right) \pm \left(\frac{1}{\sqrt{ab}}\frac{B}{j\omega(m_0,n_0}\right)\left(\frac{a}{2}\times\frac{b}{2}\right)\right)\end{aligned}$$

$$(C(m,n), D(m,n)) = \sqrt{ab}\left(\frac{A}{4}\pm\frac{j\cdot B}{4}\right)\frac{\sqrt{ab}}{4}(A\pm jB)\delta[m-m_0]\delta[n-n_0]$$

$$\psi(x,y,t) = \frac{\sqrt{ab}}{2}\mathrm{Re}(A-jB)\sin\left(\frac{m_0\pi x}{a}\right)\sin\left(\frac{n_0\pi y}{b}\right)e^{j\omega(m_0,n_0t)}$$

Hence, we complete solution of two-dimensional resonator.

$$\psi(x,y,t) = \frac{\sqrt{ab}}{2}(A\cos(\omega(m_0n_0)t) + B\sin(\omega(m_0n_0)t))\sin\left(\frac{m_0\pi x}{a}\right)\sin\left(\frac{n_0\pi y}{b}\right) \tag{A4.18}$$

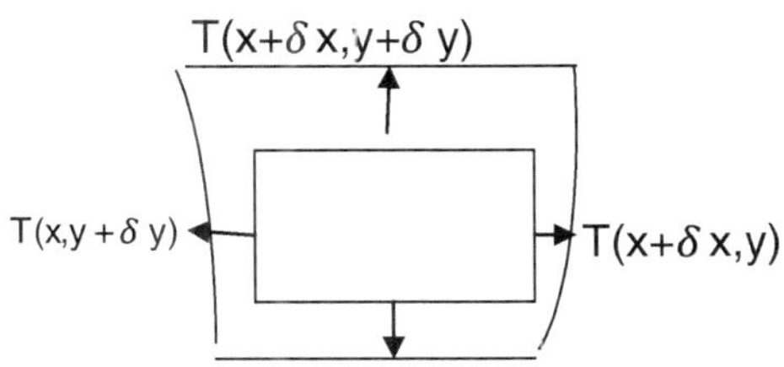

Fig. A4.3 Deformation due to excitation $T(x, y)$

Alternate method

$$m = 2b \sin \theta;$$
$$n = 2a \cos \theta;$$

Dividing both sides of above equations by $2a$ and $2b$ and adding them gives us

$$\frac{1}{\lambda_2} = \frac{n^2}{4a^2} + \frac{m^2}{4b^2}; \quad \text{where } k^2 = k_x^2 + k_y^2$$

Thus, resonant frequency of resonator can be determined.
Half-wave Fourier analysis:

$$f(x) = \frac{a_0}{2} + \sum_{n=1}^{\infty} \left[a_n \cos\left(\frac{2n\pi}{a}x\right) + b_n \sin\left(\frac{2n\pi}{a}x\right) \right]$$
$$a_n = \frac{2}{a} \int_0^a f(x) \cos\left(\frac{2n\pi}{a}x\right) \mathrm{d}x$$
$$b_n = \frac{2}{a} \int_0^a f(x) \sin\left(\frac{2n\pi}{a}x\right) \mathrm{d}x$$

Half-wave Fourier analysis will have odd or even terms, i.e., sine–sine or cosine–cosine.

If $f(x) = f(-x)$, even harmonics will take place and only cosine terms will occur, i.e.,

$$f(x) = \sum_{n=1}^{\infty} C_n \cos\left(\frac{\pi n x}{a}\right)$$

where

$$C_n = \frac{2}{a} \int_0^a f(x) \cos\left(\frac{n\pi}{a}x\right) \mathrm{d}x$$

Similarly for odd terms, $f(x) \neq f(-x)$;

$$f(x) = \sum_{n=1}^{\infty} B_n \sin\left(\frac{\pi n x}{a}\right)$$

where

$$B_n = \frac{2}{a}\int_0^a f(x)\sin\left(\frac{n\pi}{a}x\right)\mathrm{d}x$$

Spectral resolution of EM waves

Every wave can be subjected to the process of spectral resolution, i.e., can be represented as a superposition of monochromatic waves of various frequencies. The character of this expansion varies according to the character of the time dependence of the fields.

One category consists of those cases, where the expansion contains frequencies forming a discrete sequence of values. The simplest case of this type arises in the resolution of a purely periodic field. This is the usual expansion in Fourier series. It contains the frequencies which are integral multiples of the "fundamental" frequency $\omega_0 = {}^{2\pi}/_{T}$, where T is the period of the field. We therefore write it in the form as follows:

$$f = \sum_{n-\infty}^{\infty} f_{ne^{-j\omega_0 nt}}$$

(where f is any of the quantities describing the field). The quantities f_n *are* defined in terms of the function f by the integrals

$$f_n = \frac{1}{T}\int_{-T/2}^{T/2} f(t)e^{jn\omega_0 t\mathrm{d}t.}$$

Because $f(t)$ must be real

$$f_n = f_n^*.$$

in more complicated cases, the expansion may contain integral multiples of several different incommensurable fundamental frequencies. When the sum is squared and averaged over the time, the product of terms with different frequencies is given zero because they contain oscillating factors.

Only terms of the form $f_n f_{-n} = |f_n|^2$ remain. Thus, the average of the square of the field, i.e., the average intensity of the wave, is the sum of the intensities of its monochromatic components. $\overline{f^2} = \sum_{n=-\infty}^{\infty} |f_n|^2 = 2\sum_{n=1}^{\infty} |f_n|^2$; where it is assumed that the average of the function f over a period is zero. Another category consists of fields which are expandable in a Fourier integral containing a continuous distribution of different frequencies. For this to be possible, the function $f(t)$ must satisfy certain definite conditions; usually we consider functions which vanish for $t \to \pm\infty$.

Similarly, $f_{-\omega} = f_{\omega}^{*}$; Let us express the total intensity of the wave, i.e., the integrals of f^2 over all time, in terms of the intensity of the Fourier components. Now, we have

$$\int_{-\infty}^{\infty} f^2 \mathrm{d}t = \int_{-\infty}^{\infty} \left\{ f \int_{-\infty}^{\infty} f_{\omega} e^{-j\omega t} \frac{\mathrm{d}\omega}{2\pi} \right\} \mathrm{d}t = \int_{-\infty}^{\infty} \left\{ f_{\omega} \int_{-\infty}^{\infty} f e^{-j\omega t} \mathrm{d}t \right\} \frac{\mathrm{d}\omega}{2\pi}$$

$$= \int_{-\infty}^{\infty} f_{\omega} f_{-\omega} \frac{\mathrm{d}\omega}{2\pi},$$

or

$$\int_{-\infty}^{\infty} f^2 \mathrm{d}t = \int_{-\infty}^{\infty} |f_{\omega}|^2 \frac{\mathrm{d}\omega}{2\pi} = 2 \int_{0}^{\infty} |f_{\omega}|^2 \frac{\mathrm{d}\omega}{2\pi}.$$

$f(t) = \frac{1}{2\pi} \int_{-\infty}^{\infty} f_{\omega} e^{-j\omega t} \mathrm{d}\omega$; where the Fourier components are given in terms of the function

$f(t)$ by the integrals, $f_{\omega} = \int_{-\infty}^{\infty} f(t) e^{j\omega t} \mathrm{d}t$.

Power and Energy Signals:
Let $x(t)$ is the input signal, i.e., voltage signal. As per Parseval's power theorem, energy associated with this signal be

$$E = \int_{-\infty}^{\infty} |x(t)|^2 \mathrm{d}t; \text{ in time domain}$$

$$= \frac{1}{2\pi} \int_{-\infty}^{\infty} |X(\omega)|^2 \mathrm{d}\omega; \text{ in frequency domain}$$

The amount of energy radiated by this signal, when applied across Antenna having radiation resistance R_r shall be

$$E = \frac{1}{R_r} \int_{-\infty}^{\infty} |x(t)|^2 \mathrm{d}t = \frac{1}{2\pi R_r} \int_{-\infty}^{\infty} |X(\omega)|^2 \mathrm{d}\omega$$

Now if input signal is $x(t)$ having current signal

$$E = R_r \int_{-\infty}^{\infty} |x(t)|^2 \mathrm{d}t = \frac{R}{2\pi} \int_{-\infty}^{\infty} |X(\omega)|^2 \mathrm{d}\omega$$

ESD energy spectral density; energy spread per unit volume across 1 Ω resister

$$\mathrm{ESD} = |X(\omega)|^2$$

Discrete Fourier transform (DFT) in time domain into frequency domain spectral analysis

$$x(k) = \sum_{n=0}^{N-1} x(n)e\frac{-j2\pi nk}{N}; \quad k = 0, 1, \ldots, N-1.$$

$$X(n) = \frac{1}{N}\sum_{k=0}^{N-1} X(k)e\frac{-j2\pi kn}{N}; \quad n = 0, 1, 2, \ldots, N-1.$$

$X(n)$ finite sequence.

DFT has finite length N, period N

$$\psi(\theta, \phi) = k\left(\overrightarrow{r_n} \cdot r_0\right) = (n-1)kd\sin\theta$$

$$E(\sin\theta) = \sum_{n=1}^{N} e^{j(n-1)(kd\sin\theta)}$$

Annexure-5

Resonant Mode Generation and Control in RDRA

In this annexure, resonant modes TE and TM have been generated inside RDRA whose dimensions are *a*, *b*, and *d*. Two parallel plates are attached along with dielectric slab in between these plates to RDRA. This slab forms non-resonant part and RDRA is main resonant. This is shown in Fig. A5.1a, b. The resonant modes dominant and higher-order modes are being generated by maintaining appropriate aspect ratio of RDRA. Then, the non-resonant slab inductance and capacitance is introduced into main RDRA. This lumped value of inductance and capacitance is seen in the resonant frequency.

(a) The increase in the length of **internal strip** introduce shift in the **higher resonant modes** frequency, as they shift toward lower side and vice versa. Hence, **resonant frequency is reduced**.
(b) On the other side, increase in the length of **external strip** introduces **shift in the lower resonant modes** frequency **shifts toward higher side** and vice versa. Hence, frequency is increased with strip length.
(c) Increase in **spacing between parallel plates** introduces the **combined effect** of internal as well as external strip length variation, i.e., **higher- and lower-order resonant modes** shift toward the **centre frequency** which can be seen as mode-merging effect.
(d) Finally, the effect of placing a lumped varactor diode between parallel plates is seen. The increase in the capacitance value of lumped varactor diode causes **shift in the higher resonant frequency** toward **lower resonant frequency** side.

These results have been investigated using HFSS and they shown with S_{11} results along with each RDRA model. By varying length, "*a*," width "*b*," and height "*d*" of RDRA modes are generated. The internal strip, external strip, and dielectric slab and dielectric constant provided several degrees of freedom in the RDRA design. This has extended the control on the amount of coupling, hence resonant frequency. This shall have large impact on resonant modes, compactness of antenna, radiation pattern, and polarization.

R.S. Yaduvanshi and H. Parthasarathy, *Rectangular Dielectric Resonator Antennas*, DOI 10.1007/978-81-322-2500-3

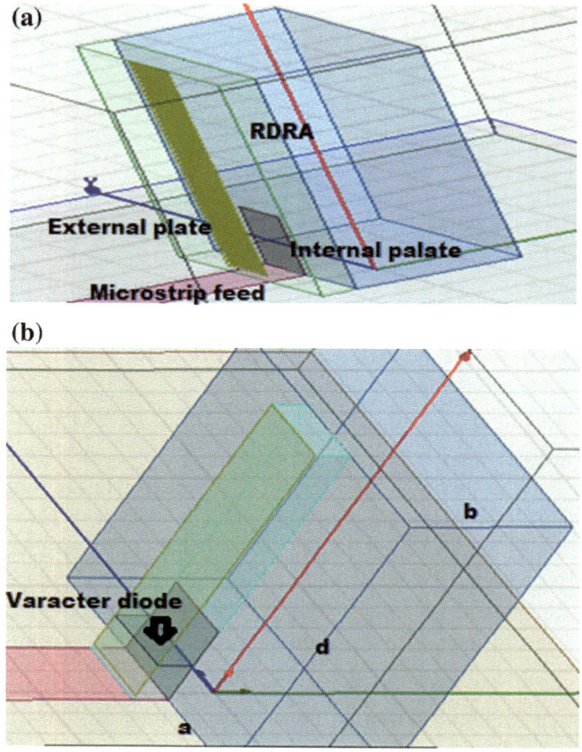

Fig. A5.1 **a** RDRA with two parallel standing strips having rectangular non-resonant slab in between. **b** RDRA with lumped varactor diode between strips

A5.1 Effect of Change of Aspect Ratio (a/b) and (a/d) of RDRA on Resonant Modes

See Fig. A5.2.

A5.2 Effect of Strip Length, Separation, ε_r on the Modes Developed Inside the RDRA

See Fig. A5.3 and Table A5.1.

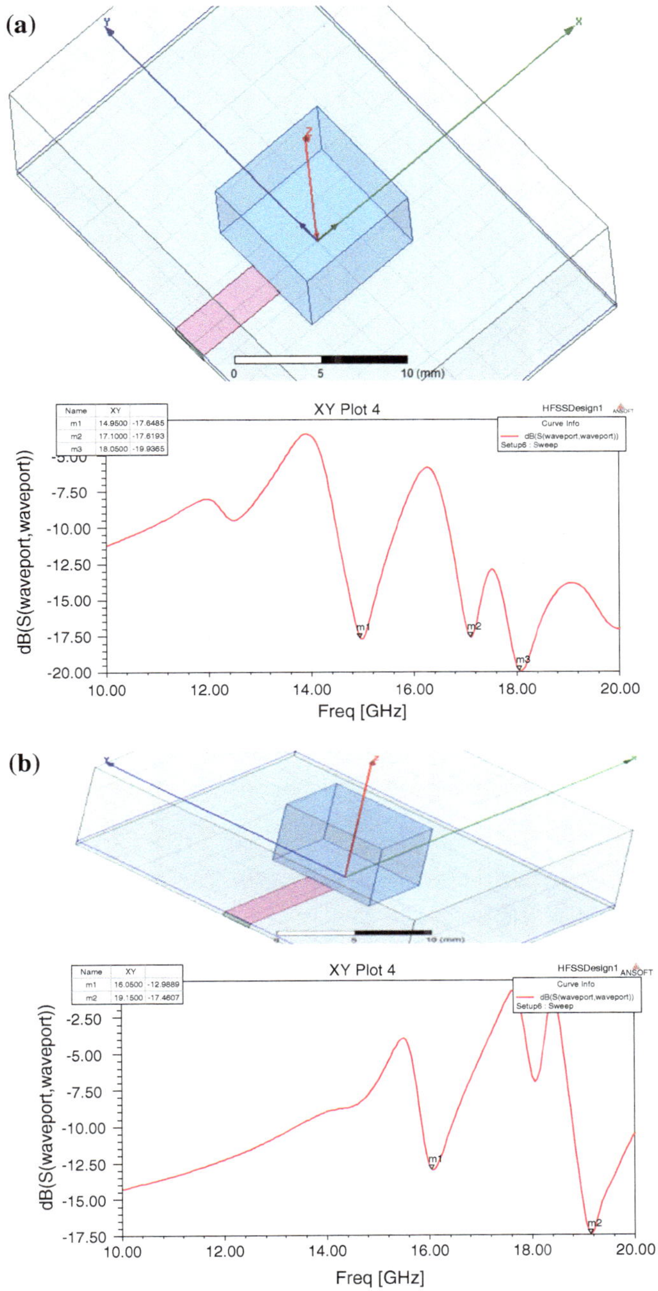

Fig. A5.2 **a** Higher-order modes generated in RDRA with square base. **b** Higher-order modes generated in the rectangular base RDRA

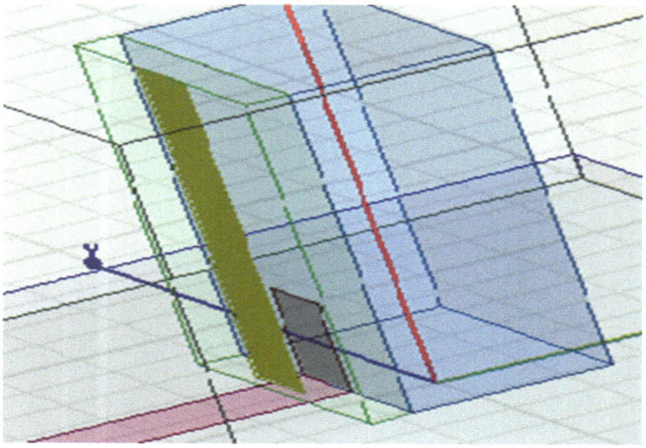

Fig. A5.3 RDRA with parallel standing strips

Table A5.1 Effect of strip length

Structure	x (mm)	y (mm)	z (mm)	ε_r
DRA	4.6	9	10.8	9.8
substrate	20	30	0.8	2.2
Micro-strip	2.4 (width)			
Rect. SLAB	1	9	10.8	1
External strip	2.4		10.5	
Internal strip	2.4		3	

A5.2(a) Effect of Internal Strip Length Variation on Resonant Modes Inside RDRA

The effect of the internal strip length is seen on resonance frequency and resonant modes of RDRA.

The reflection coefficient plot can be seen for the possible changes as given in Fig. A5.4a.

The effective electrically length of RDRA is changed by introducing change in length of internal strip as given below.

Changing the effective dimension of the dielectric resonator changes the resonant frequency.

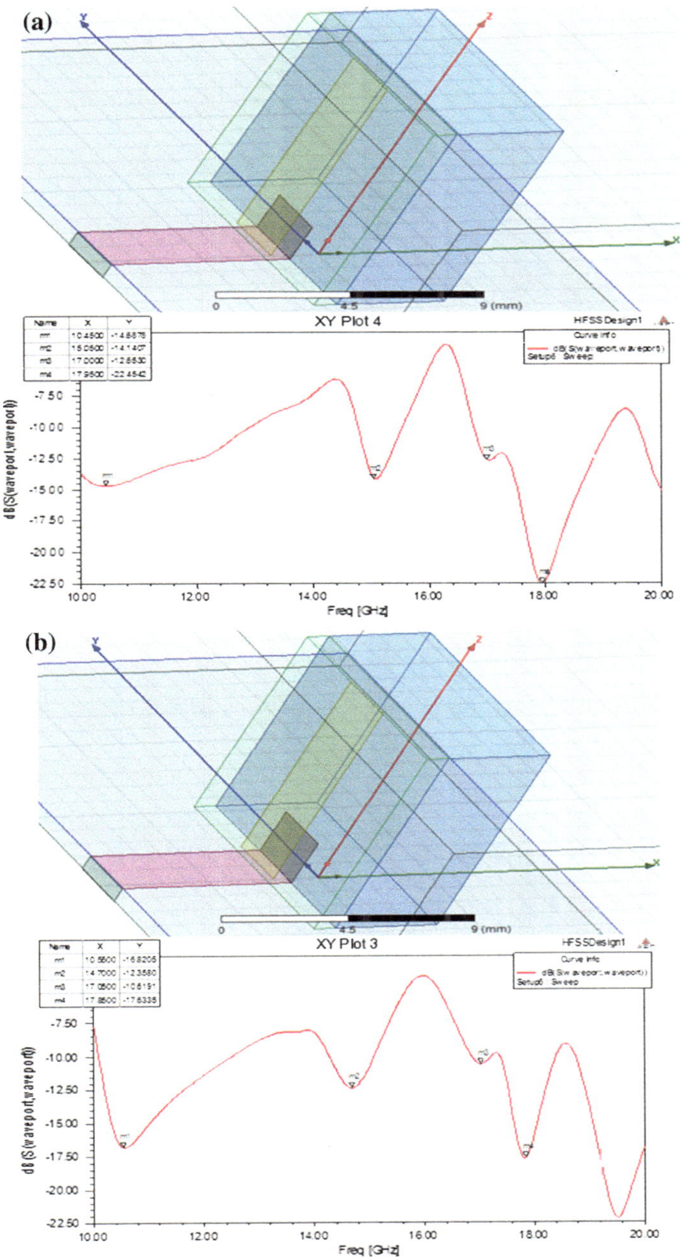

Fig. A5.4 **a** External strip (fixed) = 10.5 mm and variation in internal strip from (2 mm). **b** External strip (fixed) = 10.5 mm and variation in internal strip from (2.5 mm). **c** External strip (fixed) = 10.5 mm and variation in internal strip from (3 mm)

(c)

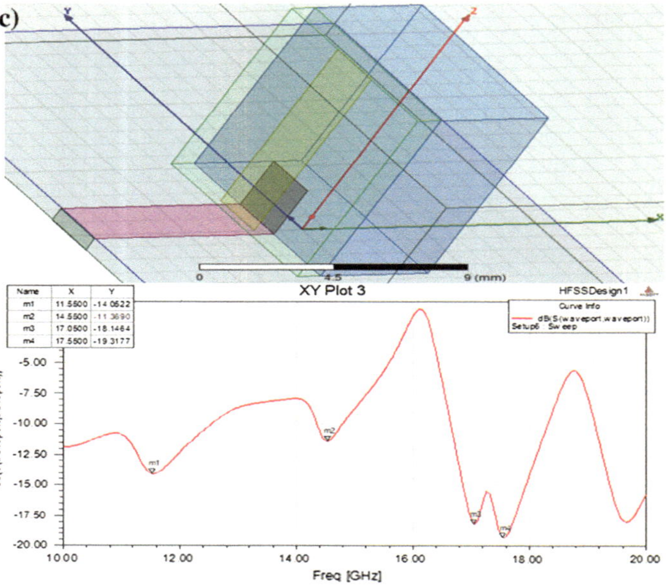

Fig. A5.4 (continued)

A5.2(b) Effect of External Strip Length Variation on Resonant Modes Inside RDRA

The effect of the external strip length on resonance frequency and resonant modes is shown in Fig. A5.5b. Internal strip (fixed) = 3 mm and variation in external strip from 10.5, 7, 0 mm is investigated. Contrary to the previous case, the third resonance stays mainly fixed at the same frequency, while the first and second resonant frequencies are considerably decreased with increasing external strip length.

A5.2(c) Effect of Separation Width Between the Two Parallel Standing Strips and ε_r

The effect of the spacing between parallel plates and permittivity of the rectangular slab between parallel plates is seen on resonance frequency and modes (Figs. A5.6 and A5.7).

- The separation width variation ranges as 0.5, 1.5, and 2.5 mm [$\varepsilon_r = 1$, external strip = 10.5 mm and inner strip = 3.5 mm (fixed)].
- we will change the variable separation Width (0.5, 1.5, 2.5) for $\varepsilon_r = 2$ keeping external strip = 10.5 mm, inner strip = 3.5 mm constant (Fig. A5.8).

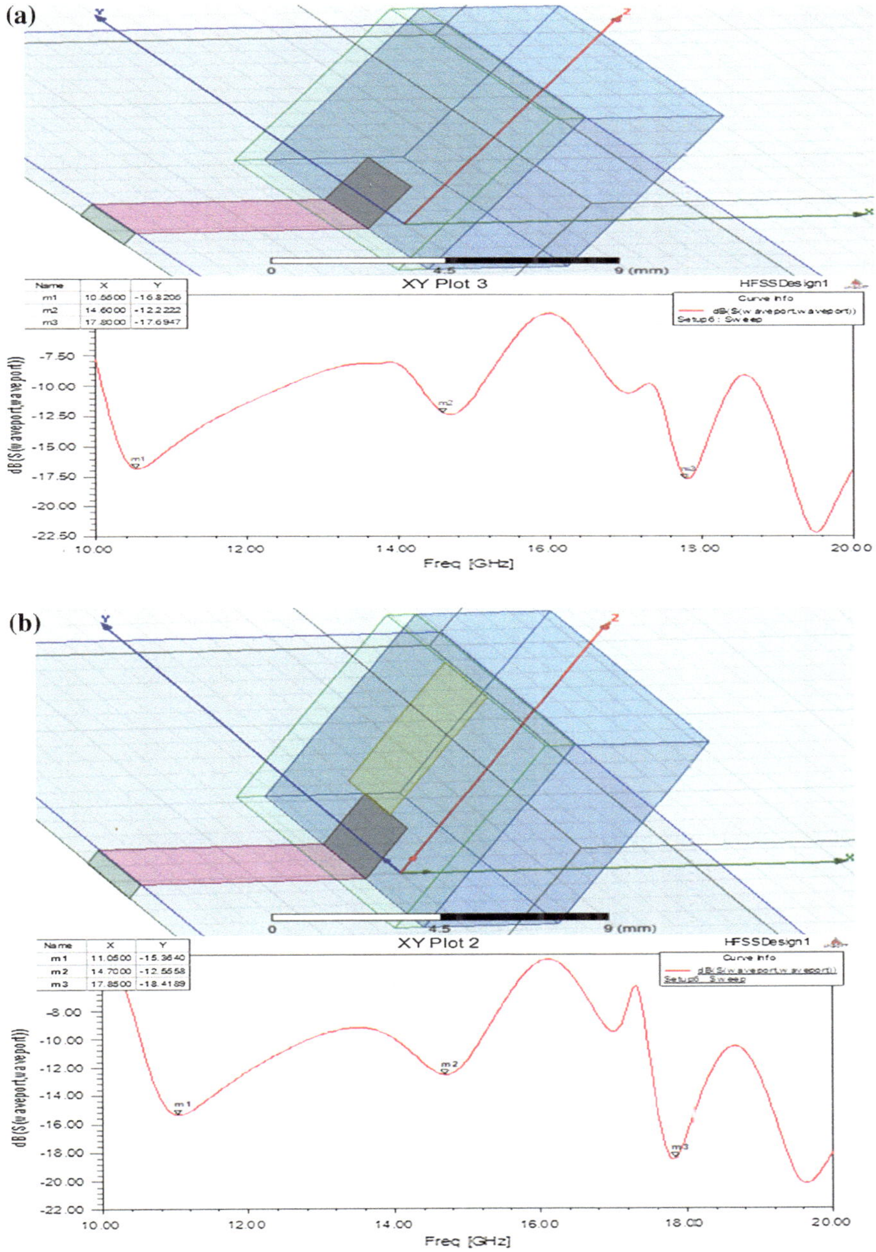

Fig. A5.5 **a** Internal strip (fixed) = 3 mm and variation in external strip from (0 mm). **b** Internal strip (fixed) = 3 mm and variation in external strip from (7 mm). **c** Internal str_p (fixed) = 3 mm and variation in external strip from (10.5 mm)

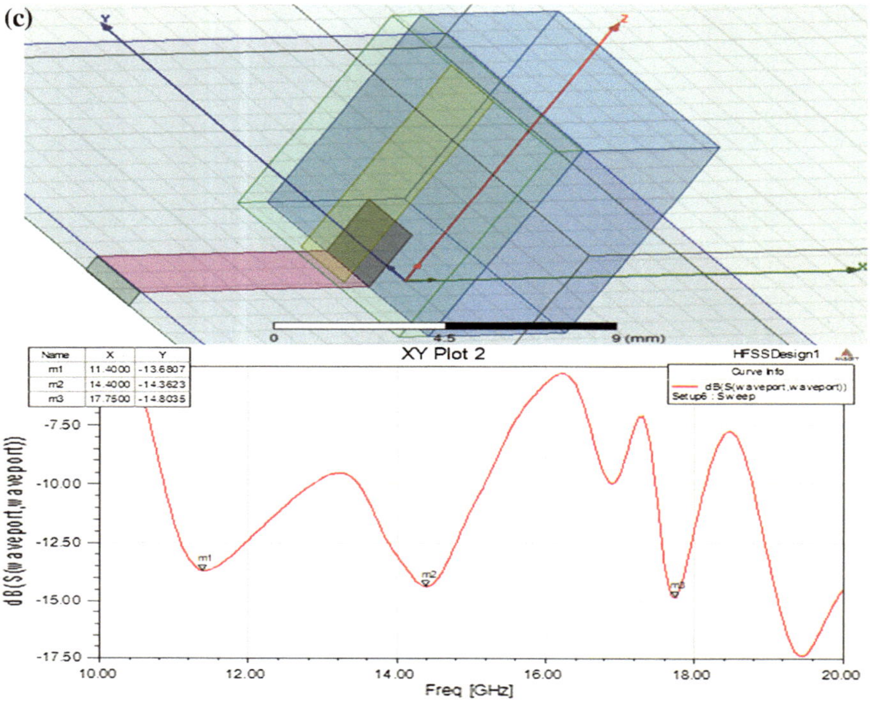

Fig. A5.5 (continued)

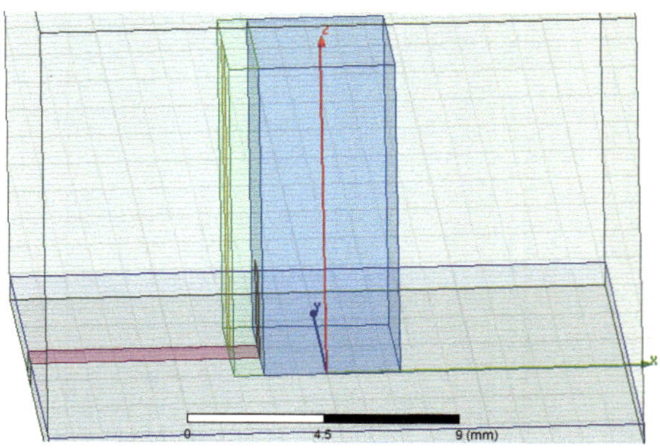

Fig. A5.6 RDRA with separated plates

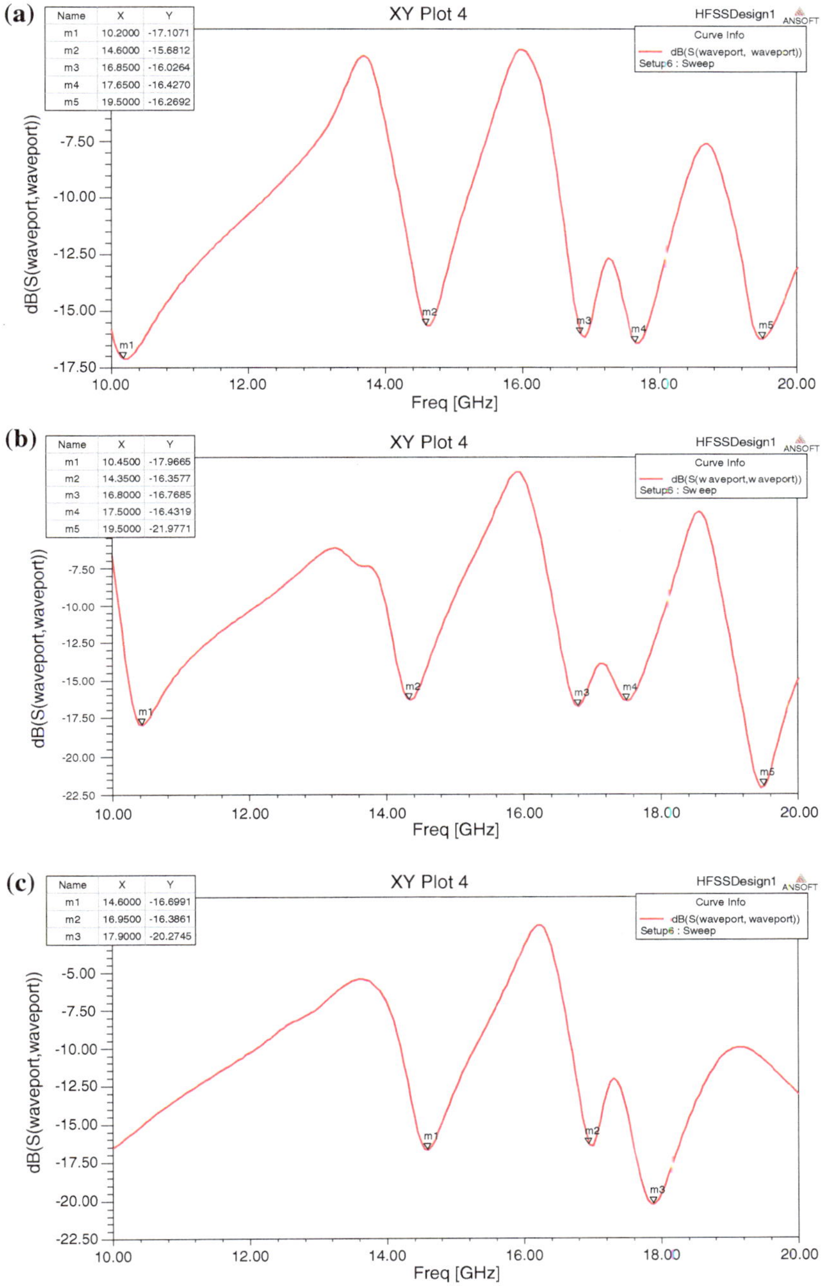

Fig. A5.7 **a** Separation width (0.5 mm) for $\varepsilon_r = 1$, external strip = 10.5 mm, inner strip = 3.5 mm. **b** Separation width (1.5 mm) for $\varepsilon_r = 1$, external strip = 10.5 mm, inner strip = 3.5 mm. **c** Separation width (2.5 mm) for $\varepsilon_r = 1$, external strip = 10.5 mm, inner strip = 3.5 mm

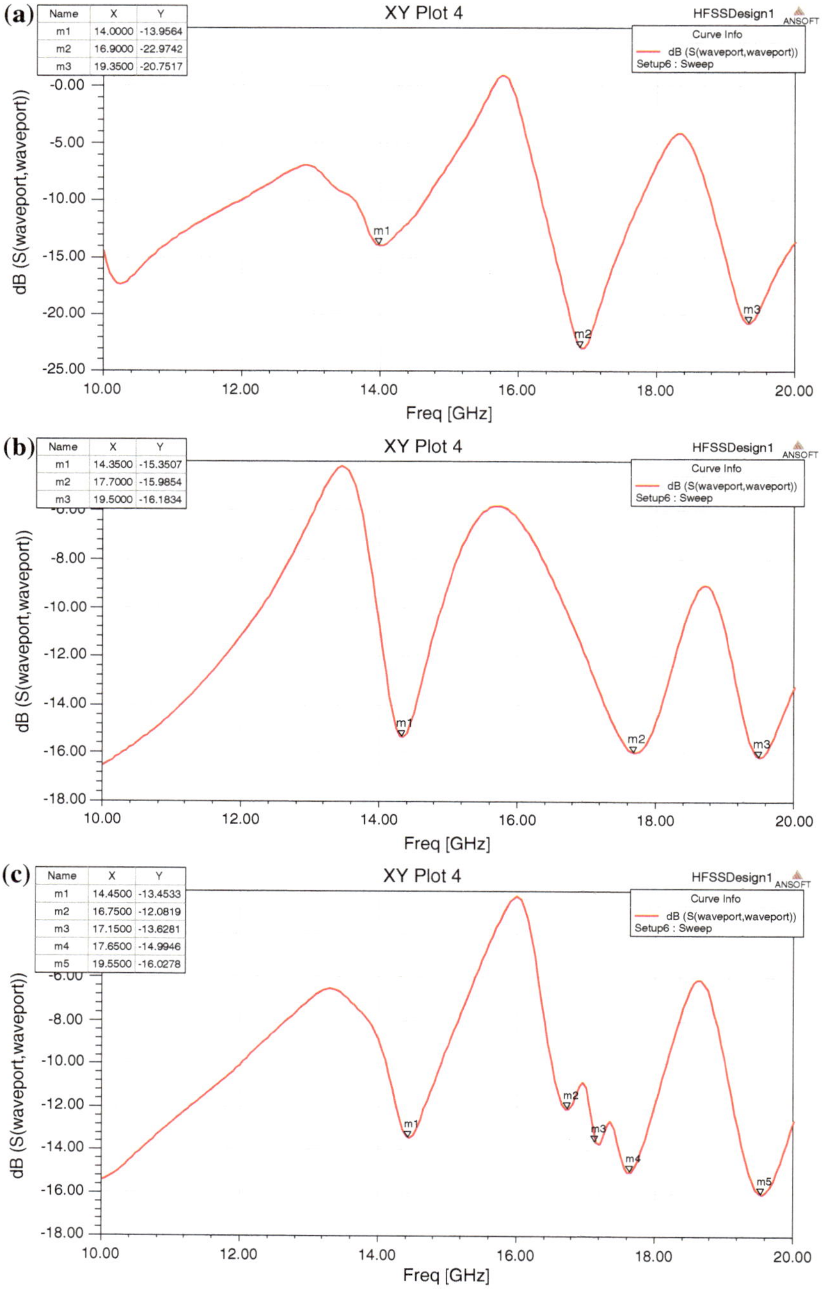

Fig. A5.8 **a** Separation width (0.5) for $\varepsilon_r = 2$, external strip = 10.5 mm, inner strip = 3.5 mm. **b** Separation width (1.5) for $\varepsilon_r = 2$, external strip = 10.5 mm, inner strip = 3.5 mm. **c** Separation width (2.5) for $\varepsilon_r = 2$, external strip = 10.5 mm, inner strip = 3.5 mm

A.5.2(d) Effect of Variable Capacitance (Varactor Diode) in Between the Plate

The effect of the varactor diode capacitance placed in between the parallel standing strips is seen. The resonant modes get shifted lower side (Fig. A5.9).

The separation width = 1.0, $\varepsilon_r = 1$, external strip = 10.6 mm, inner strip = 3.0 mm, varactor diode (variation from 1 to 5 μF with step of 1 μF) at position ($z = 2.3$) in vertical direction. The resulting effect is shown in Fig. A5.10.

A5.3 Designing Steps

HFSS steps_Project1

1. Open HFSS.
2. Create file name project1.
3. Define in the Cartesian co-ordinate system origin as ($x = 0$, $y = 0$, $z = 0$).
4. Choose 3-D rectangular box for substrate by defining the desired substrate material and its dimensions such as (RT Duroid and $x = 20$ mm, $y = 30$ mm, $z = 0.8$ mm).
5. Create DRA structure with desired material and dimensions on the substrate top surface (e.g., If substrate dimension from origin was 0.8 mm in z-direction. Then choose DRA #d dimension keeping substrate dimension as reference).

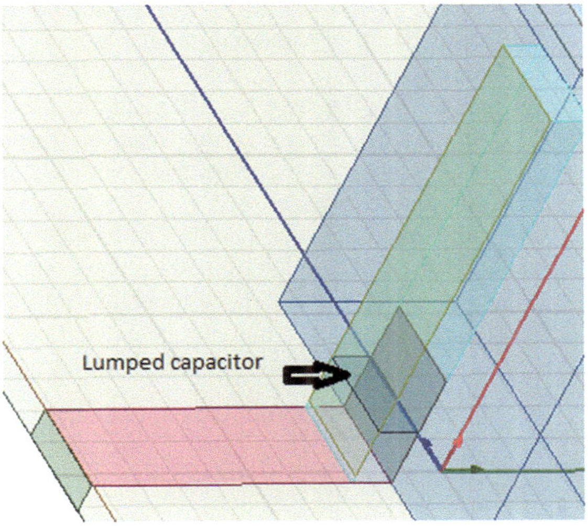

Fig. A5.9 RDRA with lumped capacitance

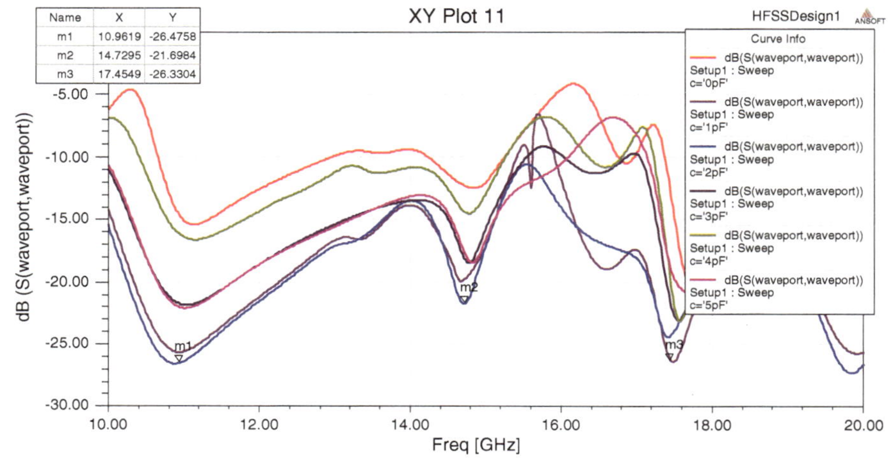

Fig. A5.10 Variation of resonant frequency with lumped capacitance

6. Create two parallel strips adjacent to DRA above the substrate surface with rectangular slab in between them keeping substrate dimension as reference.
7. Apply micro-strip feeding to the DRA structure by defining the micro-strip port with appropriate length and width for impedance matching (e.g., wave port) assigning in the desired direction of input excitation.
8. Variation in height of external strip keeping the internal strip height fixed and vice versa.
9. Effect of the permittivity of rectangular slab can be seen by varying the material property and thickness of the slab in between two fixed parallel plates.
10. Placing a lumped capacitor between two parallel standing strips with desired value (e.g., 2 μF) and perform parametric analysis for variable capacitance value of lumped element.
11. Performing the simulation for the steps 8, 9, 10 mentioned above separately and for mode analysis of DRA which give modal frequency response and effect of the variation of radiation parameters associated with DRA and non-resonant slab with parallel standing strip geometry.
12. Analysis of the simulated structure can be performed by taking various response quantities such as S11, radiation pattern, gain, and field distribution.
13. The above mechanism can also be validated in RDRA by VNA with anechoic chamber on prototype model after structure is simulated.

Annexure-6

Cartesian, Cylindrical, and Spherical Coordinate System

There are three different coordinate systems, i.e., Cartesian, cylindrical, and spherical systems. Cartesian are (x, y, z), cylindrical are (ρ, ϕ, z), and spherical are (r, θ, ϕ) representation (Figs. A6.1 and A6.2).

(a) Cylindrical to cartesian	(b) Cartesian to cylindrical
$X = \rho \cos \phi$ $Y = \rho \sin ø$ $Z = z$	$\rho = \sqrt{x^2 + y^2}$ $ø = \tan^{-1} \frac{y}{x}$ $Z = z$
(c) Spherical to cartesian	**(d) Cartesian to cylindrical**
$X = r \sin \theta \cos \phi$ $Y = r \sin \theta \sin \phi$ $Z = r \cos \theta$	$r = \sqrt{x^2 + y^2 + z^2}$ $\theta = \tan^{-1}\left(\frac{\sqrt{x^2 + y^2}}{z}\right)$ $\phi = \tan^{-1}\left(\frac{y}{x}\right)$
(e) Cylindrical to spherical	**(f) Spherical to cylindrical**
$r = \sqrt{\rho^2 + z^2}$ $\theta = \tan^{-1}\left(\frac{\rho}{z}\right)$ $\phi = \phi$	$\rho = r \sin \theta$ $\phi = \phi$ $z = r \cos \theta$

1. **DEL (∇) derivation in cylindrical system:**

   The Cartesian ∇ (Del) is given as follows:

$$\vec{\nabla} = \vec{x}\frac{\partial}{\partial x} + \vec{y}\frac{\partial}{\partial y} + \vec{z}\frac{\partial}{\partial z}$$

R.S. Yaduvanshi and H. Parthasarathy, *Rectangular Dielectric Resonator Antennas*, DOI 10.1007/978-81-322-2500-3

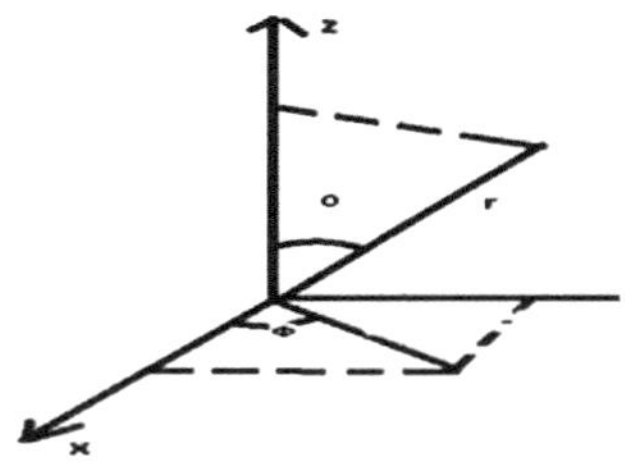

Fig. A6.1 Cartesian system

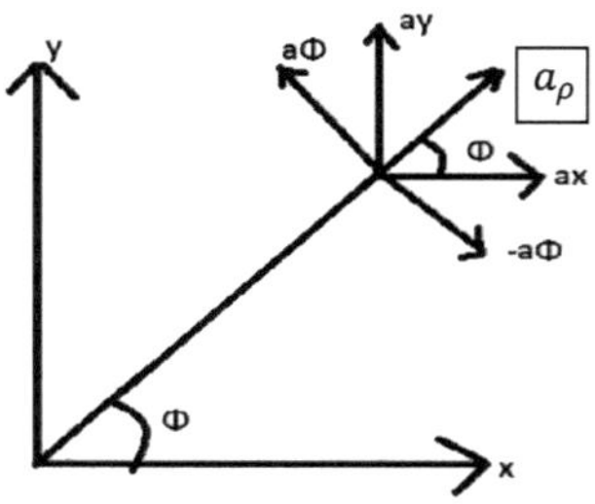

Fig. A6.2 Cylindrical components

Cylindrical ∇ (Del) is given below:

$$\vec{\nabla} = \hat{\rho}\frac{\partial}{\partial\rho} + \hat{\phi}\frac{1}{\rho}\frac{\partial}{\partial\phi} + \hat{z}\frac{\partial}{\partial z}$$

Converting differential operators in terms of the cylindrical system by chain rule:

$$\hat{x}\frac{\partial}{\partial x} = \left(\frac{\partial}{\partial\rho}\frac{\partial\rho}{\partial x} \neq \frac{\partial}{\partial\phi}\frac{\partial\phi}{\partial x} \neq \frac{\partial}{\partial z}\frac{\partial z}{\partial x}\right)\hat{x}$$

$$\frac{\partial\rho}{\partial x} = \frac{\partial}{\partial x}\left(\sqrt{x^2+y^2}\right) = \frac{1}{z}\frac{2x}{\sqrt{x^2+y^2}} = \frac{x}{\sqrt{x^2+y^2}} = \cos\phi$$

$$\left(\because \rho = \sqrt{x^2+y^2}\right)$$

$$\rho = \sqrt{x^2+y^2}$$

Hence,

$$\cos\phi = \frac{x}{\rho}$$

$$\frac{\partial\phi}{\partial x} = \frac{\partial}{\partial x}\left[\tan^{-1}\left(\frac{y}{x}\right)\right) = \frac{1}{1+\frac{y^2}{x^2}}\frac{x(0)-y1}{x^2} = \frac{x^2}{x^2+y^2} = \frac{-y}{x^2}$$

$$\frac{\partial\phi}{\partial x} = \frac{-1}{\sqrt{x^2+y^2}}\frac{y}{\sqrt{x^2+y^2}} = \frac{-1}{\rho}\cdot\sin\phi$$

$$\frac{\partial\phi}{\partial x} = 0\,(\because z \text{ is the same } z \text{ as in Cartesian system it doesn't depend on } x)$$

As per chain rule

Thus, we have

$$\hat{x}\frac{\partial}{\partial x} = \hat{x}\left(\frac{\partial}{\partial p}\frac{\partial P}{\partial x} + \frac{\partial}{\partial \phi}\frac{\partial \phi}{\partial x} + \frac{\partial}{\partial z}\frac{\partial z}{\partial x}\right)$$
$$= \hat{x}\left(\frac{\partial}{\partial \rho}\cos\phi + \frac{\partial}{\partial \phi}\left(\frac{-1}{p}\sin\phi\right) + 0\right) \tag{A6.1}$$

Using the same technique to convert the differential for y:

$$\hat{y}\frac{\partial}{\partial y} = \hat{y}\left(\frac{\partial}{\partial \rho}\frac{\partial \rho}{\partial y} + \frac{\partial}{\partial \phi}\frac{\partial \phi}{\partial y} + \frac{\partial}{\partial z}\frac{\partial z}{\partial y}\right)$$
$$\frac{\partial \rho}{\partial y} = \frac{\partial}{\partial y}\sqrt{(x^2+y^2)} = \frac{2y}{2\sqrt{x^2+y^2}} = \frac{y}{\sqrt{x^2+y^2}} = \sin\phi$$
$$\frac{\partial \phi}{\partial y}\frac{\partial}{\partial y}\left(\tan^{-1}(y)x\right) = \frac{1}{1+\frac{y^2}{x^2}}\frac{1}{x} = \frac{x}{x^2+y^2} = \frac{1}{\rho}\cos\phi$$
$$\frac{\partial z}{\partial y} = 0$$

Thus,

$$\hat{y}\frac{y}{\partial y} = \hat{y}\left(\frac{\partial}{\partial \rho}\frac{\partial \rho}{\partial y} + \frac{\partial}{\partial \phi}\frac{\partial \phi}{\partial y} + \frac{\partial}{\partial z}\frac{\partial z}{\partial y}\right)$$
$$\hat{y}\left(\sin\phi\frac{\partial}{\partial \rho} + \frac{1}{\rho}\cos\phi\frac{\partial}{\partial \phi} + 0\right) \tag{A6.2}$$

Finally, since z is not transformed between coordinate systems

$$\frac{\partial}{\partial z} = \frac{\partial}{\partial z} \tag{A6.3}$$

$$\vec{\nabla} = \hat{x}\frac{\partial}{\partial x} + \hat{y}\frac{\partial}{\partial y} + \hat{z}\frac{\partial}{\partial z}$$
$$\vec{\nabla} = \hat{x}\left(\cos\phi\frac{\partial}{\partial \rho} - \frac{1}{\rho}\sin\phi\frac{\partial}{\partial \phi}\right) + \hat{y}\left(\sin\phi\frac{\partial}{\partial \rho} + \frac{1}{\rho}\cos\phi\frac{\partial}{\partial \phi}\right) + \hat{z}\frac{\partial}{\partial z}$$

Cylindrical

$$\vec{\nabla} = (\hat{x}\cos\phi + \hat{y}\sin\phi)\frac{\partial}{\partial \rho} + \frac{1}{\rho}(\hat{y}\cos\phi - \hat{x}\sin\phi)\frac{\partial}{\partial \phi} + \hat{z}\frac{\partial}{\partial z} \tag{A6.4}$$

Hence, definition to cylindrical unit vector is given as follows:

$$\hat{p} = \hat{x}\cos\phi + \hat{y}\sin\phi = \hat{p}$$
$$\hat{\phi} = \hat{x}\sin\phi + \hat{y}\cos\phi = \hat{\phi}$$
$$\hat{z} = \hat{z} = \hat{z}$$

Thus, Del cylindrical can be written as follows:

$$\vec{\nabla} = \hat{\rho}\frac{\partial}{\partial\rho} + \frac{1}{\rho}\hat{\phi}\frac{\partial}{\partial\phi} + \hat{z}\frac{\partial}{\partial z}$$

which is the desired solution of ∇ in cylindrical coordinates.

2. **DEL (∇) expression as spherical system** (Figs. A6.3, A6.4 and A6.5):

Spherical to Cartesian	Cartesian to Spherical
$X = r\sin\theta\cos\phi$ $Y = r\sin\theta\sin\phi$ $Z = r\cos\theta$	$r = \sqrt{x^2+y^2+z^2}$ $\theta = \tan^{-1}\left(\frac{\sqrt{x^2+y^2}}{z}\right)$ $\phi = \tan^{-1}\left(\frac{y}{x}\right)$

$$\vec{\nabla} = \frac{\partial}{\partial x}\hat{x} + \frac{\partial}{\partial y}\hat{y} + \frac{\partial}{\partial z}\hat{z} \tag{A6.5}$$

$$\hat{x}\frac{\partial}{\partial x} = \hat{x}\left[\frac{\partial}{\partial r}\frac{\partial r}{\partial x} + \frac{\partial}{\partial\theta}\frac{\partial\theta}{\partial x} + \frac{\partial}{\partial\phi}\frac{\partial\phi}{\partial x}\right] \tag{A6.6}$$

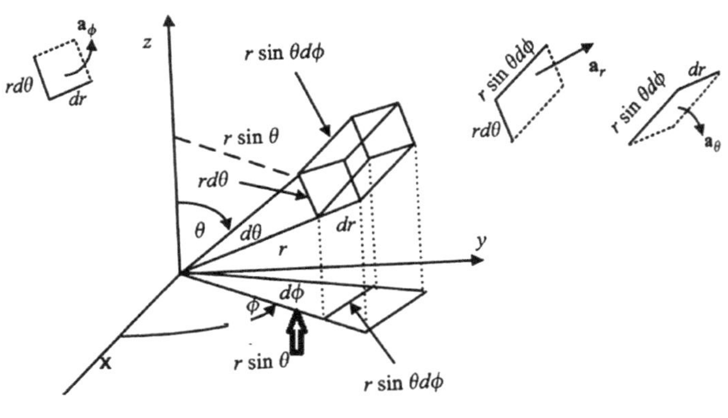

Fig. A6.3 Spherical system

Fig A6.4 Spherical components

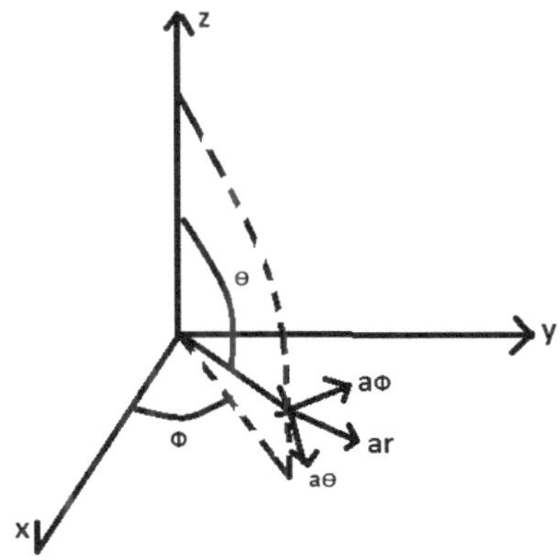

Fig. A6.5 Spherical subcomponents

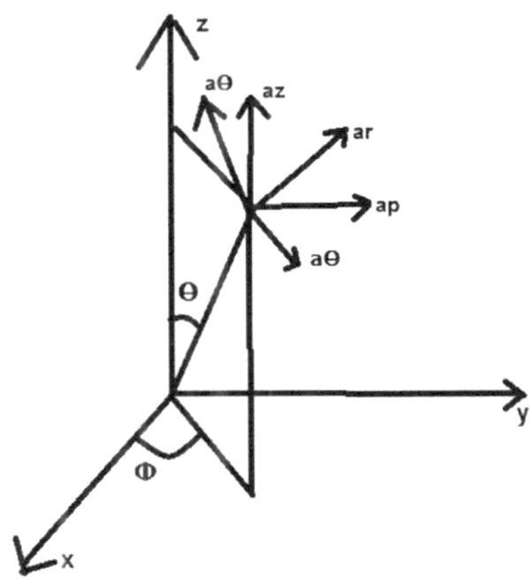

$$\hat{y}\frac{\partial}{\partial y} = y\left[\frac{\partial}{\partial r}\frac{\partial r}{\partial y} + \frac{\partial}{\partial \theta}\frac{\partial \theta}{\partial y} + \frac{\partial}{\partial \phi}\frac{\partial \phi}{\partial y}\right] \tag{A6.7}$$

$$\hat{z}\frac{\partial}{\partial z} = z\left[\frac{\partial}{\partial r}\frac{\partial r}{\partial z} + \frac{\partial}{\partial \theta}\frac{\partial \theta}{\partial z} + \frac{\partial}{\partial \phi}\frac{\partial \phi}{\partial z}\right] \tag{A6.8}$$

Now, partially differentiate r with respect to x

$$\begin{aligned}\frac{\partial r}{\partial x} &= \frac{\partial}{\partial x}\sqrt{x^2+y^2+z^2} \\ &= \frac{2x}{2\sqrt{x^2+y^2+z^2}} \\ &= \frac{x}{\sqrt{x^2+y^2+z^2}} \\ &= \frac{r\sin\theta\cos\phi}{r} \\ &= \sin\theta\cos\phi\end{aligned} \tag{A6.9}$$

Similarly partially differentiate r with respect to y

$$\begin{aligned}\frac{\partial r}{\partial y} &= \frac{\partial}{\partial y}\sqrt{x^2+y^2+z^2} \\ &= \frac{2y}{2\sqrt{x^2+y^2+z^2}} \\ &= \frac{y}{\sqrt{x^2+y^2+z^2}} \\ &= \frac{r\sin\theta\sin\phi}{r} \\ &= \sin\theta\sin\phi\end{aligned} \tag{A6.10}$$

Partially differentiate r with respect to z

$$\begin{aligned}\frac{\partial r}{\partial z} &= \frac{\partial}{\partial z}\sqrt{x^2+y^2+z^2} \\ &= \frac{2z}{2\sqrt{x^2+y^2+z^2}} \\ &= \frac{z}{\sqrt{x^2+y^2+z^2}} \\ &= \frac{r\cos\theta}{r} \\ &= \cos\theta\end{aligned} \tag{A6.11}$$

Partially differentiate θ with respect to x

$$\begin{aligned}
\frac{\partial \theta}{\partial x} &= \frac{\partial}{\partial x} \tan^{-1}\left(\frac{\sqrt{x^2+y^2}}{z}\right) \\
&= \frac{1}{1+\frac{x^2+y^2}{z^2}} \frac{1}{z} \frac{2x}{2\sqrt{x^2+y^2}} \\
&= \frac{z^2}{z^2+x^2+y^2} \frac{x}{z\sqrt{x^2+y^2}} \\
&= \frac{x}{\frac{(\sqrt{x^2+y^2})}{z}(x^2+y^2+z^2)} \\
&= \frac{r\sin\theta\cos\phi}{r^2\tan\theta} \\
&= \frac{\cos\theta\cos\phi}{r}
\end{aligned} \tag{A6.12}$$

Partially differentiate θ with respect to y

$$\begin{aligned}
\frac{\partial \theta}{\partial y} &= \frac{\partial}{\partial y} \tan^{-1}\left(\frac{\sqrt{x^2+y^2}}{z}\right) \\
&= \frac{1}{1+\frac{x^2+y^2}{z^2}} \frac{1}{z} \frac{2y}{2\sqrt{x^2+y^2}} \\
&= \frac{z^2}{z^2+x^2+y^2} \frac{y}{z\sqrt{x^2+y^2}} \\
&= \frac{y}{\frac{(\sqrt{x^2+y^2})}{z}(x^2+y^2+z^2)} \\
&= \frac{r\sin\theta\sin\phi}{r^2\tan\theta} \\
&= \frac{\cos\theta\sin\phi}{r}
\end{aligned} \tag{A6.13}$$

Partially differentiate θ with respect to z

$$\begin{aligned}\frac{\partial\theta}{\partial z} &= \frac{\partial}{\partial z}\tan^{-1}\left(\frac{\sqrt{x^2+y^2}}{z}\right)\\ &= \frac{-1}{1+\frac{x^2+y^2}{z^2}}\frac{\sqrt{x^2+y^2}}{z^2}\\ &= \frac{-z}{(x^2+y^2+z^2)}\left(\frac{\sqrt{x^2+y^2}}{z}\right)\\ &= -\frac{\sin\theta}{r}\end{aligned} \tag{A6.14}$$

Partially differentiate ϕ with respect to x

$$\begin{aligned}\frac{\partial\phi}{\partial x} &= \frac{\partial}{\partial x}\tan^{-1}\left(\frac{y}{x}\right)\\ &= \frac{1}{1+\frac{y^2}{x^2}}\left[\frac{x(0)-y(1)}{x^2}\right]\\ &= \frac{x^2}{x^2+y^2}\left[-\frac{y}{x^2}\right]\\ &= -\frac{y}{x^2+y^2}\\ &= \frac{-\sin\phi}{r\sin\theta}\end{aligned} \tag{A6.15}$$

Partially differentiate ϕ with respect to y

$$\begin{aligned}\frac{\partial\phi}{\partial y} &= \frac{\partial}{\partial y}\tan^{-1}\left(\frac{y}{x}\right)\\ &= \frac{1}{1+\frac{y^2}{x^2}}\left[\frac{1}{x}\right]\\ &= \frac{x}{x^2+y^2}\\ &= \frac{\cos\phi}{r\sin\theta}\end{aligned} \tag{A6.16}$$

Partially differentiate ϕ with respect to z

$$\frac{\partial \phi}{\partial z} = \frac{\partial}{\partial z}\tan^{-1}\left(\frac{y}{x}\right) = 0 \tag{A6.17}$$

Put Eqs. (A6.5), (A6.8), (A6.11) in Eq. (A6.2), Put Eqs. (A6.6), (A6.9), (A6.12) in Eq. (A6.3) and Put Eqs. (A6.7), (A6.10), (A6.13) in Eq. (A6.4).

$$\hat{x}\frac{\partial}{\partial x} = \hat{x}\left[\frac{\partial}{\partial r}\sin\theta\cos\phi + \frac{\partial}{\partial \theta}\frac{\cos\theta\cos\phi}{r} - \frac{\sin\phi}{r\sin\theta}\frac{\partial}{\partial\phi}\right] \tag{A6.18}$$

$$\hat{y}\frac{\partial}{\partial y} = \hat{y}\left[\frac{\partial}{\partial r}\sin\theta\sin\phi + \frac{\partial}{\partial \theta}\frac{\cos\theta\sin\phi}{r} + \frac{\cos\phi}{r\sin\theta}\frac{\partial}{\partial\phi}\right] \tag{A6.19}$$

$$\hat{z}\frac{\partial}{\partial z} = \hat{z}\left[\frac{\partial}{\partial r}\cos\theta - \frac{\partial}{\partial \theta}\frac{\sin\theta}{r}\right] \tag{A6.20}$$

Put Eqs. (A6.13), (A6.14) and (A6.15) in Eq. (A6.1).
And by using original definition to Spherical unit vector,

$$\hat{r} = \hat{x}\sin\theta\cos\phi + \hat{y}\sin\theta\sin\phi + \hat{z}\cos\theta$$
$$\hat{\theta} = \hat{x}\cos\theta\cos\phi + \hat{y}\cos\theta\sin\phi - \hat{z}\sin\theta$$
$$\hat{\phi} = -\hat{x}\sin\phi + \hat{y}\cos\phi$$

We get

$$\vec{\nabla} = \hat{r}\frac{\partial}{\partial r} + \hat{\theta}\frac{1}{r}\frac{\partial}{\partial\theta} + \hat{\phi}\frac{1}{r\,sin\,\theta}\frac{\partial}{\partial\phi}$$

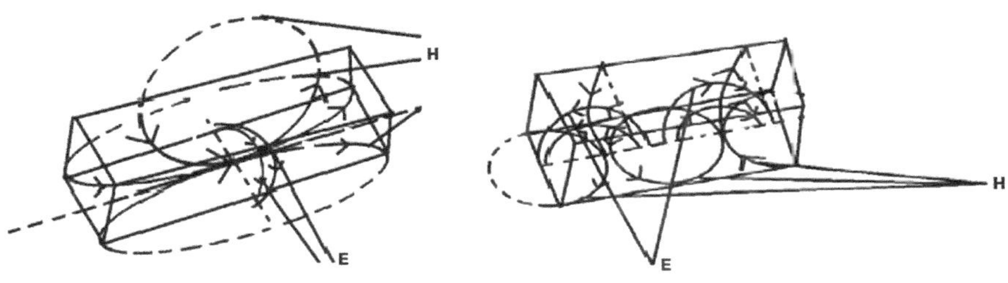

Fig. A6.6 *E* and *H* fields pattern in RDRA

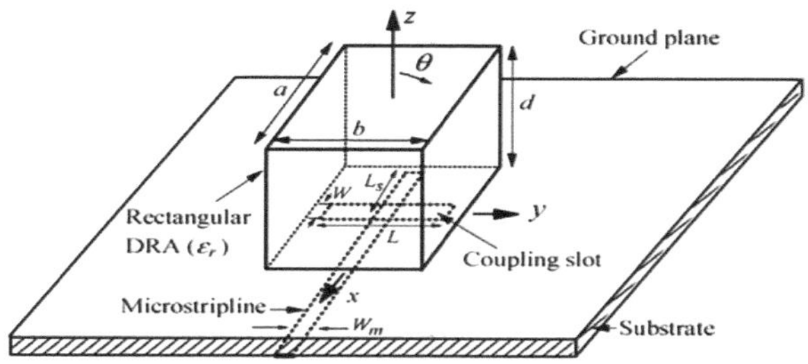

Fig. A6.7 Rectangular DRA

Table A6.1 Transcendental equation solution

S. No.	Permittivity	Dimension (a(length) × b (width) × d(depth)) (mm)	Resonant frequency	Effective width (b')	Multiple factor	% change in width
1.	10.0	14.3 × 25.4 × 26.1	3.5	34.22	1.3474	34.7381
2.	10.0	14 × 8 × 8	5.5	14.13	1.7665	76.6535
3.	10.0	15.24 × 3.1 × 7.62	6.21	8.33	2.8872	168.7230
4.	20.0	10.2 × 10.2 × 7.89	4.635	15.31	1.5014	50.1419
5.	20.0	10.16 × 10.2 × 7.11	4.71	15.15	1.4858	48.5797
6.	35.0	18 × 18 × 6	2.532	24.12	1.34	33.9973
7.	35.0	18 × 18 × 9	2.45	25.64	1.4244	42.4423
8.	100.0	10 × 10 × 1	7.97	11.24	1.1242	12.4237

Hence $\vec{\nabla}$ from Cartesian to spherical converted.

3. ***E* and *H* fields in RDRA**

   Fields converting into TE and TM modes inside rectangular DRA (Fig. A6.6).

4. **Transcendental equation solution using MATLAB programs (simulated rectangular DRA)** (Fig. A6.7; Table A6.1).

Program 1

```
%%Dimensions of DRA
%%length
d=[14.3,14.0,15.24,10.2,10.16,18,18,10];
%%width
w=[25.4,8,3.1,10.2,10.2,18,18,10];
%%height
h=[26.1,8,7.62,7.89,7.11,6,9,1];
%%Mode
m=1;
n=1;
p=1;
c=3e8;
cons=[10.0,10.0,10,20,20,35,35,100];
syms y real
for i=drange(1:8)
kx(i)=pi/d(i);
kz(i)=pi/2/h(i);
ko=sqrt((kx(i).^2+y.^2+kz(i).^2)/cons(i));
f=real(y.*tan(y*w(i)/2)-sqrt((cons(i)-1)*ko.^2-y.^2));
ky(i)=fzero(inline(f),[0,(pi/w(i))-0.01]);
%%Resonant frequency
fre(i)=c/2/pi*sqrt((kx(i).^2+ky(i).^2+kz(i).^2)/cons(i))*1e3;
Effwidth(i)=pi/ky(i);
factor(i)=Effwidth(i)./w(i);
perchangwidth(i)=((Effwidth(i)-w(i))/w(i))*100;
end
```

Results:

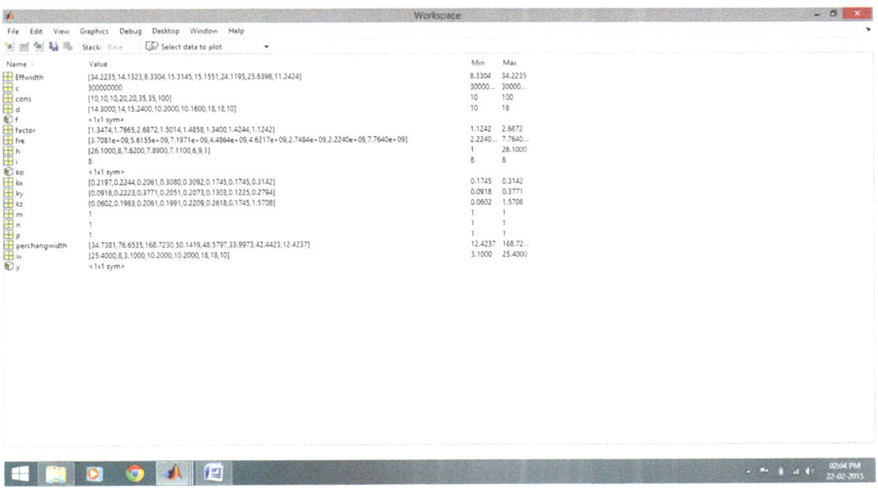
Workspace

File Edit View Graphics Debug Desktop Window Help

Stack: Base Select data to plot

Name	Value	Min	Max
Effwidth	[34.2235,14.1323,8.3304,15.3145,15.1551,24.1195,25.6396,11.2424]	8.3304	34.2235
c	300000000	30000...	30000...
cons	[10,10,10,20,20,35,35,100]	10	100
d	[14.3000,14,15.2400,10.2000,10.1600,18,18,10]	10	18
f	<1x1 sym>		
factor	[1.3474,1.7665,2.6872,1.5014,1.4858,1.3400,1.4244,1.1242]	1.1242	2.6872
fre	[3.7081e+09,5.6155e+09,7.1971e+09,4.4864e+09,4.6217e+09,2.7484e+09,2.2240e+09,7.7640e+09]	2.2240...	7.7640...
h	[26.1000,8,7.6200,7.8900,7.1100,6,9,1]	1	26.1000
i	8	8	8
ko	<1x1 sym>		
kx	[0.2197,0.2244,0.2061,0.3080,0.3092,0.1745,0.1745,0.3142]	0.1745	0.3142
ky	[0.0918,0.2223,0.3771,0.2051,0.2073,0.1303,0.1225,0.2794]	0.0918	0.3771
kz	[0.0602,0.1963,0.2061,0.1991,0.2209,0.2618,0.1745,1.5708]	0.0602	1.5708
m	1	1	1
n	1	1	1
p	1	1	1
perchangwidth	[34.7381,76.6535,168.7230,50.1419,48.5797,33.9973,42.4423,12.4237]	12.4237	168.72...
w	[25.4000,8,3.1000,10.2000,10.2000,18,18,10]	3.1000	25.4000
y	<1x1 sym>		

Program2

```
m=1;
n=1;
p=1;
E_r=10;
a=15.24e-03;
b=3.1e-03;
d=7.62e-03;
c=3e+08;
kx=m*pi/a;
ky=n*pi/b;
kz=p*(pi/d)/2;
ko=sqrt(kx^2+ky^2+kz^2)/sqrt(E_r);
fo=(c*ko/pi)/2;
foghz=fo/(1e+09);
```

Results:

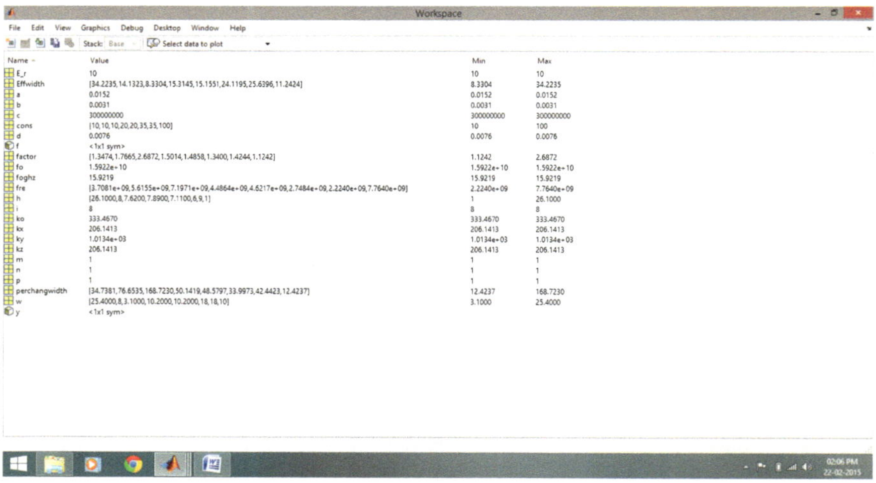
Workspace

File Edit View Graphics Debug Desktop Window Help

Stack: Base Select data to plot

Name	Value	Min	Max
E_r	10	10	10
Effwidth	[34.2235,14.1323,8.3304,15.3145,15.1551,24.1195,25.6396,11.2424]	8.3304	34.2235
a	0.0152	0.0152	0.0152
b	0.0031	0.0031	0.0031
c	300000000	300000000	300000000
cons	[10,10,10,20,20,35,35,100]	10	100
d	0.0076	0.0076	0.0076
f	<1x1 sym>		
factor	[1.3474,1.7665,2.6872,1.5014,1.4858,1.3400,1.4244,1.1242]	1.1242	2.6872
fo	1.5922e+10	1.5922e+10	1.5922e+10
foghz	15.9219	15.9219	15.9219
fre	[3.7081e+09,5.6155e+09,7.1971e+09,4.4864e+09,4.6217e+09,2.7484e+09,2.2240e+09,7.7640e+09]	2.2240e+09	7.7640e+09
h	[26.1000,8,7.6200,7.8900,7.1100,6,9,1]	1	26.1000
i	8	8	8
ko	333.4670	333.4670	333.4670
kx	206.1413	206.1413	206.1413
ky	1.0134e+03	1.0134e+03	1.0134e+03
kz	206.1413	206.1413	206.1413
m	1	1	1
n	1	1	1
p	1	1	1
perchangwidth	[34.7381,76.6535,168.7230,50.1419,48.5797,33.9973,42.4423,12.4237]	12.4237	168.7230
w	[25.4000,8,3.1000,10.2000,10.2000,18,18,10]	3.1000	25.4000
y	<1x1 sym>		

Program 3

MATLAB programs taking parameters a,b,d same and comparing frequency using :

Program 1 : Characteristic Equation

```
m=1
n=1
p=1
E_r=10
a=14.3e-03
b=25.4e-03
d=26.1e-03
c=3e+08
k_x=m*pi/a
k_y=n*pi/b
k_z=p*(pi/d)/2
k_o=sqrt(k_x^2+k_y^2+k_z^2)/sqrt(E_r)
f_o=(c*k_o/pi)/2
f_oGHz=f_o/1e+09
```

output:

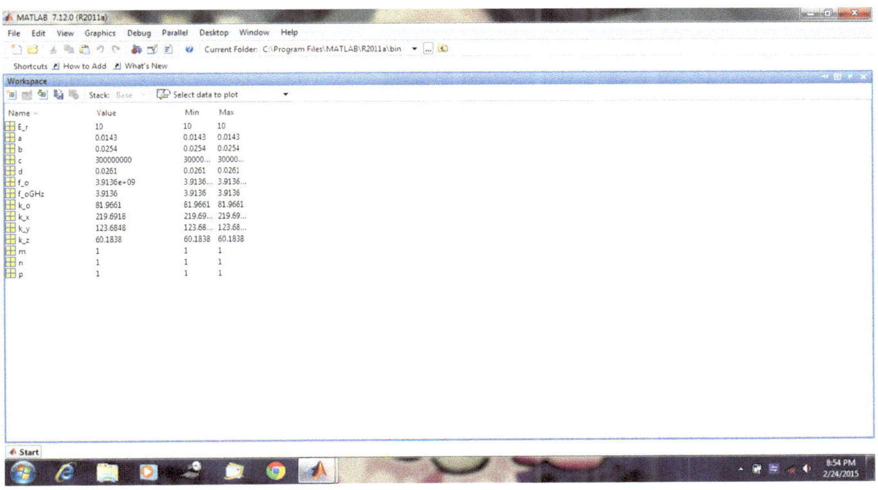

Program 4

Transcendental Equation for same dimensions:

```
m=1;
n=1;
p=1;
E_r=10;
a=14.3e-03;
b=25.4e-03;
d=26.1e-03;
c=3e+08;
syms y real
kx=pi/a;
kz=pi/d/2;
ko=sqrt(kx^2+y^2+kz^2)/sqrt(E_r);
f=real(y*tan(y*b/2)-sqrt((E_r-1)*ko^2-y^2));
ky=fzero(inline(f),[0,(pi/b)-0.01]);
fre=c/2/pi*sqrt((kx^2+ky^2+kz^2)/E_r)*1e3;
effwidth=pi/ky;
factor=effwidth/b;
perchangwidth=((effwidth-b)/b)*100;
```

output:

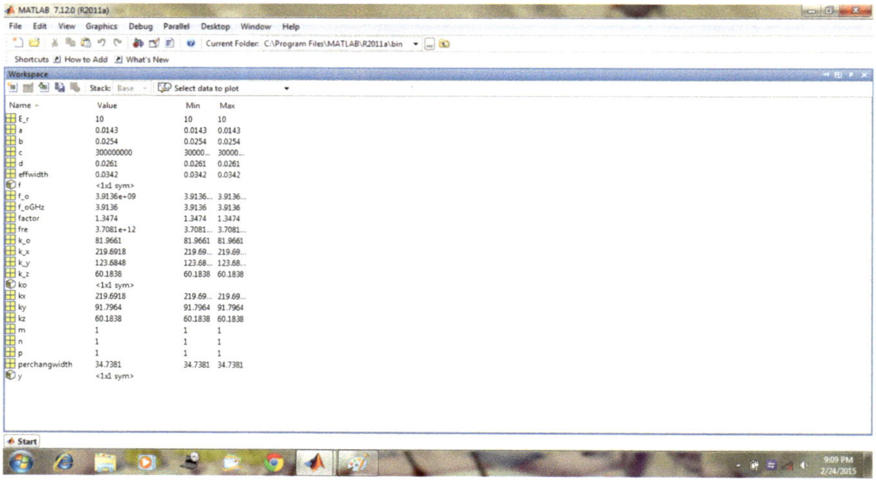

Program5

MATLAB programs taking parameters a,b,d same and comparing frequency using :

: Characteristic Equation

```
    Where a=17mm
          b=25mm
          c=10mm
m=1;
n=1;
p=1;
E_r=10;
a=17e-03;
b=25e-03;
d=10e-03;
c=3e+08;
k_x=m*pi/a;
k_y=n*pi/b;
k_z=p*(pi/d)/2;
k_o=sqrt(k_x^2+k_y^2+k_z^2)/sqrt(E_r);
f_o=(c*k_o/pi)/2;
f_oGHz=f_o/1e+09;
```

Output

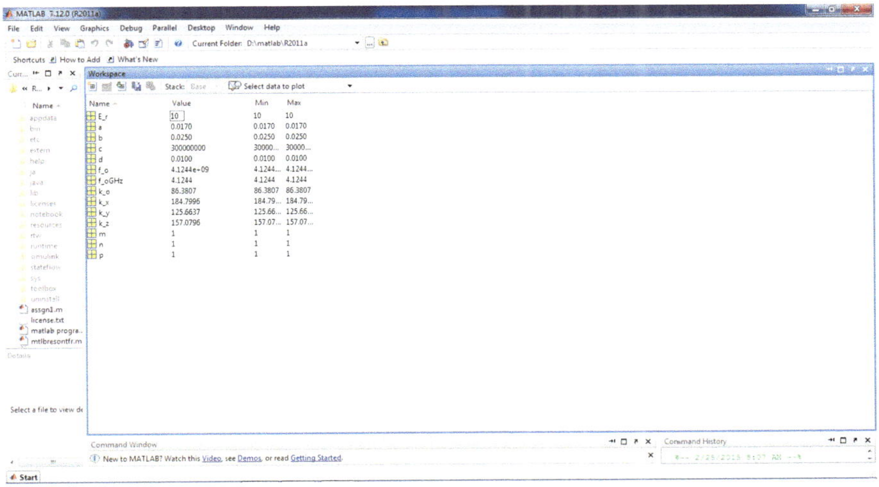

Program 6

Transcendental Equation

```
m=1;
n=1;
p=1;
E_r=10;
a=17e-03;
b=25e-03;
d=10e-03;
c=3e+08;
syms y real
kx=pi/a;
kz=pi/d/2;
ko=sqrt(kx^2+y^2+kz^2)/sqrt(E_r);
f=real(y*tan(y*b/2)-sqrt((E_r-1)*ko^2-y^2));
ky=fzero(inline(f),[0,(pi/b)-0.01]);
fre=c/2/pi*sqrt((kx^2+ky^2+kz^2)/E_r)*1e3;
effwidth=pi/ky;
factor=effwidth/b;
perchangwidth=((effwidth-b)/b)*100;
```

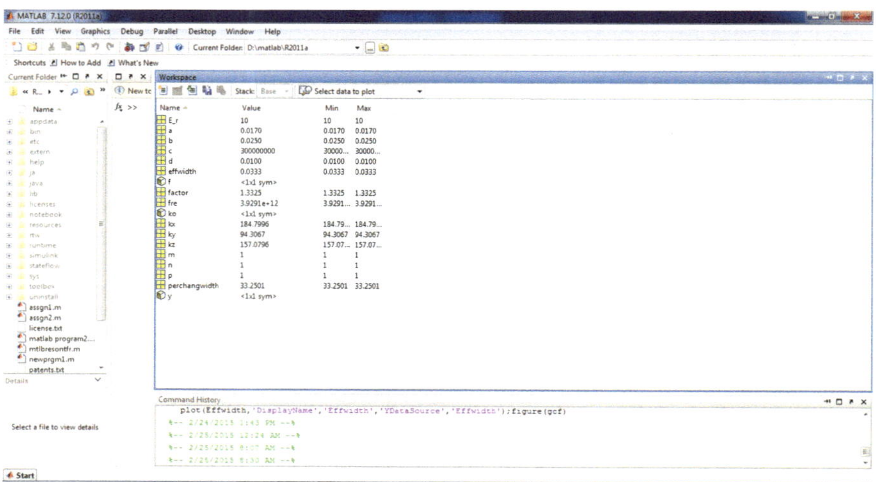

Program 7

```
MATLAB programs taking parameters a,b,d same and comparing frequency using :
 Characteristic Equation

m=1
n=1
p=1
E_r=10
a=14.3e-03
b=25.4e-03
d=26.1e-03
c=3e+08
k_x=m*pi/a
k_y=n*pi/b
k_z=p*(pi/d)/2
k_o=sqrt(k_x^2+k_y^2+k_z^2)/sqrt(E_r)
f_o=(c*k_o/pi)/2
f_oGHz=f_o/1e+09
```

Output:

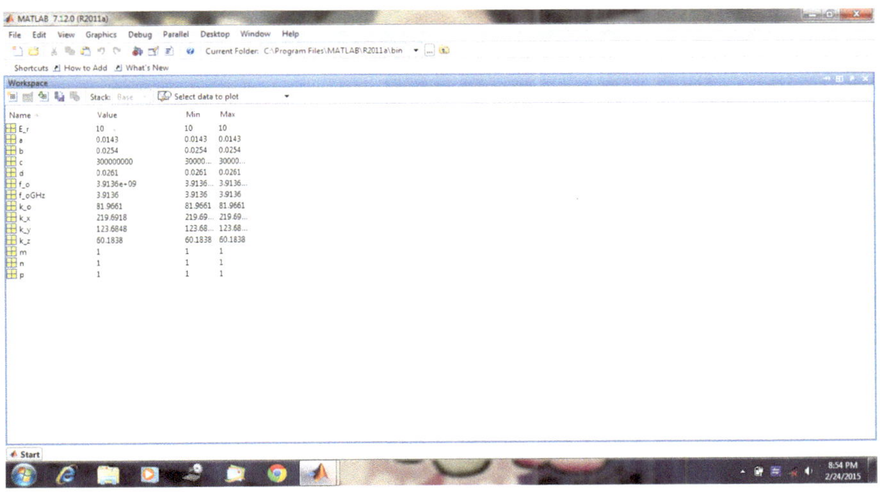

Program 8

Tanscandental Equation

```
m=1;
n=1;
p=1;
E_r=10;
a=14.3e-03;
b=25.4e-03;
d=26.1e-03;
c=3e+08;
syms y real
kx=pi/a;
kz=pi/d/2;
ko=sqrt(kx^2+y^2+kz^2)/sqrt(E_r);
f=real(y*tan(y*b/2)-sqrt((E_r-1)*ko^2-y^2));
ky=fzero(inline(f),[0,(pi/b)-0.01]);
fre=c/2/pi*sqrt((kx^2+ky^2+kz^2)/E_r)*1e3;
effwidth=pi/ky;
factor=effwidth/b;
perchangwidth=((effwidth-b)/b)*100;
```

Output:

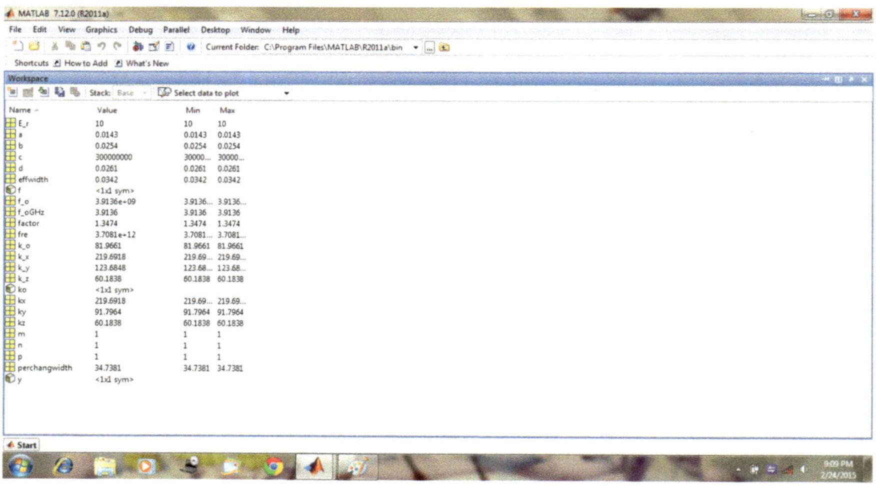

Program 9

```
m=3;
n=3;
a=10;
b=5;
x=linspace(-5,5,51);
y=linspace(-2.5,2.5,51);
[xi,yi]=meshgrid(x,y);
Ez=cos(m*pi*xi/a).*cos(n*pi*yi/b);
Ez=Ez.^2;
Ez=sqrt(Ez);
surf(xi,yi,Ez)
view([-45,60])
%%view([180,0])
drawnow
```

Workspace

Stack: Base Select data to plot

Name	Value	Min	Max
Ez	<51x51 double>	3.37...	1
a	10	10	10
b	5	5	5
m	3	3	3
n	3	3	3
x	<1x51 double>	-5	5
xi	<51x51 double>	-5	5
y	<1x51 double>	-2.50...	2.50...
yi	<51x51 double>	-2.50...	2.50...

Program 10:

```
d=[14.3,14.0,15.24,10.2,10.16,18,18,10];
w=[25.4,8,3.1,10.2,10.2,18,18,10];
h=[26.1,8,7.62,7.89,7.11,6,9,1];
m=1;
n=1;
p=1;
c=3e8;
cons=[10.0,10.0,10,20,20,35,35,100];
syms y real
for i=drange(1:8)
kx(i)=pi/d(i);
kz(i)=pi/2/h(i);
ko=sqrt((kx(i).^2+y.^2+kz(i).^2)/cons(i));
f=real(y.*tan(y*w(i)/2)-sqrt((cons(i)-1)*ko.^2-y.^2));
ky(i)=fzero(inline(f),[0,(pi/w(i))-0.01]);
fre(i)=c/2/pi*sqrt((kx(i).^2+ky(i).^2+kz(i).^2)/cons(i))*1e3;
Effwidth(i)=pi/ky(i);
factor(i)=Effwidth(i)./w(i);
perchangwidth(i)=((Effwidth(i)-w(i))/w(i))*100;
end
```

Workspace

Stack: Base   Select data to plot

Name	Value	Min	Max
Effwidth	[34.2235,14.1323,...	8.33...	34.2...
Ez	<51x51 double>	3.37...	1
a	10	10	10
b	5	5	5
c	300000000	3000...	3000...
cons	[10,10,10,20,20,3...	10	100
d	[14.3000,14,15.24...	10	18
f	<1x1 sym>		
factor	[1.3474,1.7665,2....	1.12...	2.68...
fre	[3.7081e+09,5.61...	2.22...	7.76...
h	[26.1000,8,7.6200...	1	26.1...
i	8	8	8
ko	<1x1 sym>		
kx	[0.2197,0.2244,0....	0.17...	0.31...
ky	[0.0918,0.2223,0....	0.09...	0.37...
kz	[0.0602,0.1963,0....	0.06...	1.57...
m	1	1	1
n	1	1	1
p	1	1	1
perchangwidth	[34.7381,76.6535,...	12.4...	168....
w	[25.4000,8,3.1000...	3.10...	25.4...
x	<1x51 double>	-5	5
xi	<51x51 double>	-5	5
y	<1x1 sym>		
yi	<51x51 double>	-2.50...	2.50...

Bibliography

1. Long SA, Mcallister MW, Shen LC (1983) The resonant cylindrical dielectric cavity antenna. IEEE Trans Antennas Propag 31(3):406–412
2. Kajfez D, Guillon P (eds) (1986) Dielectric resonators. Artech House, Dedham
3. Kranenburg RA, Long SA, Williams JT (1991) Coplanar wave guide excitation of resonator antennas. IEEE Trans Antennas Propag 29(1):119–122
4. Lee RQ, Simons RN (1994) Bandwidth of dielectric resonator antennas. In: IEEE antennas propagate society of international symposium. Seattle, pp 1500–1503
5. Wu JY, Hung CY, Wong KL (1994) Low profile very high permittivity DRA excited by coplanar waveguide. Microw Opt Technol Lett 22(2):96–97
6. Yung EKN, Lee WWS, Luk KM (1994) A Dielectric resonator on a micro strip antennas. In: IEEE antennas propagate society of international symposium serial. Washington, pp 1504–1507
7. Mongia RK, Luk KM, Yung EKN (1994) Low profile dielectric resonator antennas using a very high permittivity material. Electron Lett 30(17):1362–1363
8. Oliver MB, Antar YMM, Mongia RK, Iittipiloon A (1995) Circularly polarized rectangular dielectric resonator antenna. Electron Lett 31(6):418–419
9. Pozar DM, Schubert DH (1995) Micro strip antennas: the analysis of micros trip antennas and arrays. IEEE Press, New York
10. Keller MG, Roscone DJ, Oliver MB, Mongia RK, Antar YMM, Iittipiloon A (1995) Active aperture coupled rectangular dielectric resonator antenna. IEEE Microw Guided Wave Lett 5 (11):376–378
11. Esselle KP (1996) A low profile rectangular dielectric resonator antennas. Electron Lett 44 (9):1296–1297
12. Hwang Y, Zhang YP, Mulk K, Yung ENK (1997) Gain enhanced miniaturized rectangular dielectric resonator antennas. Electron Lett 33(5):350–352
13. Leung KW, To MW (1997) Slot coupled dielectric resonator antennas using a proximity feed on a perpendicular substrate. Electron Lett 33(20):1665–1666
14. Hwang Y, Zhang YP, Luk KM, Yung EKN (1997) Gain-enhanced miniaturised rectangular dielectric resonator antenna. IEEE Electron Lett 33(5):350–352
15. Mongia RK, Ittipiboon A (1997) Theoretical and experimental investigation on rectangular dielectric resonator antennas. IEEE Trans Antenna Propag 45:1348–1356
16. Mongia RK, Iittipiboon A (1997) Theoretical and experimental investigation on rectangular dielectric resonator antennas. IEEE Trans Antennas Propag 45(9):1348–1356
17. Mongia RK, Ittipiboon A (1997) Theoretical and experimental investigations on rectangular dielectric resonator antennas. IEEE Trans Antennas Propag 45(9):1348–1356
18. Junkar GP, Kishk AA, Glisson AW (1998) Multiport network description and radiation characteristics of coupled dielectric resonator antennas. IEEE Trans Antennas Propag 46 (8):1242–1243

R.S. Yaduvanshi and H. Parthasarathy, *Rectangular Dielectric Resonator Antennas*, DOI 10.1007/978-81-322-2500-3

19. Kranenburg RA, Long SA (1998) Micro strip transmission line excitation of dielectric resonator antennas. Electron Lett 24(18):1156–1157
20. Wu JY, Hung CY, Wong KL (1999) Low profile very high permittivity DRA excited by a coplanar wave guide. Microw Opt Technol Lett 22(2):96–97
21. Hong CS (1999) Adjustable frequency dielectric resonator antennas. Proc Natl Sci Counc. ROC (A) 23(6):736–738
22. Henry B, Petosa A, Antar YMM, Morin GA (1999) Mutal coupling. Opt Technol Lett 21 (1):44–48
23. Luk KM, Lee MT, Leung KW, Yung EVN (1999) Technique for improving coupling between micro strip line and dielectric resonator antennas. Electron Lett 35(5):357–358
24. Sangiovanni A, Garel PY, Dauvigenc JY, Pichor Ch (2000) Numerical analysis of dielectric resonator antennas. Int J Numer Model Electr Netw Devices Fields 13:199–215
25. Leung KW (2000) Conformal strip excitation of dielectric resonator antennas. IEEE Trans Antennas Propag 48(6):961–967
26. Ge Y, Esselle KP (2000) Computation of radiation pattern of a rectangular dielectric resonator antenna using the method of moments. Microw Opt Technol Lett 27(6):382–384
27. Hang CY, Kuo JS (2000) Frequency adjustable circularly polarized dielectric resonator antennas. Microw Opt Technol Lett 34(3):211–213
28. Sangiovanni A, Dauvignac JY, Pichod C (2001) Embedded dielectric resonator antenna for bandwidth enhancement. Electron Lett 33(7):406–408
29. Laisne A, Gillard R, Piton G (2001) Robust slot fed dielectric resonator antennas using an intermediate. Electron Lett 37(25):1497–1498
30. Jerry Fiedziuszko S, Home S (2001) Dielectric resonator—rise your high SQ. IEEE Microw Mag 51–60
31. Deng SM, Chen TW, Kan HH (2001) A CPW fed rectangular dielectric resonator antennas. In: Proceedings of IEEE APMC, Taipei, pp 954–956
32. Al Salmeh MS, Antar YMM, Seguin G (2002) Coplanar wave guide fed coupled rectangular dielectric resonator antennas. IEEE Trans Antennas Propag 50(10):11415–1419
33. Fiedziuszko SJ, Hunter IC, Itoh T, Kobyashi Y, Nishokawa T, Suitzer SN, Wakino K (2002) Dielectric materials, devices and circuits. IEEE Trans Microw Theory Tech 50(3):706–720
34. Petosa A, Littipiboon A, Thirkoune S (2002) Perforated dielectric resonator antennas. Electron Lett 38(24):1493–1495
35. Kishk AA, Yan Y, Glisson AW (2002) Conical dielectric resonator antennas for wide band applications. IEEE Trans Antenna Propag 50(4):469–474
36. So KK, Leung KW (2002) Annular-slot excited dielectric resonator antenna with a backing cavity. IEEE Antenna Propagation Society international symposium, San Antonio, Texas, June 2002, pp 470–473
37. Ferrando-Bataller M, Valero-Nogueira A, Cabedo-Fabrés M, Antonino Daviu E (2003) Design of ultra-wideband antennas using characteristic modes. INICA/COST 284 workshop, Berlin, Germany
38. Li B, So KK, Leung KW (2003) A circular polarized dielectric resonator antenna excited by asymmetrical U-slot with a backing cavity. IEEE Antenna Wirel Propag Lett. 2:133–135
39. Guo YX, Luk KM (2003) Dual polarized dielectric resonator antennas. IEEE Trans Antennas Propag 51(5):1120–1123
40. Kishk AA (2003) Wideband truncated tetrahedron dielectric resonator antenna excited by a coaxal probe. IEEE Trans Antenna Propag 51(10):2913–2917
41. Gentilli GB, Morni M, Selleri S (2003) Relevance of coupling effects on DRA array design. IEEE Antenna Propaga 51(3):399–404
42. Kishk AA, Zhang X, Glission AW, Kajfez D (2003) Numerical analysis of stacked dielectric resonator excited by a coaxial probe for wideband application. IEEE Trans Antenna Propaga 51(8):1996–2006
43. Deng SM, Tsai CL, Chiu CW, Chnag SF (2003) CPW fed rectangular ceramic dielectric resonator antennas 51(8):2142–2144

44. Petosa A, Ittipiboon A, Antaar Y (2003) Broadband dielectric resonator antennas. In: Dielectric resonator antennas. Research Studies Press Ltd., UK
45. Ghosh B, Antar YMM, Petosa A, Littipiboon A (2004) Feed configuration of CPW fed DRA. IEEE Antennas Propagation Society international symposium, Monterrey, CA, June 2004
46. Bit-Babik G, DiNallo C, Faraone A (2004) Multimode dielectric resonator antenna of very high permittivity. In: Proceedings of IEEE antennas and propagation symposium, Monterrey, CA, June 2004, pp 1383–1386
47. Bit Babik B, Di Nallo C, Fraone A (2004) Multimode dielectric resonator antennas of very high permittivity. IEEE Antennas Propagation Society international symposium, Monterrey, CA, June 2004
48. Sung Y, Ahn CS, Kim YS (2004) Micro strip fed dual frequency dielectric resonator antennas. Microw Opt Technol 42(5):388–390
49. Littipiboon A, Mongia RK, Antar YMM, Bhartia P, Cuhaci M (2004) An integrated rectangular dielectric resonator antenna. IEEE Antennas Propagation Society international symposium, Michigan, June 2004, pp 604–607
50. Petosa A, Thirkoune S, Lttipiboon A (2004) Array of performed dielectric resonator antennas. IEEE Antennas Propagation Society international symposium, Monterrey, CA, June 2004
51. Li B, Leung KW (2004) A wideband strip fed rectangular dielectric resonator antenna. In: Proceedings of IEEE antennas and propagation symposium, DC, Washington, July 2004, pp 1383–1386
52. Sheng XQ, Leung KW, Yung EKN (2004) Analysis of waveguide fed dielectric resonator antenna using a hybrid infinite element method moment method. IEEE Proc Microw Antenna Propag 151:91–95
53. Eshrah IA, Kishk AA, Yakovlev AB, Gilsson AW (2004) Modeling of a waveguide probe excitation for dielectric resonator antenna. IEEE Antennas Propagation Society international symposium, Monterrey, CA, June 2004
54. Nannini C, Ribero JM, Dauvignac JY, Pichot C (2004) Bifrequency behavior and bandwidth enhancement of a dielectric resonator antenna. Microw Opt Technol Lett 42(5):432–434
55. Nannini C, Ribero JM, Dauvignac JY, Pichot C (2004) Bifrequency and bandwidth enhancement of a dielectric resonator antennas. Microw Opt Technol Lett 42(5):432–434
56. Joliveau E, Smith J, Wolfe J (2004) Tuning of vocal tract resonances by sopranos. Nature 427:116
57. Li B, Leung KW (2005) Strip-fed rectangular dielectric resonator antennas with/without a parasitic patch. IEEE Trans Antenna Propag 53(7):2200–2207
58. de Young CS, Long SA (2006) Wideband cylindrical and rectangular dielectric resonator antenna. IEEE Antennas Wirel Propag Lett 5:426–429
59. Petosa A, Thirakoune S, Littipiboon A (2009) Higher order modes in rectangular DRAs for gain enhancement. Presented at the ANTEM 2009, Banff, Canada, Feb 2009
60. Chang T-H, Kiang J-F (2009) Bandwidth broadening of dielectric resonator antenna by merging adjacent bands. IEEE Trans Antennas Propag 57(10):3316–3320
61. Sebastian ML (2009) Materials for wireless communication. Elsevier, Amsterdam
62. Kong LB (2010) Electrically tunable materials. Elsevier, Amsterdam
63. Huff GH, Rolando DL, Walters P, McDonald J (2010) A frequency reconfigurable dielectric resonator antenna using colloidal dispersions. IEEE Antenna Wirel Propag Lett 9
64. Huff GH, Rolando DL, Walters P, McDonald J (2010) A frequency reconfigurable dielectric resonator antenna using colloidal dispersions. IEEE Antennas Wirel Propag Lett 9
65. Petosa A, Thirakoune S (2011) Rectangular dielectric resonator antennas with enhanced gain. IEEE Trans Antennas Propag 59(4):1385–1389
66. Yaduvanshi RS (2012) Hybrid MHD antenna. Intersci J IJUWBCS 2(4):201–206
67. Wang YF, Denidni TA, Zeng QS, Wei G (2014) IEEE Lett
68. Khan QU Higher mode excitation for high gain micro strip antenna. AIU-Int J Electron Commun

69. Guha D, Antaar YMM (2012) Higher order mode excitation for high-gain broadside radiation from cylindrical dielectric resonator antennas. IEEE Antenna Propag
70. Popov A (1999) Multi-mode dielectric resonator antenna with controllable radiation pattern. IEEE Trans Antenna Propag
71. Kreyszig E (2002) Advanced engineering mathematics. Wiley for Half wave Fourier analysis
72. Jackson JD (1962) Classical electrodynamics. Wiley, New York
73. Cheng X (2011) Hybrid mode radiation in patch antenna loaded with corrugated electric arrays. Department of Electrical and Computer Engineering, University of Florida, Gainesville, FL, USA
74. Chang K, Hsieh LH (2004) Microwave ring circuits and related structures. Wiley Interscience, New York
75. Adam JJ, Bernhard JT (2013) Broadband equivalent circuit models for antenna impedances and fields using characteristic modes. IEEE Trans Antenna Propag
76. Jaggi VP, Mathur AB (1985) Advanced engineering mathematics. Khanna publisher (Chapter 13 Series solutions and special functions I.e. Bessel function)
77. Neshati MH, Wu Z (2001) The determination of the resonant frequency of the TE^{y}_{111} mode in a rectangular dielectric resonator for antenna application. 11th international
78. Colline BS, Nahar V, Kingsley SP, Zhang SQ, Krupa S (2004) A dual band hybrid dielectric antenna for laptop computer. IEEE Antenna Propagation Society international symposium
79. Jordan EC, Balmain KG (1968) Electromagnetic and radiating systems. Wiley, New York
80. Landau LD, Liftshiz EM (1984) Electrodynamics of continuous media. Elsevier, Amsterdam
81. Jacksion JD (1962) Classical electrodynamics. Wiley, New York
82. Balanis CA (2005) Antenna theory analysis and design. Wiley, New York
83. Harrington RF (1961) Time harmonic electromagnetic fields. IEEE Press series on electromagnetic wave theory
84. Sadiku MNO (2009) Principles of electromagnetics. Oxford University Press, Oxford
85. Feynman RP (2013) Lecture series on physics volume II. Pearson Publishing
86. Guha D, Antar YMM (2011) Microstrip and printed antennas. Wiley, New York

Index

R.S. Yaduvanshi and H. Parthasarathy, *Rectangular Dielectric Resonator Antennas*, DOI 10.1007/978-81-322-2500-3